Water Governance

Water Governance

An Evaluation of Alternative Architectures

Edited by

Asanga Gunawansa

Attorney-at-Law, Supreme Court of Sri Lanka, Adjunct Assistant Professor, School of Design and Environment and International Research Associate, Institute of Water Policy, National University of Singapore

Lovleen Bhullar

Former Research Associate, Institute of Water Policy, National University of Singapore and Researcher, Environmental Law Research Society, New Delhi, India

Edward Elgar

Cheltenham, UK • Northampton, MA, USA

Published by
Edward Elgar Publishing Limited
The Lypiatts
15 Lansdown Road
Cheltenham
Glos GL50 2JA
UK

Edward Elgar Publishing, Inc.
William Pratt House
9 Dewey Court
Northampton
Massachusetts 01060
USA

A catalogue record for this book
is available from the British Library

Library of Congress Control Number: 2012941548

This book is available electronically in the ElgarOnline.com
Social and Political Science Subject Collection, E-ISBN 978 1 78100 642 9

FSC
www.fsc.org
MIX
Paper from
responsible sources
FSC® C018575

ISBN 978 1 78100 641 2

Typeset by Servis Filmsetting Ltd, Stockport, Cheshire
Printed and bound by MPG PRINTGROUP, UK

Contents

Contributors

Akintola Akintoye, PhD in Construction Economics and Management (1991). Head of School of Built and Natural Environment, University of Central Lancashire, United Kingdom.

Lovleen Bhullar, LLM (Environmental Law) (2007), School of Oriental and African Studies, University of London, London, United Kingdom. Researcher, Environmental Law Research Society, New Delhi, India.

Eshien Chong, PhD in Economics (2006), University Paris Sud 11, France. Assistant Professor in Economics, Sorbonne Business School, University of Paris 1 Panthéon-Sorbonne, France.

Abu Naser Chowdhury, PhD Candidate (2012), Nanyang Technological University, Singapore. Sub-Divisional Engineer, Public Works Department (PWD), Ministry of Housing and Public Works, Bangladesh.

Camelia Dewan, MSc. Development Studies (2010), London School of Economics and Political Science, London, United Kingdom. Research Manager/Consultant, International Water Management Institute, Sweden.

Vanessa Garcia, Master of Public Policy (2012), Lee Kuan Yew School of Public Policy, National University of Singapore.

Belen Olmos Giupponi, PhD in Law (2004), University Carlos III, Madrid, Spain. Assistant Professor, International Law and International Relations, King Juan Carlos University, Madrid, Spain.

Asanga Gunawansa, PhD in Law (2007), National University of Singapore; LLM (1997), University of Warwick, United Kingdom. Attorney-at-Law, Supreme Court of Sri Lanka; International Research Associate, Institute of Water Policy, Lee Kuan Yew School of Public Policy, National University of Singapore.

Sonia Ferdous Hoque, MSc. Environmental Management (2012), National University of Singapore, Singapore. Research Associate, Institute of Water Policy, Lee Kuan Yew School of Public Policy, National University of Singapore.

Arani Kajenthira, D.Phil Engineering Science (2010), University of Oxford, United Kingdom. Research Fellow, Belfer Center for Science and International Affairs, Harvard Kennedy School of Government, USA.

Ashwin Mahalingam, PhD in Civil Engineering (Construction Engineering and Management) (2006), Stanford University, USA. Assistant Professor, Indian Institute of Technology Madras, India.

Claude Ménard, PhD in Economics (1980), University of Paris (Panthéon-Sorbonne), France. Full Professor of Economics, University of Paris (Panthéon-Sorbonne).

Sharmila Murthy, Juris Doctorate (2003), Harvard Law School; Masters in Public Administration (2003), Harvard Kennedy School of Government, USA. Fellow, Carr Center for Human Rights Policy, Harvard Kennedy School of Government.

Michael Paddon, MA Economics (1973), University of Cambridge, United Kingdom; MA Sociology (1973), University of Essex, United Kingdom. Associate Professor, Institute for Sustainable Futures, University of Technology Sydney, Australia.

Edouard Pérard, PhD in Economics (2009), Institut d'Etudes Politiques de Paris, France. Regional Coordinator for Asia and Global Water Coordinator, Public–Private Infrastructure Advisory Facility, The World Bank.

Suresh Renukappa, PhD (2008), Glasgow Caledonian University, United Kingdom. Research Associate, Centre for Sustainable Development, School of Built and Natural Environment, University of Central Lancashire, United Kingdom.

Zheng Sha, PhD Candidate (2012), Nanyang Technological University, Singapore. Senior Associate, Infrastructure Government & Utilities, PricewaterhouseCoopers, Corporate Finance Pte Ltd, Singapore.

Robert L.K. Tiong, PhD (1994), Nanyang Technological University, Singapore. Associate Professor, School of Civil and Environmental Engineering, Nanyang Technological University.

Håkan Tropp, PhD in Water and Environmental Studies (1999), Linköping University, Sweden. Director, Stockholm International Water Institute, Sweden.

Alex Wong, Bachelor of Laws (1996), University of London, United Kingdom. Partner, Hogan Lovells Lee & Lee, Singapore.

1. Introduction

Asanga Gunawansa, Lovleen Bhullar and Sonia Ferdous Hoque

1.1 WATER SUPPLY CHALLENGES – A CRISIS OF GOVERNANCE

Water is essential to life and while access to improved water supply and sanitation services are vital to human health and well-being, it also plays a crucial role in determining the development of a country or region. According to the World Health Organization and United Nations Children's Fund's report on 'Progress on Sanitation and Drinking Water: 2010 Update', in 2010 about 87 percent of the global population (about 5.9 billion people) and 84 percent of the people in developing regions had access to piped water supply through house connections or to an improved water source, including standpipes, water kiosks, protected springs and protected wells. However, about 14 percent (884 million people) did not have access to an improved water source and had to use unprotected wells or springs, canals, lakes or rivers for their water needs.

Over the years, access to water as a basic human right has been mentioned in a number of international documents of varied normative value. It was first implicitly mentioned in the Universal Declaration of Human Rights (United Nations, 1948), which stated in Article 25 that 'Everyone has the right to a standard of living adequate for the health and well-being of himself and of his family, including food, clothing, housing and medical care'. Although not legally binding, it has been considered as customary international law and has been reasserted in many international agreements. Article 12(1) of the 1966 International Covenant on Economic, Social and Cultural Rights (ICESCR) stated that 'The States Parties to the present Covenant recognize the right of everyone to the enjoyment of the highest attainable standard of physical and mental health' and Article 6(1) of the 1966 International Covenant on Civil and Political Rights (ICCPR) mentioned that 'Every human being has the inherent right to life. This right shall be protected by law. No one shall be arbitrarily deprived of his life'. These two legally binding covenants have once again mentioned water implicitly, as it is assumed that water is a fundamental necessity for enjoying good health.

Beginning in the 1970s, a series of international environmental or water conferences have taken on the issue of access to basic resource needs and marked the transition towards a more explicit recognition of the right to water. Resolution II on 'Community Water Supply' of the 1977 United Nations Water Conference, held in Mar del Plata in Argentina, recognized water as a right for the first time, declaring that 'All peoples, whatever their stage of development and social and economic conditions, have the right to have access to drinking water in quantities and of a quality equal to their basic needs'. The 1979 Convention on the Elimination of Discrimination against Women and the 1989 Convention on the Rights of the Child serve as two binding instruments that oblige national governments, as signatories to the conventions, to protect and realize the rights of women and children to water.

The challenges of water provision and pricing started to gain increased attention during the 1990s, when the focus of the international community shifted to sustainable development with greater emphasis on the economic and environmental aspects of growth. In 1992, the International Conference on Water and Sustainable Development, held in Dublin, Ireland, recognized the increasing scarcity of water and stated in Principle 4 that: 'Water has an economic value in all its competing uses and should be recognized as an economic good. Within this principle, it is vital to recognize first the basic right of all human beings to have access to clean water and sanitation at an affordable price' (Dublin Statement, 1992).

In the same year, Chapter 18 of Agenda 21, an outcome of the United Nations Conference on Environment and Development (UNCED), held in Rio de Janeiro, endorsed the resolution of the Mar del Plata Water Conference and called this 'the commonly agreed premise'. These statements, followed by others, gradually affirmed the role of national governments and international organizations, to take responsibility to fulfill the stated 'right to water' (United Nations, 1992). The 1999 UN General Assembly Resolution (A/Res/54/175) 'The Right to Development' stated in Article 12 that 'in the full realization of the right to development, inter alia: (a) The rights to food and clean water are fundamental human rights and their promotion constitutes a moral imperative both for national Governments and for the international community'. In 2002, the General Comment 15 of ICESCR provided guidelines for the interpretation of the right to water, framing it within two articles, Article 11, the right to an adequate standard of living, and Article 12, the right to the highest attainable standard of health. In paragraph 1 it stated that 'The human right to water is indispensable for leading a life in human dignity. It is a pre-requisite for the realization of other human rights', and in paragraph 2 it mentioned that 'The human right to water entitles everyone to sufficient, safe, acceptable, physically accessible and affordable water for personal and domestic uses.'

The various international declarations made after the Dublin Statement and UNCED confirm that water has an economic value and when pricing water, factors such as public right to access to water, the need to maintain affordability, and reduction of wastage, need to be taken into consideration. For example, the Ministerial Declaration of the 2nd World Water Forum organized by the World Water Council, held in The Hague in 2000, declared *inter alia* that:

> To manage water in a way that reflects its economic, social, environmental and cultural values for all its uses, and to move towards pricing water services to reflect the cost of their provision. This approach should take account of the need for equity and the basic needs of the poor and the vulnerable. (World Water Council, 2000)

The Ministerial Declaration of the 3rd World Water Forum, held in Kyoto in 2003, declared *inter alia* that:

> Funds should be raised by adopting cost recovery approaches which suit local climatic, environmental and social conditions and the 'polluter-pays' principle, with due consideration to the poor. All sources of financing, both public and private, national and international, must be mobilized and used in the most efficient and effective way. (World Water Council, 2003)

All the conventions, conferences, declarations and action plans discussed so far aimed to address the problems ensuring safe water supply to all. The most notable of these was the United Nations Millennium Declaration 2000, which addressed eight Millennium Development Goals (MDGs) that all 193 United Nations member states and at least 23 international organizations have agreed to achieve by the year 2015. Target 7C under Goal 7 aims to halve, by 2015, the proportion of the population without sustainable access to safe drinking water and basic sanitation. Moreover, to emphasize the urgency of dealing with water security issues, the UN General Assembly issued a number of resolutions. In 1992, it declared March 22 of each year as 'World Water Day', in 2000 it proclaimed the year 2003 as the 'International Year of Freshwater' and in 2003 it decided to mark 2005–15 as the 'International Decade for Action – Water for Life'. The goal of the decade has been to provide greater focus on water-related programs and projects to achieve the internationally agreed, water-related goals. Finally, in July 2010, the UN General Assembly adopted a resolution that 'recognized the right to safe and clean drinking water and sanitation as a human right that is essential for the full enjoyment of life and all human rights' (A/RES/64/292). Following this resolution, in September 2010, the Human Rights Council took a further step by specifying that the right to water and sanitation is legally binding.

The progress towards meeting the MDG target in developing countries in terms of access to safe drinking water is uneven. Most of the improvements are in China and India, while sub-Saharan Africa is lagging behind, with only 60 percent of the population using improved sources of drinking water. There is also wide disparity between urban and rural water supply, and currently while 94 percent of the urban population of developing countries uses improved sources of water, it is only 76 percent of the rural population. Since 1990, 1051 million people in urban areas have gained access to improved drinking water sources. However, during the same period, the world's urban population rose by 1090 million people; resulting in a net 39 million increase in the urban population that rely on unimproved drinking water sources, most of whom live in the cities and towns of developing countries (WHO/UNICEF, 2010).

In the document 'Towards Water Security: A Framework for Action', prepared by the Global Water Partnership (GWP) for presentation at the Second World Water Forum, it was stated:

> The water crisis is mainly a crisis of governance. Working towards effective governance requires an enabling environment and appropriate institutional structures that allow stakeholders to work together for the effective water management. Financial practices must be realigned to support the sustainable use of water resources (GWP, 2000).

The term 'water governance' has no universally accepted definition. According to several agencies of the United Nations (UNDESA/UNDP/UNECE, 2003) and the Global Water Partnership (Rogers and Hall, 2003), 'water governance' refers to 'the range of political, social, economic and administrative systems that are in place to develop and manage water resources, and the delivery of water services, at different levels of society'. This definition appears broad enough to encompass the various relevant systems, the role of such systems and the end-users. However, it does not adequately address issues such as governmental obligations to the people, setting priorities for water policy, identifying minimum water requirements and allocation of water, and sourcing the necessary financial and technical inputs for development and management of water infrastructure facilities.

Given the modern forms of government, increasing population and the need for decentralized systems of administration, public participation and decentralization are integral elements of governance of public resources such as water. Bakker (2003) defines 'water governance' to include 'the range of political, organizational and administrative processes through which communities articulate their interests, their input is absorbed, decisions are made and implemented, and decision makers are held accountable in the development and management of water resources and delivery of water services'. According to Lewis (2004), 'adequate governance can decrease political and social risks, as well as institutional failures and rigidity. It can also improve capacities to cope with shared problems'. In these circumstances, water governance is no longer considered from the narrow perspective of water resource management through the political and administrative instruments of a government.

Recent research on water governance has observed a direct correlation between the lack of access to water services and poor governance structures, especially in developing countries where water is a scarce resource (Lewis, 2004; Lenton et al., 2005; WSP, 2007). Camdessus and Winpenny (2003) argue that even if more financial assistance is given to developing countries, it is unlikely that the funding will translate into improved service outcomes unless wider governance issues are also simultaneously addressed. In many developing countries, the governance of the water sector as a whole is in a state of confusion and dysfunction with little responsiveness or accountability to citizens (Tropp, 2005). The key reasons are lack of institutional clarity, uncoordinated approaches towards water management by different state entities, and lack of proper legal regulatory framework to provide a sound foundation for water governance. According to the UN (2006), the key reasons for poor water governance in most countries are fragmented institutional structures, lack of clarity of roles and responsibilities, inefficient resource allocation, ineffective financial management, lack of management know-how, weak accountability of policymakers and implementing agencies, unclear or non-existent regulatory environments, and unpredictability in the investment climate that discourages private sector investment.

Hence, capturing all the different dimensions of the concept, the term 'water governance' refers to the power and authority of public sector entities as well as cooperation and collaboration between public sector, private sector and the community, and the various legal, institutional, administrative and technical mechanisms established to facilitate water resource management, delivery of water to people and sustainable water use, which together encompass the water governance architecture. Successful and sustainable water governance architectures require the participation of the public sector, the private sector and the community. This does not necessarily mean that only market-based mechanisms with private sector participation and competitive pricing policies will be successful. Water governance under strict public administration regimes can also succeed if the support of end-users as well as private sector participation in some form, for example, the provision of technology, is available.

This book is an outcome of the research project titled 'Water Governance: An Evaluation of Alternative Architectures' conducted by the editors for the Institute of Water Policy at the Lee Kuan Yew School of Public Policy in the National University of Singapore, in collaboration with the Public Utilities Board, Singapore. It compiles the work of academics, researchers and experts, working in various sectors of water management, from different institutions and universities across the world. Other than this introductory chapter and the concluding chapter, the book consists of 13 chapters revolving around the basic theme of urban water governance, where the authors discuss the various types of water governance structures being used across the world as well as identifying trends, challenges, gaps and opportunities, and the recent shifts in management and perceptions. Many chapters are specifically dedicated to outlining and analyzing the urban water supply scenarios in selected countries, with specific focus on legal, policy and institutional frameworks governing water supply provision. The book comprises separate chapters on water governance in the United Kingdom, France, Spain, India, China and Australia as well as the Middle East and North African region as a whole. Section 1.2 of this chapter provides a comprehensive overview of the various water governance mechanisms in place and thus provides an understanding of the basic terms and types of contracts that have been dealt with in this book. A detailed discussion on the evolution of public and private water governance architectures over the last two centuries is provided in section 1.3. Section 1.4 provides an outline of each of the chapters, including their main focus and findings.

1.2 OVERVIEW OF ALTERNATIVE WATER GOVERNANCE ARCHITECTURES

Most urban water supply services around the world are provided by public entities, although there is no agreement as to the exact division of public governed services, public–private partnerships (PPP) and totally privatized entities. Approximately 90 percent of the global water supply is in the hands of public sector entities and the balance is provided by private or mixed public–private companies under various contractual and procurement arrangements (UNICEF, 2010; Wolff and Palaniappan, 2004; Kishimoto, 2009; Marin, 2009). According to the *Pinsent Mason's*

Table 1.1 Governance architectures for urban water supply

Governance Mechanism	Description	Examples
Public	Public water utilities	Porto Alegre (Brazil)
	Public–public partnerships	UN's Water Operators' Partnership and Waternet (Amsterdam, The Netherlands)
	Small-scale water cooperatives	Gramastetten (Austria)
	Corporatized water utilities	Singapore, Saudi Arabia and Egypt
	Autonomization	Uganda and Zambia
Public–private	Public–private partnership	Several countries, such as The Philippines
Private	Privatization	United Kingdom (nationwide), Chile, and the Czech Republic

Water Yearbook 2011–2012 (Pinsent Masons LLP, 2011), in 2011 about 909 million people or about 13 percent of the global population was provided with water supply and sanitation services either entirely by the private sector or by some form of PPP. As far as developing countries are concerned, according to the Third World Centre for Water Management (TWCWM), the private sector currently serves about 3–6 percent (estimates vary) of urban water consumers. This percentage is likely to increase in the coming years, but not rapidly. TWCWM predicts that under all foreseeable conditions, a vast majority of domestic consumers in developing countries are likely to continue to be served by publicly run water companies in the next decade.

In urban areas, the water supply service is provided by water utilities, which often differ from each other in terms of their geographical coverage relative to administrative boundaries, sectoral coverage, ownership structure and governance arrangements. Many water utilities provide services in a single city, town or municipality, while in some federal countries a single utility is often responsible for providing water supply services to an entire state. In case of some smaller developed countries, the entire country or at least all the large urban areas are provided with water and sanitation services by one service provider. Water supply providers can be public, private, mixed or cooperative and there can be many forms of governance arrangements for both public and private utilities. Table 1.1 summarizes the different forms of water governance architectures currently in place, with totally publicly owned and managed water utilities and totally privatized utilities at two ends of the continuum. Sections 1.2.1 to 1.2.7 below present a brief analysis of each of the seven alternative water governance architectures described in the table under three main governance mechanisms.

1.2.1 Public Water Utilities

Traditionally, the government has been responsible, partly or fully, for the provision of water and the development and management of related water infrastructure facilities. As noted above, even today, public entities continue to provide most of the water and wastewater services worldwide, although the number of end-users served by the private sector entities has increased.

Under the totally public managed water governance model, a formal public authority or authorities (local, regional or national) is responsible for the provision of water services (including infrastructure development and funding, operation of the supply system, regulating tariffs, billing and collection of tariffs (if they are raised), and system management and maintenance). The public sector also retains full ownership of the related water infrastructure. Where this model thrives or is strongly supported due to political reasons such as the fear of public criticism for removing heavy state subsidies for water, private sector involvement has been considered inappropriate given the public good and basic need (or merit or beneficial good) characteristics of water supply, and the inherent monopolistic tendency of water systems due to economies of scale in service provision (Johnstone and Wood, 2001).

There are successful public water utilities. In Porto Alegre, the capital of the south-most state of Brazil, Rop Grande do Sul, the Municipal Department of Water and Sewerage (DMAE), a public-owned water utility, has become a model for successful public management of water for those who are opposed to private participation in water services (Maltz, 2005). The Transnational Institute (The Netherlands), Our Water Commons (United States), Food and Water Watch (United States), Blue Planet Project (Canada), Centre for Human Rights and Law (India), and the Red Vida (Latin American water justice network) are some organizations that strongly advocate retention of public management of water. These entities, and scholars such as David Hall of the Public Services International Research Unit (PSIRU) of the University of Greenwich, United Kingdom, advocate public sector control of water on several grounds, the main ones being that water is a public right, public sector management is capable of preventing exploitation of consumers by profit-driven private entities, and the wealth of experience and knowledge that exists in public sector given that the vast majority of water operators in the world are in the public sector.

Despite the strong case made for the continuation and/or reclaiming of public management of water, there are many cases where public management has been unsuccessful. Urban water supply in particular requires the adoption of expensive engineering solutions due to reasons such as high population, high concentration of industrial and business entities, and lack of clean and adequate water resources within close proximity to urban settlements. Sourcing sufficient financial and technological resources for developing and maintaining urban water services thus becomes a major challenge for those in charge of urban governance. Additionally, the demand for large government subsidies renders investments in the development and management of urban water facilities unprofitable public burdens. Furthermore, if the public entity integrates the responsibility for water supply into its administration of an urban area, city, town or municipality there is also the risk of water tariff revenue being diverted for other purposes. Other concerns relate to high levels of inefficiency, low levels of coverage (and equity concerns because richer neighborhoods are the primary recipients of subsidized water services), faulty incentive structures, politicization of personal appointments and management, and other bureaucratic weaknesses. These risks are particularly high in developing countries. As a result, the effectiveness of public management has been questioned.

Based on lessons learnt from successful public utilities across the world, Chapter 15 in this volume provides a detailed analysis of the factors that determine the ability of a public utility to excel.

1.2.2 Public–Public Partnerships

A public–public partnership is a collaboration between two or more public authorities or organizations (international or national), based on solidarity, to improve the capacity and effectiveness of one partner in providing public water services, which encourages sharing of experiences and knowledge, cross-subsidizing, and twinning and other not-for-profit cooperation (Hachfeld et al., 2009). Public–public partnerships have often been described as 'a peer relationship forged around common values and objectives, which exclude profit-seeking' (Hall, 2010). Their key functions include training and developing human resources, providing technical support on a wide range of issues, improving efficiency and building institutional capacity, financing water services and improving community participation (Hall et al., 2009).

According to Hall et al. (2009), public–public partnerships avoid the risks of public–private partnerships, such as transaction costs, contract failure, renegotiation, and the complexities of regulation, commercial opportunism, monopoly pricing, commercial secrecy, currency risk and lack of public legitimacy. Thus, financial and logistical support comes mainly from mutual arrangements between public entities with the same country or between public entities of different countries. For example, in 2010, the European Union earmarked funding, under the ACP-EU Water Facility, to support public–public partnerships between public water operators in the European Union, and Africa, Caribbean and the Pacific. The water partnership aims to transfer 'expertise, knowledge and learning' from water and sanitation utilities, local authorities and other water sector organizations (PSIRU, 2010).

Hall et al. (2009) documented more than 130 public–public partnerships in around 70 countries. Their success is demonstrated by Hall (2010) pointing to the fact that over 70 countries have implemented public–public partnerships in the water sector, compared to the 44 countries which have introduced private sector participation in the water sector. Since the 1980s, K-Water, a state-owned bulk water supplier in South Korea, has assisted local governments and municipalities with capacity building and technical consultancy for their water supply systems. After the termination of the Azurix-led concession contract in Greater Buenos Aires, Argentina, the provincial government set up Aguas Bonaerense SA (ABSA) as a public sector water company, which is co-owned and operated by the workers' cooperative '5 de setiembre S.A.'. In Peru, the social movement organization FREDEAJUN (Frente de Defensa del Agua de la Region Junín) and one of its members, the local sector trade union SUTAPAH (Sindicato Único de Trabajadores de Agua Potable de Huancayo), established a public–public partnership between SEDAM, the public utility for Huancayo in the department of Junín, and the Argentinian public water operator ABSA. In The Netherlands, Waternet has been active in twinning and other international not-for-profit cooperation, through WereldWaternet. The company currently has projects aimed at sharing expertise and experience to improve drinking water supplies in Suriname, Egypt, Indonesia, Palestine and several other countries (Hachfeld et al., 2009).

However, the number of countries which have implemented public–public partnerships alone is insufficient evidence to demonstrate its success. This is because they are soft, non-binding understandings between public entities to assist public management of water services, thus vulnerable to changing government policy and economic situations. In other words, their greatest weakness lies in the fact that they are non-contractual and thus non-binding. In the circumstances, unlike in the case of PPPs, the parties involved in public–public partnerships would lack the legitimate right to demand the performance of obligations during the agreed period of partnership and press for legal remedies in the event of failure to do so by either party.

It should be noted, however, that the creation of public–public partnerships does not necessarily have to rule out private sector participation. Concerned about the risk of missing the water and sanitation targets in the MDGs, the Water Operators' Partnership was established in November 2010 under the mandate of UN-HABITAT to enable water operators (including private operators) to systematically communicate amongst themselves and to share their experiences and learn from each other on a not-for-profit basis. It is based on the premise that establishing direct and effective partnerships and networking among operators at global, regional, subregional and national levels and facilitating a process of 'learning by doing' amongst them is an urgent necessity for better equipping the key players in the water sector to meet the challenges of attaining the MDGs on water supply (UN-HABITAT, 2008).

For example, Empresa de Servicios Sanitarios del Paraguay (ESSAP) – Paraguay's state-owned water operator – signed an agreement with Brazilian operator Companhia de Saneamento de Minas Gerais (COPASA) in September 2007 for the provision of technical assistance on a not-for-profit basis, which was aimed at helping ESSAP reduce water leakage in the Paraguayan capital Asunción and its metropolitan area, to increase ESSAP's water provision capacity. However, water operators' partnerships may negate the goal of public–public partnerships to protect against privatization of public services and dilute the initially progressive alternative concept.

1.2.3 Small-scale Water Co-operatives

Small-scale water co-operatives are special forms of public water companies, which are licensed by the government and are allotted a limit as to the amount of water they are allowed to extract and supply to the community. The cost and investment of running small-scale water co-operatives is low (but the unit cost is high). In some cases, the municipalities may support their establishment and provide operational assistance. Rural areas in Austria and Finland are serviced by a large number of small-scale water cooperatives. In the federal district of Oberösterreich (Austria), where the government supports decentralized water services, more than 10 percent of the population is served by around 1500 co-operatives. One example is the Wassergenossenschaft Gramastetten, which was founded in 1947 and provides drinking water to about 2000 people. It was one of the largest water cooperatives with 569 members in 2008. Membership is connected to the ownership of real estate and apartments and a connection fee has to be paid to access the water network of the co-operative. All relevant information is available to everyone and important decisions are taken by the general assembly of all members. The administrative work and most of the technical work are done on a voluntary basis. The regional association of water co-operatives provides expertise, quality control, and training. They charge low prices for water, follow the principle of strict non-profit management, use local water sources and their administrative costs are low due to voluntary work by the members. However, while they have been successful on a small scale, it may not be possible to replicate this model on a larger scale and therefore, water co-operatives have limited application for urban water supply (Hachfeld et al., 2009).

1.2.4 Corporatized Public Water Utilities

Corporatization refers to the transformation of state assets or agencies into state-owned corporations in order to introduce corporate management techniques to their administration. It is one form of new public management initiatives implemented in some countries to improve the efficiency of water governance by transforming a public water agency or utility into an independent corporation. The key aim of corporatization is to mimic the structure and efficiency of private corporations while assuring that social objectives are still emphasized through public ownership (Harding and Preker, 2000). It is, therefore, a suitable model for increasing efficiency of governance in public water utilities and a viable alternative to total privatization, which often attracts public criticism and political opposition.

Under corporatization, managers are given virtually complete control over all inputs and issues related to production of services and thus are not subject to day-to-day management controls by the state. Efficiency is achieved by accountability over a range of service-related and economic targets, which in effect mimic the effective governance structures that are associated with private companies. However, the risk of political interference in tariffs and employment, as well as incentive problems, cannot be ruled out.

Several African countries have implemented a hybrid solution of public ownership and private sector efficiency by transforming the public entities in charge of water provision into state-owned corporate business undertakings. A good example of this model can be found in Burkina Faso, formally known as the Republic of Upper Volta, a small landlocked country in West Africa, which is considered to be the third poorest country in Africa. After the country was established in 1958 as a self-governing French colony, for almost two decades, water services were provided by a private operator, which focused on a few rich neighborhoods of the capital, Ouagadougou (Marin et al., 2010). The contract was terminated in 1977 and the responsibility for water services was transferred to municipalities until 1985, when the Office national de l'Eau et de l'Assainissement (ONEA) was established as the new public sector entity responsible for provision of urban water supply services (Marin et al., 2010). Early performance of ONEA was typical of inefficient public enterprises in charge of public utilities in most developing nations. As a result, by the early 1990s, when the subject of water governance was evaluated, it was evident that the urban water sector had made little progress as more than one-third of the urban population lacked access to piped water and household connection coverage stood at only 24 percent in urban areas (Marin et al., 2010).

In order to introduce reforms to improve water management, ONEA was transformed from a quasi-public agency into a 100 percent government-owned limited liability company in 1994, converting it into a state-owned business entity governed by private law (Marin et al., 2010). In the year 2001, ONEA entered into an innovative performance-based service contract with Veolia, to provide more efficient services to the urban households (Marin, 2009). ONEA's management now operates based on an arm's-length relationship with the government, which enables it to appoint, promote and terminate its employees based on performance. Further, it can disconnect water services for non-payment of bills without being hindered by typical restraints on public sector-run utility providers such as the lack of legal authority to terminate water connections as a result of applicable laws on public rights. All these changes have helped ONEA to become one of the few well-managed public water utilities in sub-Saharan Africa. In December 2008, ONEA became the first public water utility in sub-Saharan Africa to be IS0 9001 certified (Jagannathan et al., 2009).

Public water supply institutions in Arab countries have also undergone various reforms, shifting more towards client-oriented service provision with improved institutional capacity to pursue the principles of efficiency, transparency, accountability and equity in delivering services to the people. According to the World Bank (2007), these transformations have been designed to address the problems afflicting these public-run water utilities, 'such as unclear lines of responsibility for operations, low tariffs, difficulties retaining qualified personnel, and political interference in staffing policies and other aspects of operations'. A good example of such reforms can be seen in Egypt, where the government has separated service provision from regulation in the water supply and sanitation sector. The holding company (HC), established in 2004, now has autonomous authority and is managed by a board consisting of a wide range of stakeholders. It operates at the local level through companies responsible for service provision and day-to-day operation and maintenance of the networks. The work of these companies is monitored monthly against a set of performance indicators to ensure that they strive to achieve excellence and efficiency. Along with the HC, Egypt has also established a Water Regulatory Agency to provide economic regulation for the sector, which links the government, society, and the HC to ensure the effective implementation of national policies and regulations concerning water (Jagannathan et al., 2009).

Other examples of corporatized public water utilities include Aqaba Water Company (Jordan), AQUA (Poland), COPASA (Brazil), Johannesburg Water (South Africa), National Water and Sewerage Company (Uganda), Public Utilities Board (PUB) (Singapore) and Sydney Water (Australia). PUB in Singapore, which has successfully managed the water services in the city-state with approximately 4.5 million people but with scarce freshwater resources, is considered a role model for sustainable water management by the international community. In Saudi Arabia, all key sector functions were formerly the direct responsibility of the Ministry of Water and Electricity. There was no separation between institutions in charge of policy and regulation and service provision. In 2008, the National Water Company was established to take over service provision, and it has contracted private companies to operate the water infrastructure in large cities (Riyadh, Jeddah, Mecca and Taif) under management contracts.

It should be noted that there is also a growing interest in India towards corporatization of public water entities. The Center of Excellence for Change, a think tank established as a result of collaborative effort of officials across hierarchies from the drinking water utilities, water resource organizations, agricultural engineering and livelihood organizations, and academia across India, is currently spearheading the efforts to promote reforms in the public sector water governance through 'change management', a structured approach to transitioning public organizations from the current state to a desired future state by empowering employees to accept and embrace changes in their current approaches to management. In other words, this is a process by which public entities plan to be as efficient as private sector entities.

1.2.5 Autonomization

Autonomization refers to the empowering of local governments and other public sector water operators' statutory autonomy to function independently from strict public sector control with the objective of allowing them more autonomy and flexibility in managing some services, as well as to raise resources through tariffs and fares (Kuhlmann and Fedele, 2010). Unlike in the case of corporatization, the intention is not to transform a typical public sector entity into a government-owned business undertaking that operates almost like a private sector corporation, but to increase the level of autonomy of the public sector water operators by clarifying and strengthening the legal framework within which they function.

Autonomized public water utilities operate according to institutional arrangements and management practices associated with the private sector. This model was adopted by the Dar es Salaam Water and Sewerage Corporation in Tanzania following the termination of a lease contract. Namibia, Uganda and Zambia have adopted the model without any previous experience of private sector participation in urban water supply. These reforms have been promoted by sector professionals, donors and lending agencies in an environment marked by the declining interest of the large international water companies to venture into these countries. Table 1.2, which is based on Schwartz (2008), evaluates the characteristics of the model in Uganda and Zambia.

The prerequisites for success of the autonomized model include: leadership of top management, strong institutional culture, degree of professionalism of the staff, and the level of support from the Ministries and the international donors and lending agencies. However, the model has been criticized for reshaping public water utilities into profit seeking entities. The success of the model is also heavily dependent on government grants and donor funds, which may be withdrawn or reduced, and therefore the long-term sustainability of the model is questionable (Schwartz, 2008).

1.2.6 Public–Private Partnerships

Until about the mid-1980s, the provision of water-related services to the people remained strictly within the domain of the public sector because of the public good aspect and the natural monopoly inherent in water supply infrastructure. There was a change in water governance when countries began to feel the need to overcome the inefficiency of public management and the inadequacy of public funds and technology to develop and maintain facilities to cater to the growing demand for new and efficient services. Various forms of private sector participation in the development, maintenance and operation of water services were pursued as a solution. The reluctance of public sector entities to totally relinquish control over a sector in which they had absolute monopoly, and the public and political opposition in many countries to total privatization, led to the development

Table 1.2 Characteristics of autonomization in Uganda and Zambia

Characteristics	Uganda	Zambia
Increase in autonomy	The legal framework under which the National Water and Sewerage Corporation (NWSC) operates was strengthened. The NWSC was reorganized into an autonomous corporate body with institutional and financial autonomy.	A new institutional framework for the water sector was adopted, which included the concept of commercially operating public utilities.
Separation of service provision and regulatory tasks	The Board of Directors of the NWSC holds the management accountable using the performance targets set in the performance contracts between the NWSC and the Government of Uganda.	The National Water Supply and Sanitation Council (NWASCO), an independent regulatory agency, is responsible for the regulation of all water utilities.
Creation of quasi-competition	Internally Delegated Area Management Contracts (IDMACs) between the NWSC Head Office and the Area service providers reward good performance.	NWASCO facilitates a benchmarking exercise.
Increase tariffs to cost recovery levels and increase customer orientation	Customer care and service training; Customer charter; Annual customer surveys; Strategic alliance meetings (involving customers).	Consumer water watch groups
Increase accountability for results	Performance contracts; IDMACs	NWASCO develops guidelines for service levels and is authorized to enforce them. It also implements a Performance Incentive Scheme.

of public–private partnership (PPP), an arrangement under which public sector entities could continue to retain the permanent ownership of facilities but share the financing, development, maintenance and revenue responsibilities with the private sector.

The term 'public–private partnership' is used to describe a variety of contractual arrangements between public sector and private sector entities to develop and/or deliver public services in partnership, a model that is widely used for developing infrastructure facilities, including water. PPPs are based on the idea that the private sector is better positioned to generate the capital investment required to undertake network rehabilitation or maintenance, and expansion. The private sector's potential for increased efficiency is also emphasized. In practice, however, there may be other reasons for the introduction of PPPs, including loan conditionalities imposed by international development banks and a government's ability to dismiss public opinion, such as in the case of a country under military rule.

However, from a legal and project procurement perspective, a public–private partnership is seen as an altogether different concept from outright privatization, discussed below. In fact, it provides a structure which exploits some of the lessons countries have learned from privatization. Compared to total privatization, the slower, balanced process of PPP is felt to enhance government value and provide long-term continuing government participation in functions of public interest. Furthermore, PPP provides the flexibility that governments lack in privatization as it allows them to engage the private sector to develop and operate facilities whilst having some control in projects as partners. Moreover, if structured properly, PPPs can create a fair distribution of wealth, avoiding the financial windfalls often criticised in privatization (Lovells Lee & Lee, 2008).

There exists a spectrum of PPP models for urban water supply depending on several factors, including the distribution of decision and property rights and risks and incentives between the public and private entities (Finger and Allouche, 2001).

Service contract

A private entity provides specific services, such as leak detection, meter reading, billing or collecting invoices, and water quality measurements, for a short time period. The fees are fixed per unit of work. The private entity is required to make very limited capital investment, and these are short-term contracts. In this situation, the private sector bears very little risk and there is very little uncertainty around the expected outputs (Kauffmann, 2007). This form of PPP allocates the least responsibility to the private operator. The government retains ownership, control and responsibility (and risk).

Management contract

A private operator manages and maintains the water facility for the contract period without any investment obligations. A management contract can be used to bring in new management systems, organizations and skills, or as a preliminary step to restructure a dilapidated utility before a concession. The government compensates the private operator (costs-plus-fee). The government retains most of the operational and commercial risks, though some risk-sharing may be built into the contract using performance bonuses or contingent fees. The success of this model largely depends on the manner in which the commercial risk of the operation is allocated. For example, where the management fee to the private operator is fixed at the outset, the commercial risks of the operation are borne entirely by the public sector. However, if the management fee is linked to the performance of the managed utility, the private operator bears the commercial risks relating to operations.

Lease contract

The government leases the right to operate and maintain a water system, and to collect user charges to a private operator, and the latter is compensated with an agreed portion of the revenues. The private operator takes on the operational risk but the public authority retains ownership and responsibility for system finance and expansion, and replacement of major assets, and it recovers part or all of its costs from its own share of user charges. The lease holder may also administer investment funds as an agent of the municipality, without taking related risks. If the payments from users recovered by the private sector operator cover more than the operator's agreed remuneration, the operator is generally supposed to return the difference to the lessor (public entity) in order to cover the cost of the investments under the latter's responsibility (Kauffmann, 2007).

Greenfield contracts (BOT/BOOT/BOTT contracts)

A greenfield contract is similar to a lease except that the private operator is responsible for investment and network expansion as well. Under this approach, the public entity delegates the right to provide water services to the private sector, yet retains some control over the sector by incorporating terms and conditions, including the rights and obligations of the service provider and the term and scope of private sector operation, in a concession contract or license. These contracts often involve take-or-pay provisions, that is, revenue guarantees that subject governments to contingent liabilities. The terms BOT (Build–Operate–Transfer), BOOT (Build–Own–Operate–Transfer) and BOTT (Build–Operate–Train–Transfer) are project financing and procurement mechanisms often associated with greenfield contracts. Under these schemes, the private sector operator obtains the capital needed for construction, builds and operates the infrastructure for an agreed period of time (anywhere between 15 and 30 years) and then transfers ownership back to the relevant public entity. Under the BOTT model, the private sector has the additional specific responsibility of training local staff before handing over the project to the public entity at the conclusion of the concession period.

These contracts are also often called concession contracts. However, in some cases, a concession describes a contract where the public authority transfers ownership and control of the entire water system, which is already constructed, to a private operator for a given period. The private operator assumes responsibilities for operation and maintenance as well as investment and service obligations. The operator bills and retains user charges for the concession period and the government retains ownership of the assets.

Joint venture

The private company forms a legal entity with the public sector, and both parties share responsibilities and investment obligations. Thus the public sector entity can share ownership with private shareholders. The joint venture company itself may either own the assets or (most often) be given a franchise by the local government. Compared to other PPP models, joint ventures are seen as project development and operation vehicles in which both the public and private sectors are expected to play an active joint role in project management.

Table 1.3 describes the allocation of key responsibilities between public and private entities, in the different models of PPPs for urban water supply, and their duration (World Bank, 1997).

Several examples and cases of PPPs in the water sector have been provided throughout this volume, and Chapter 15 contains a detailed analysis of the factors which determine the extent of success of a PPP.

1.2.7 Privatization

In a general sense, 'privatization' is said to have taken place when a specific function is handed over to the private sector and regulatory control remains a public sector responsibility. Therefore, in some cases, the terms 'privatization' and 'PPP' are used interchangeably. The term 'private sector participation' is also used sometimes instead of PPP but it does not highlight the important role of the public authorities.

In the strict sense, privatization implies a full divestiture or the sale of public assets to a private operator, and therefore the main property and

Table 1.3 Allocation of key responsibilities between the public and private entities in different types of PPPs

Type of PPP	Asset Ownership	Operation and Maintenance	Capital Investment	Commercial Risk	Duration
Service contract	Public	Shared	Public	Public	1–2 years
Management contract	Public	Private	Public	Public	3–5 years
Lease/ *affermage*	Public	Private	Public	Shared	8–15 years
Build–operate–transfer (BOT)	Shared	Private	Private	Private	20–30 years
Concession	Public	Private	Private	Private	25–30 years
Joint venture	Shared	Private	Private	Private	Indefinite

decision rights are transferred from the public entity. This is rare in the water supply sector (except in the United Kingdom and Chile). The privatization of water supply in the United Kingdom in 1989 is discussed in detail in Chapter 4.

In Chile, the privatization of the water companies was preceded by decades of institutional and policy reforms in the water sector. Under the military rule from 1973 to 1989, the existing economic paradigm changed from one where the state must ensure the optimum allocation of resources to one where market forces dictated efficient allocation. During this period, the national agency, Servicio Nacional de Obras Sanitarias (SENDOS), was responsible for providing water and sanitation services to most parts of Chile as well as for regulatory and supervisory functions. In 1981, the enactment of the Water Code separated water rights from land ownership to make the former fully tradeable, and the water market flourished as a result in large urban and agricultural areas. In 1990, the sector was transformed with the establishment of 13 regional water companies and the creation of a separate regulatory agency, Superintendency of Sanitary Services (SISS). The ownership of the regional water companies was transferred to Corporación de Fomento de la Producción de Chile (CORFO), a government organization established in 1939 to promote economic growth, and they were authorized to function as private enterprises with concessions to service particular communities and urban areas. An innovative model of tariff regulation was borrowed from the Chilean electricity and telecommunications sector; efficient cost levels were estimated for an imaginary model company and used as a benchmark to set tariffs for the utilities. Means-tested subsidies (that is, subsidies that are granted only to those that have demonstrably limited means) were also introduced at the same time to cushion the effect of the tariff increase on the poor (Hearne and Donoso, 2005; Bitran and Arellano, 2005).

Privatization of water companies in Chile was brought about in stages between 1998 and 2005 under democratically elected governments. During 1998–2001, five of the 13 companies, serving more than 75 percent of all users, including those in the three large urban centres (Santiago, Valparaíso and Concepción), were privatized by selling 51 percent of their shares to private stakeholders. After 2001, instead of selling the assets to private stakeholders, the government awarded concession contracts to the private companies for management of water services in the remaining seven regional companies. All seven companies that were privatized in this second way merged in 2005, assuming the name ESSAN. Unlike the massive failures of PPP projects in other Latin American countries, privatization has been quite successful in Chile because of the reforms preceding the privatization, which allowed the public companies to improve efficiency and financial sustainability before being handed over to the private sector.

In most other cases, however, total privatization of water has met with stiff resistance in both developed and developing countries. Privatization was seen as too blunt and abrupt a technique to involve the private sector in the provision of public services that were previously strictly within state control (Lovells Lee & Lee, 2008). The difficulty of selling the idea of privatization to the public who have got used to free or heavily subsidized services from the public sector is one of the key reasons for the failure of privatization in many countries. International financial institutions (IFIs), which used to support water privatization, have now adopted a more cautious approach.

1.3 EVOLUTION OF WATER GOVERNANCE ARCHITECTURES

An analysis of the historical and current patterns of financing and managing urban water supply services reveals that both public and private players have important roles to play. The major challenge is allocating these responsibilities in ways that best address the needs of the people in the local context (Gentry and Auyuan, 2000). While private sector participation in the water industry can be traced back to the late eighteenth century, there have been fluctuations in the level of involvement of private actors and these variations have largely been shaped by the international ideologies dominant during each period of time. Moreover, the nature of private sector involvement has evolved in terms of sharing of risks and responsibilities as determined by the different contractual arrangements and also in terms of market share in developed and developing countries. The following section outlines these changes in water governance over time with reference to major events in the water industry and the global development agenda. A timeline of the events is also summarized in Figure 1.1.

1.3.1 Late Eighteenth to Early Nineteenth Century: Emergence of Privatization in the North

In the eighteenth century, rapid urbanization in North American and European cities created demand for establishment of water networks, mainly driven by the need for fire-fighting and public health concerns (Prasad, 2007; Bakker, 2010, p. 83). In New York in 1795, poor quality of water sold by vendors and polluted public wells led to a major epidemic of yellow fever, which pressurized the city authorities to seek private sector investment for providing clean drinking water. However, poor performance and corruption by the private operator, the Manhattan Company, and a series of mishaps (large fire and cholera outbreaks) between 1828 and 1835, forced the government to take over the water supply system (Prasad, 2007). Private companies also supplied water to Boston, Philadelphia, New Orleans and other big cities in the United States until the mid-nineteenth century. However, most of them were eventually returned to municipal ownership as private operators, driven by the profit motive, were reluctant to increase coverage outside wealthier communities and also failed to provide water for cleaning streets, public fountains and fire hydrants (Hanson, 1959). Hence, compared to 93.8 percent of waterworks (one in 16 waterworks) being owned by the private sector in 1800 in the United States, the proportion of privately owned waterworks fell to about 30 percent (2950 out of 9850 waterworks) in 1924 (Masten, 2004; Baker, 1899).

Similarly, by the beginning of the nineteenth century, private water companies had already been serving the city of London for more than two centuries (Gentry and Auyuan, 2000). However, after the major cholera outbreak in 1840, these became regulated entities and, over time, the provision of water and sewerage services in the United Kingdom (UK) came to be dominated by the public sector. In 1861, the private sector served about 60 percent of the population in larger towns and the share decreased to 20 percent in 1881 and 10 percent in 1901 (Prasad, 2007). During the period from 1900 to 1974, municipalities were mostly in charge of water

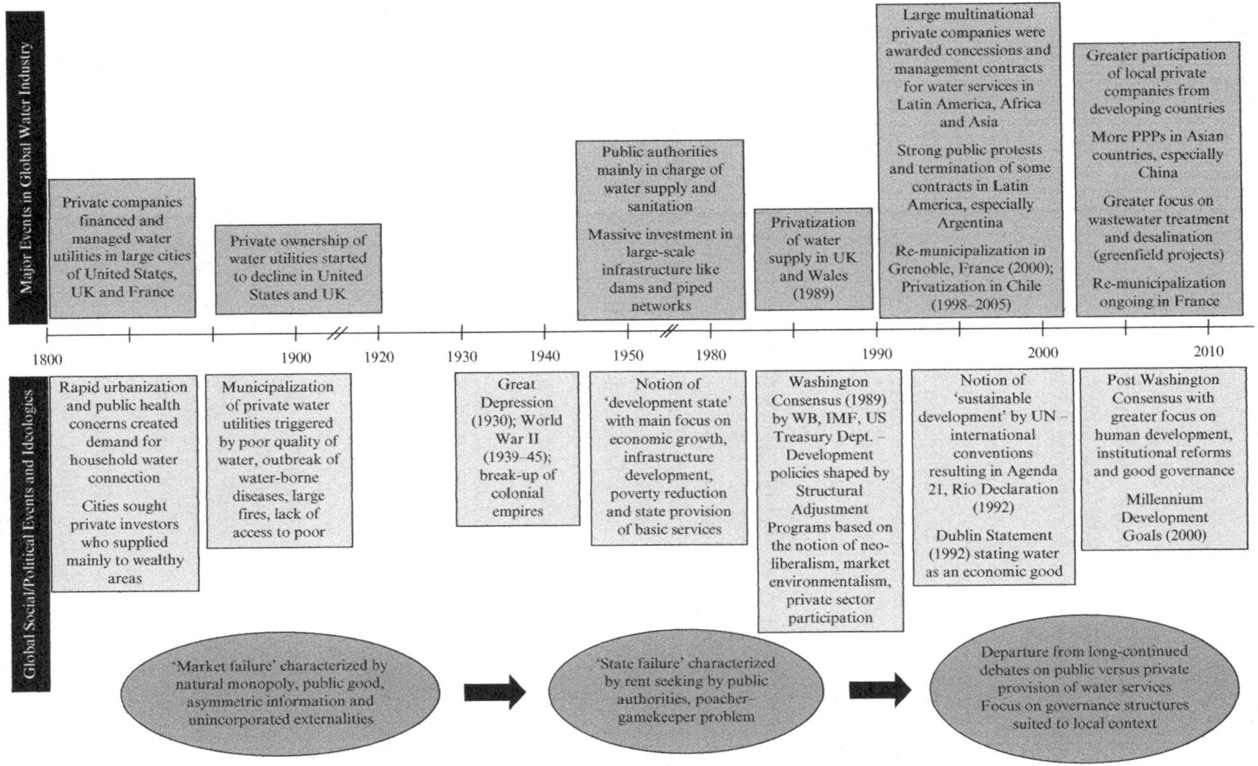

Figure 1.1 Timeline illustrating major events in the global water industry along with changing international perceptions

supply. There were ten government-owned regional water authorities supplying water and sewerage services, and 29 statutory water companies supplying water only. Following the introduction of the Water Act in 1989, the UK privatized its ten regional water and sewerage authorities by selling their assets and liabilities (OFWAT, 1993).

In France, private sector participation in the operations of water systems started in 1782 when a company founded by the brothers Perrier was granted a license to supply piped water in an area in Paris for 15 years (Roth, 1987). This was followed by the first municipal contract won by Generale des Eaux (currently Veolia) in 1853 under the reign of Napoleon III (*Financial Times*, 1999). Unlike the UK, which opted for complete divestiture, France used the 'delegated management' model, under which municipalities retain ownership of infrastructure and the right to impose tariffs, while private companies bid for long-term contracts to provide the services (*Financial Times*, 1999).

1.3.2 Post-World War II: Emergence of Development States

As discussed above, by the late nineteenth century and early twentieth century, many privately owned utilities in the United States and the UK had been municipalized as the governments were thought to be more judicious in providing water. The Great Depression in the 1930s, the end of World War II and the break-up of colonial empires pushed many governments in both developed and developing countries to play a more active role in providing utility services (Gunatilake and San Jose, 2008). The post-World War II period was characterized by the notion of the 'development state' where governments were considered to be prime movers of economic development. The central philosophy was that rapid economic growth can only be fostered by diverting factors of production from primary sectors typified by traditional technology, low productivity and decreasing returns to industrial sectors that would yield higher productivity and increasing returns (Adelman and Yeldan, 1999). Hence the 'big push' strategy, based on the assumption that developing countries could achieve rapid economic development by following the path taken by the West, was adopted. Under this model, development organizations such as the World Bank and governments focused on financing large-scale infrastructure to stimulate production (Bakker, 2010).

During this time the World Bank, as the largest multilateral lender, played a leadership role in shaping water policy and lending, and targeted investment towards large water dams that could support agricultural modernization, urbanization and industrialization. This move away from the private sector towards greater state involvement was justified by the supposed existence of 'market failures', such as natural monopoly and environmental externalities, along with the reluctance of the private sector to invest in these highly capital-intensive and less profitable projects. Moreover, it was expected that governments could use cross-subsidies to expand water services to the poorer customers (Bakker, 2010, p. 85).

1.3.3 Washington Consensus: Emergence of Neoliberalism

Although large sums of international aid and multilateral loans were granted since the 1950s, public authorities only focused on central affluent urban areas, leaving the peripheral and rural areas underserved (Winpenny, 2003). Moreover, it was realized that governments acting as both owners and regulators of water utilities were prone to the 'poacher–gamekeeper' problem, leading to deteriorating performance, corruption and inefficiency due to lack of incentives (Bakker, 2010, p. 43). These 'state failures' were accompanied by the public sector's indebtedness and lack of access to commercial finance.

By the end of the 1980s, a decade designated as the International Drinking Water and Sanitation Decade, international actors in the water sector started to reach a new consensus that private sector participation could address the problems of water supply and sanitation in developing countries (Budds and McGranahan, 2003). During this period, the earlier view of governments as the prime movers in economic development was replaced by the perception of governments as 'the problem' or cause of underdevelopment (Adelman and Yeldan, 1999). Using their leverage as creditors, the International Financial Institutions, especially institutions such as the World Bank and the Asian Development Bank, aggressively promoted neoliberal reforms to indebted governments of developing countries, through structural adjustment policies that advocated reduction of state spending. These reforms, termed the 'Washington Consensus' in 1989, emphasized that economic development should be undertaken by the private sector within free markets, with the state playing a facilitating and regulatory role without direct involvement (Budds and McGranahan, 2003). Moreover, the official development assistance (ODA) for water supply and sanitation started to decline in the 1990s, preceded by three decades of steady increase (Bakker, 2010, p. 73).

1.3.4 Post-Washington Consensus: Emergence of Sustainable Development

Meanwhile, in the 1980s, the environmental myopia of the mainstream multilateral and bilateral institutions met fierce criticism, which led to the 'greening' of these organizations. The response to these criticisms and the resulting changes to the discourse and priorities of multilateral agencies are discussed in Chapter 13 in this volume. As mentioned in the latter, these reforms were linked to the ascendance of the notion of sustainable development in environmental policy (Bakker, 2010, p. 88). Besides the World Bank, other international institutions such as the United Nations (UN) and the Organisation for Economic Co-operation and Development (OECD) began to have substantial impact on the water sector. As discussed earlier in section 1.1, a series of international conferences, declarations, resolutions and action plans all focused on the need to address sustainable development of water resources and to ensure access to clean water for all.

1.3.5 Resurgence of the Private Sector: Emergence of Concessions and Management Contracts

The privatization of the water industry in the United Kingdom in 1989, which is discussed in detail in Chapter 4 in this volume, was of strategic importance as it convinced many governments that private sector financing for water services was a viable option. Having gained experience in the domestic market, the large French and British companies such as Generale des Eaux (Vivendi/Veolia), Suez Lyonnaise des Eaux, Thames Water, Anglian Water, United Water and Biwater were willing to enter the international market (Gentry and Auyuan, 2000). Moreover, the water crisis in the low- and middle-income countries, which presented a huge market and hence high profit potential, acted as a pull factor for these large companies.

The formation of PPPs for the provision of water supply and sanitation services gained initial momentum in the late 1980s in Latin America, as the region was undergoing widespread economic liberalization and was in dire need of massive investments for infrastructure development. According to Marin (2009), between 1991 and 2000, the number of developing countries with PPP projects increased from four to 38, and by the end of 2000 about 93 million people were being served by private operators across the world. Latin America itself accounted for 45 percent of the population served by private operators and Argentina became the largest market for PPPs, with more than 18 million people served by private operators. Most of the contracts in Latin America have been concession agreements in which private operators were required to make large capital investments, and there was almost no government or donor funding. These private operators were foreign companies who invested in United States (US) dollars, but could only earn revenue in the local currency from consumer tariffs. Hence, devaluation of the local currencies led to serious debts for these operators, which in return limited their ability to make further investments. They had to periodically request the government to raise the tariffs to cover their costs, and whenever this happened there were widespread protests from the local people. Moreover, while the contract demanded expansion of service coverage to poor areas at low costs, the private operators had little incentive to do so, as poor households would be paying lower tariffs. When the fees for new connections were increased, in order to earn revenues, this again resulted in social resistance.

In response to growing public unrest and criticism of private sector-operated water utilities, the governments decided to terminate the contracts. The private companies which had invested and developed or operated such facilities went for dispute settlement before forums such as the International Centre for Settlement of Investment Disputes (ICSID), claiming compensation (see Table 14.2 in Chapter 14 in this volume for outcomes of some of these cases). This was followed by mass exit of the multinationals, which either tried to sell their shares to local private operators or had to hand over control to the public sector. Chile is the only country in Latin America that had privatized its entire water supply and sanitation sector during the period from 1998 to 2005 (see section 2.7).

Between 2000 and 2007, although the population served by PPPs fell sharply in Latin America, it rose steadily in other regions such as Eastern and Central Asia, sub-Saharan Africa, the Middle East and Eastern Europe (Marin, 2009). PPPs have managed to increase water supply and sewerage facilities in many of the poor urban areas of Africa. Unlike in Latin America, the PPPs in Africa were mainly management contracts and leases in which the ownership of capital assets remained with the government, with significant capital input from foreign donors. The private operators were thus not required to make significant investments and hence faced lower risks. Tariff reforms were initiated before signing the contract and the private operators received a fixed remuneration for each cubic meter of water sold, depending only on the average tariff, not the social tariff. In most cases, the governments conducted preparatory workshops and promoted public participation to convince stakeholders that private sector participation was the right approach, although these involved large transaction costs. In cases where there were combined power and water supply concessions, private operators were able to finance a large portion of their investment programs for water by reinvesting, each year, a portion of the cash flows collected from tariff revenues from the power sector. Following a specific model that evolved from the French standard *affermage* approach, private operators were awarded penalties or incentives for failing or achieving their targets, respectively. Some *affermages* relied heavily on public borrowing through an asset-holding company, until the company becomes self-financed with customer tariffs supporting both the operation and maintenance (O&M) cost and the government's debt service (Gunawansa and Hoque, 2012).

1.3.6 Re-municipalization of Water Utilities

In certain cases, failure of the private sector to meet contractual obligations, in terms of expanding coverage, increasing investment and enhancing efficiency, coupled with rise in water tariffs and popular discontent with the private sector, have led to re-municipalization of water utilities, sometimes through early termination or upon expiry of contracts. A detailed discussion of the key reasons and outcomes of re-municipalization is provided in Chapter 14 in this volume. According to the Water Re-municipalization Tracker, this change in ownership and management from the private sector back to the public sector is occuring at the municipal and community level (as in France or the United States), at the regional level (as in Buenos Aires and the Santa Fe provinces in Argentina) and also at the national level (as in Uruguay and Mali). Based on their analysis of five examples of re-municipalization – in Paris (France), Dar es Salaam (Tanzania), Buenos Aires (Argentina), Hamilton (Canada) and a series of Malaysian municipalities – Pigeon et al. (2012) concluded that re-municipalization can yield positive results in terms of equity, transparency, efficiency and sustainability. However, the transition to public hands is fraught with difficulties, even in the presence of strong political will and financial and technical capacity. This is due to loss of institutional memory, degraded assets, political and legal difficulties created by private operators, and the need to bring change in service delivery cultures and ideological reforms.

As discussed in Chapter 14 in this volume, in Paris the decision to regain management and ownership of the water utilities from the two powerful water companies (Suez and Veolia) upon contract expiry was purely a political one, with little pressure from the public. Massive tariff increases of about 265 percent between 1985 and 2009, which enabled the private companies to earn huge profits, coupled with audit reports showing potential of financial savings, led to this decision. In the very first year, the city had savings of €35 million and an 8 percent reduction in tariff (Pigeon et al., 2012). According to Anne Le Strat, the Deputy Mayor of Paris, the successful re-municipalization initiative in Paris is providing food for thought for other municipalities as well, and with 75 percent of delegated water services contracts in France up for renewal by 2015, councils are mulling whether to consider in-house management (Le Strat, 2011).

In the case of Buenos Aires, the decision to terminate the contract half way led to legal consequences, as Suez claimed compensation before the ICSID. This decision was the result of the failure of the company to meet its contractual targets for expanding coverage and improving quality, accompanied by repeated calls for renegotiations to increase tariffs. Since 2006, the new public company, Aguas y Saneamientos Argentinos, owned 90 percent by the Argentine government and 10 percent by the workers' union, has produced impressive results, particularly in terms of increasing access to water and sanitation in poorer neighborhoods. In Dar es Salaam, Tanzania's high dependence on international donors forced the government to award a ten-year lease to City Water Services (CWS) in 2003; however, due to the private company's poor preparation, difficult relations with staff and shareholder in-fighting, the government abruptly terminated the contract in 2005. The new public company Dar es Salaam Water and Sewerage Corporation (DAWASCO) has managed to reduce water leakages and to extend coverage, although only partially. As the transition was highly controlled by donors promoting neoliberal policies, the previous institutional set-up is still there and the World Bank continues to determine DAWASCO's priorities (Pigeon et al., 2012).

1.3.7 New Generation PPPs: Emergence of New Markets, New Contract Types and Innovative Financing Mechanisms

Learning from experiences with PPPs in the 1990s and 2000s, recent contracts are taking more innovative, context-specific approaches that are better suited to developing-country circumstances. Moving beyond the sterile debates about public versus private ownership and management, the water sector is now focusing on more concrete discussions about how to ensure the best possible service under a given set of resource limitations (Anderson and Janssens, 2011). Over the last decade, private sector activity has boomed in the water treatment market, and of the total 337 projects, most have been implemented in China (286), followed by Latin America (20) and the Middle East and North Africa (MENA) (12) (Perard, 2012). Reforms in the Chinese water sector over the last decade, which are discussed in detail in Chapter 9 in this volume, resulted in the implementation of 318 PPP projects in the country, of which most were sewerage treatment plants (224), followed by potable water treatment plants (55) and water utility without sewerage (22) (Perard, 2012). Moreover, India is showing an interest in private sector participation in water supply with a number of new projects in development (Perard, 2012). Chapter 2 in this volume provides a detailed region-wise overview of the extent of PPPs in the water sector across the world. At the same time, there has been increased activity in the water desalination market, with projects being implemented or under development in some cities. Most of these projects are greenfield projects, particularly build–operate–transfer (BOT) contracts, since water treatment or desalination plants require significant capital investment, but present a narrower range of risks for private companies as the projects are focused on a single client rather than utility operations that require ongoing engagement with multiple consumers and stakeholders (Anderson and Janssens, 2011).

Recent press releases are indicating that public authorities are exploring new types of partnerships with private companies, through which they can benefit from the experience of the private sector without handing over assets or management responsibilities. For example, the New York City Department of Environmental Protection, the largest municipal water and wastewater utility in the United States, has signed a US$36 million 'partnership contract' with Veolia Water, under which the latter will evaluate the performance of the city's existing water and wastewater systems and identify opportunities to reduce costs. Under this project, named 'Operational Excellence', the city aims to achieve annual savings of between US$100 million and US$200 million from its operations and maintenance budget. According to the contract terms, upon completion of that evaluation, the city will decide whether or not to appoint Veolia to oversee the implementation of those efficiency and cost-reduction measures (Veolia Water, 2012). This partnership is similar to the 30-year 'Design Build Assist' contract signed between the city of Winnipeg and Veolia in 2011, under which both parties will collaborate on capital improvements and provide ongoing strategic advice and guidance on design, construction, technology and operational needs for three wastewater treatment and biosolids facilities in Winnipeg (Veolia Water, 2011).

One of the main rationales for involving private companies in the water sector was to mobilize funds to secure financial investment. However, according to Marin (2009), although the private sector has contributed to improved service delivery, there is general acknowledgment that the financial investments were overstated and not fully realized. Consequently, the sector now requires more creative financing options to overcome low tariff levels and a backlog in infrastructure development and maintenance. Some reports (Vaidya and Vaidya, 2008; Ministry of Urban Development, Government of India, 2008) provide a useful analysis of these emerging options and outline opportunities to mobilize market-based repayable financing mechanisms such as loans, bonds and equity. In fact, pooled financing has already been used by Indian cities in the states of Tamil Nadu and Karnataka to expand water services (Hoque, 2012).

According to the *Pinsent Mason Water Yearbook 2010–2011* (Pinsent Masons LLP, 2011), the last few years have been marked by an increase in the number and quality of local companies. In 2007, it was estimated that 42 percent of the population in developing countries was being served by 'home-grown' companies rather than foreign companies that were required to establish local companies as their subsidiaries (Marin, 2009). Hall (2010) criticizes this report and states that the emergence of 'local companies' is a myth. According to him, apart from a few cases of actual local or national companies (such as Tata 'company town' in Jamshedpur, India, or WSSA, the company set up in South Africa under the apartheid regime), the rest are private companies which are descendants of the multinationals, with similar performance problems.

As mentioned in Chapter 9 in this volume, domestic water companies have significantly expanded their activities in the Chinese water market. Being either former state-owned enterprises with former government officials holding managerial positions, or having the state as the major shareholder, these domestic companies have close association with water-related bureaus of local governments, which facilitates access to market information. The increasing role of national local private sector as well as informal water vendors has been discussed in Chapter 13 in this volume.

1.4 OVERVIEW OF THE CHAPTERS

This section provides a brief synopsis of all the chapters in the book, highlighting the main areas of focus, the major topics of discussion and the key findings and recommendations.

Chapter 2, titled 'Private sector participation in water supply: a global perspective' by Edouard Pérard, provides an in-depth overview of the level of engagement of the private sector in providing water services in high-, middle- and low-income countries across the world. The author mentions that contrary to the popular belief that private sector participation (PSP) is a new phenomenon that gained momentum during the 1990s, it is a governance mechanism that has been in place for more than 200 years. Moreover, during the last two centuries, the level of PSP in water supply in countries has been fluctuating with periodic upward and downward trends, showing that the choice between private and public delivery of services is not a one-time decision. Unlike other sectors like energy and telecommunications, where high-income countries predominantly opt for privatization, the scope of PSP is highly heterogeneous among countries, irrespective of their income levels. Of the 84 countries in the world which have introduced PSP in the water sector, only 16 are high-income countries while the rest are middle- or low-income countries. Overall, the chapter concludes on the note that despite the highly publicized termination of several contracts, the review of the evolution of private sector participation over the recent years clearly indicates that the level of private sector participation has been expanding and not declining. However, it is true that most of the new private activity concerns mostly water treatment and wastewater treatment facilities rather than water supply.

There has been long-continued debate regarding the effectiveness and appropriateness of public versus private sector participation in the provision of water supply services in urban areas. In Chapter 3, titled 'Good governance: the key to solving Asia's water supply challenges', Alex Wong argues that the solution to the water crisis faced by the growing cities of Asia does not depend on a particular type of management structure; rather, it can be achieved through either public or private sector participation, provided the system of governance is not weak and flawed. The chapter explores a diverse range of successful urban water supply cases under both public and private management and contrasts them with relatively unsuccessful cases in similar situations. The relatively successful concession for East Manila granted to the Manila Water Corporation by the Metropolitan Waterworks and Sewerage System (MWSS) in the Philippines is compared to a similar concession granted in West Manila to Maynilad. Moreover, the relative success of the public water utility, Phnom Penh Water Supply Authority (PPWSA), in Phnom Penh in Cambodia presents a further contrast against a similar public utility in neighboring Ho Chi Minh City in Vietnam. Finally, the chapter discusses the case of Singapore, which involves a water strategy that revolves around both tight central public utility-type control and the leasing of water delivery infrastructure to the private sector through long-term PPPs. The chapter concludes with the thought that although there are political, cultural and financial barriers, it is not impossible to achieve the provision of adequate safe water for Asia's urban population. The examples of Singapore, Phnom Penh and Manila illustrate that setting the right policies and strategies coupled with changes in behavior (in terms of valuing water both physically and fiscally) and financial resources to implement these goals are the tripartite keys to success – but the final piece that is always needed to complete the puzzle is, undoubtedly, good management.

Chapter 4 on 'The UK water industry: infrastructure, governance and procurement' by Akintola Akintoye and Suresh Renukappa deals with the status and the different business models that are used for the provision of water and wastewater services in the United Kingdom (UK). Water and wastewater services in England and Wales are fully privatized, compared with water and wastewater services in Scotland and Northern Ireland, which are provided by public operators, albeit controlled by private law. Although the UK water industry has been generally considered a success, it faces a number of challenges including: rising demand and increased pressure on supply in the South and East due to climate change causing drier summers, decreased river flows and less predictable supply; maintenance and renewal of ageing network infrastructure combined with the need to meet European Union (EU) environmental standards; and increasing need for expenditure on flood defence infrastructure. The chapter proposes three different procurement models for water-related infrastructure development, that is: a pure public sector model, a public–private partnership model and a pure private sector model. Given the current challenges, the water industry needs to shift from its traditional governance mechanisms and develop solutions with the public sector to create new physical infrastructure procurement models that will underpin mutual goals and deliver products and/or services in an economically viable, socially responsible and environmentally sustainable manner.

Chapter 5, titled 'Performance of urban water governance in France' by Eshien Chong, provides an empirical and statistical assessment of the performance of PPPs and in-house provision for delivering water services. The chapter particularly looks into the organization of water supply services in France, since the country has a long experience with PSP in public services and a large variety of organizational structures for water services, ranging from in-house provision or direct public management to concession contracts. Using data collected by the French Institute of Environment from 3600 municipalities, a simple econometric analysis has been carried out. When only price is considered as a measure of performance, the results indicate that on average consumers pay a higher price when water services are organized as a PPP in France, and it is shown that, controlling for specific characteristics of a municipality's water services and situation, the higher price premium associated with services run under a PPP still persists. Interestingly, the estimations show that by making call for tenders compulsory when a municipality seeks to outsource its water service through a PPP, the Sapin Law has a moderating effect on the size of this premium.

In Chapter 6, titled 'Water governance in Spain: a dynamic system in transition', Belen Olmos Giupponi provides an overview of the Spanish water policy, underlying the main turning points in its evolution and the features of the current water governance system. In Spain, the water sector has undergone a deep-rooted change in its management paradigm and the water laws have adjusted to the new water policies. However, this transition has not been an even process, especially when the implementation of the EU Water Framework Directive raised various complex issues relating to the limited capacity of the institutional set-up to face the new challenges. Spain exhibits a long experience in increasing the participation of stakeholders, users and regional and local government agencies in water institutions (basin authorities), taking into account their essential role in the process. With respect to new challenges, water management in Spain faces problems such as environmental degradation, growing water demand, climate change and agricultural policies, and the importance of regional initiatives in the area of water pricing and environmental policies have been underlined in this regard. After the 'privatization wave', there have been significant changes in urban water supply, in particular concerning the emergence of various PPPs. Nevertheless, there is a need to provide information about the performance of private corporations in charge of water supply.

In Chapter 7, titled 'Urban water challenges in the MENA region: integrating Islamic principles with demand management strategies', Arani Kajenthira and Sharmila Murthy explore the evolution of urban water governance in the Middle East and North Africa (MENA) region and specifically consider the historical role that Islam has played in developing state-centered, centralized water ministries in the region. Based on the nature of water availability across the MENA region, the chapter examines how water scarcity has fostered supply-centered policies, the role of new technologies for augmenting supply and the potential impacts of climate change. Moreover, the recent trends in urban water utility reforms that rely on demand management principles, which promote greater decentralization, private sector participation and increasing tariffs as a way to promote financial viability and foster water conservation, have been highlighted. It has been suggested that urban water reform cannot be understood without considering the role of agricultural water management. Finally, the chapter presents a case study of Saudi Arabia, a hyper-arid country that has made significant progress in reforming its urban water governance framework, but still faces significant challenges in developing a sustainable water management strategy.

Chapter 8, titled 'Implementing PPP programs in the urban water and sanitation sector: some insights from the Indian experience in selected states' by Ashwin Mahalingam, first aims to illustrate the tight linkage between project delivery mode choice and the robustness of the institutional environment to support PPPs. The choice of project delivery mode is not based on the features of the project alone. Rather, the ability of the institutional environment to support the delivery mode is a key consideration. The author uses examples of PPPs from Tirupur and Alandur in Tamil Nadu to show that although the projects themselves were not unlikely candidates for BOTs, the inability of the institutional environment to support the complexity of the project structure led to their downfall. In Tirupur, an inability to enforce public sector commitment, poor project structuring capacity and a lack of stakeholder awareness were some of the environmental capability gaps that led to project failure. In Alandur, the inability of the project sponsor to monitor and govern a concession led to turbulences during the operations phase of the project. Then, the chapter suggests that governments should select PPP modes that can survive within their environments and 'upgrade' these modes in step with measures to strengthen the institutional environment. For instance, the DBO model requires a less mature institutional environment as compared to a BOT project, and thus fits well within Tamil Nadu. These lessons are particularly important in the context of developing countries with growing economies. The rate of infrastructure growth that is required to sustain economic growth in these countries is so large that governments are likely to have no choice but to turn towards the private sector to deliver infrastructure services. However, if governments choose forms of PPP that cannot be supported within their environment, they risk jeopardizing the overall service delivery goals.

Chapter 9, titled 'Urban water governance in China: legal and policy framework' by Lovleen Bhullar, provides an outline of the legal framework in China related to water infrastructure, together with a discussion on the various institutional and regulatory problems with regard to private sector participation. China has adopted a problem-oriented and 'develop first, control later' approach towards the urban water supply sector, which has resulted in the lack of a robust regulatory regime that is suited to China's specific institutional characteristics. Hence, interested private investors will have to adjust their expectations concerning China's urban water supply sector. Moreover, as the sector is driven by local provincial policies and approvals rather than the decisions of the central government, it is important to abandon the focus on the water sector as a whole in favor of a local or regional approach. The scope of PSP is restricted and in order to enter this lucrative market, the water multinationals have entered into joint ventures with domestic or regional private investors in the urban water sector in China. The government possesses limited financial capacity and know-how to proceed with the required investment to modernize and develop its water supply plants and networks and it is technologically backward in water treatment and production. It relies on international companies to provide the best and most up-to-date equipment and the latest innovations for a competitive price. This is expected to lead to a shift in the nature of private sector participation, from water supply and wastewater to development of advanced water technologies.

Chapter 10, titled 'Development of public–private partnerships in the water sector in Singapore' by Robert L.K. Tiong, Zheng Sha and Abu

Naser Chowdhury, presents the development and procurement of PPP in the water sector in Singapore. The chapter provides an overview of four PPP water projects: SingSpring desalination project, Second Tuas desalination project, Ulu Pandan NEWater project and Changi NEWater project. Based on the analysis of the time taken from the issuance of Request for Proposal (RFP) to reaching financial closure and commencing construction, the authors conclude that the procurement duration has decreased with each subsequent project as a result of experience gained from previous ones. The chapter also includes a case study on the Tuas Desalination Plant, which has been contracted out on a design–build–own–operate (DBOO) model, and details of the project features, organization structure and risk allocation are discussed. Based on the analysis of these PPP projects, the authors conclude that the successful implementation of PPPs in Singapore results from: (1) a clear policy framework and strong institutional capacity that results in a transparent and fair tendering process; (2) strong government commitment in terms of water policy and planning and implementing PPP projects; and (3) the presence of local private companies with expertise in the water sector.

Chapter 11, titled 'Urban water governance in Australia: the private sector at the margins' by Michael Paddon, provides an overview of the involvement of the private sector in the provision of water supply services in Australia, a country in which government provision remains dominant. While the major responsibilities for water service delivery have not been contracted out or privatized, contracting out of specific services by urban water utilities is widespread, particularly in the metropolitan areas, and there has been extensive engagement between public and private sectors in schemes to build infrastructure to meet requirements at certain points in the supply chain. The chapter discusses the concession arrangements for the city of Adelaide, which have been in operation since the mid-1990s. While public policy has largely rejected a model of extensive private provision of water or funding of water services, there have been a range of reforms since the 1990s intended to refocus the ways in which public water utilities operate through a 'corporatization model' (under the National Competition Policy), which are reviewed critically in the chapter. It also suggests that the drivers for change may have been modified in water governance, with less emphasis on investment in supply, with more extensive demand side considerations.

Chapter 12, 'Risks in urban water reforms: a challenge to public–private partnerships' by Claude Ménard, deals with the multiple risks faced by reforms in the urban water sector, with a particular emphasis on institutional risks. Indeed, important fluctuations in investment as well as protective contractual clauses suggest that innovative solutions such as private sector participation and their accompanying regulation might also generate new sources of uncertainty and vulnerability for the management of water utilities. The chapter identifies different types of risks involved with particular attention to the specific risks faced by private operators and the resulting factors of vulnerability plaguing contractual agreements. It emphasizes the difficulties in evaluating these risks *ex ante*, although such evaluation is central to legitimize the choice of public–private partnership over alternative arrangements. It also examines the sensitive issue of allocation of these risks between public authorities and private operators, and the key role of enforcement in determining if PPP was the right choice. The chapter concludes with some insights on the ambiguous role of regulation as an institutional response to the vulnerability of parties facing these risks.

Chapter 13, titled 'Public and private interfaces: changing international perceptions of public–private partnerships in water services' by Håkan Tropp and Camelia Dewan, explores the gradual change in perceptions among bilateral and multilateral organizations regarding the need for and efficacy of the private sector in the development process. In the 1980s, organizations such as the World Bank and International Monetary Fund began to promote a set of structural adjustment policies, known as the Washington Consensus, that prescribed liberalization, privatization and deregulation to bring about reform in crisis-wrecked developing states. This spurred the momentum of privatizing the water supply sector in many countries, characterized by long-term concessions and divestitures. However, the focus on sustainability brought about by the multilateral conventions of the 1990s, and the increasing criticism of the Washington Consensus because of its relative neglect of the provision of production-oriented infrastructure and services needed to supply profitable work opportunities for poor people, led to a contrasting paradigm called the Post-Washington Consensus. The latter emphasized the need for different institutions in different economies and recognized cases in which governments had to intervene to control the market interventions. It focused on poverty reduction, emphasizing the need for delivery to the poor, of social services such as education and health care, by government and civil society. Moreover, the failure of PPPs in terms of increasing tariffs, lack of service extension to poor areas and mass exit of multinationals resulting from loss due to local currency devaluation, in many countries, has led to a pragmatic shift regarding the role of the private sector. The authors have identified three emerging trends with regard to private sector involvement in the water supply sector. Firstly, there is now a trend towards short-term management and service contracts, where the private sector is responsible for operation and management while the investment risks lie with the public sector. Secondly, national companies and informal water vendors are now playing increasing roles as private players. Finally, private sector water users, such as large corporations, are perceiving water as a strategic resource whose management is critical, both as part of their corporate social responsibility (CSR) policies and, more importantly, as part of their long-term business plans.

Chapter 14, titled 'Re-municipalization of water utilities: back to the public fold' by Asanga Gunawansa and Vanessa Garcia, provides a detailed insight into the emerging trend of re-municipalizing water utilities that were previously managed by the private sector via different types of PPPs. Using relevant examples from various countries, the authors have highlighted some of the key reasons that have compelled governments in both developing and developed countries to take the responsibility of providing water and sanitation services back to the public sector. Large-scale public protests arising from poor quality of service and increased tariffs and failure of the private companies to meet contractual obligations in terms of improving technical efficiency, expanding network coverage and increasing investments are often the main reasons for early termination of contracts. However, changes in government, often accompanied by changes in ideological notions and sectoral reforms brought about by legal and policy changes, are also crucial factors stimulating the transition from the private to the public fold. Moreover, lack of good governance practices, including lack of transparency and accountability by the private companies, corruption by both public and private authorities, and lack of effective regulation, are also evident in many contracts. Finally, the authors use three case studies from France, Malaysia, and Trinidad and Tobago to highlight different scenarios of re-municipalization. This is followed by a discussion on the scope of achieving better management of water utilities after returning to public hands.

Chapter 15, the concluding chapter, titled 'Good governance of water: the final analysis' by Sonia Ferdous Hoque and Asanga Gunawansa, aims to articulate the discussions and analysis made in each of the chapters of this book, supported by evidence provided in other studies. Section 15.2 of the chapter examines the ideological notions usually set forward by proponents of public or private sector in relation to the outcomes perceived so far. Further, in this final chapter, an attempt has been made to outline the crucial parameters required for successful management both by the public and the private sector, respectively, using examples of good practices from selected utilities across the world.

REFERENCES

Adelman, I. and A.E. Yeldan (1999), 'The role of government in economic development', Working Paper No. 890, University of California at Berkeley.

Anderson, A. and J.G. Janssens (2011), 'Emerging PPP trends in the water and sanitation sector', London: Building Partnerships for Development in Water and Sanitation.

Baker, M.N. (1899), 'Water-works', in E.W. Bemis (ed.), *Municipal Monopolies*, New York: Thomas Y. Crowell & Company, pp. 3–52.

Bakker, K. (2003), *Good Governance in Restructuring Water Supply: A Handbook*, Toronto: Federation of Canadian Municipalities.

Bakker, K. (2010), *Privatizing Water: Governance Failure and the World's Urban Water Crisis*, Singapore: National University of Singapore Press.

Bitran, G. and P. Arellano (2005), 'Regulating water services – sending the right signals to utilities in Chile', World Bank, available at http://www-wds.worldbank.org/external/default/WDSContentServer/WDSP/IB/2005/04/18/000112742_20050418123921/Rendered/PDF/321260286Bitran1Arellano.pdf?pagePK=64193027&piPK=64187937&theSitePK=523679&menuPK=64187510&searchMenuPK=64187283&theSitePK=523679&entityID=000112742_20050418123921&searchMenuPK=64187283&theSitePK=523679 (accessed 15 April 2012).

Budds, J. and G. McGranahan (2003), 'Are the debates on water privatization missing the point? Experiences from Africa, Asia and Latin America', *Environment and Urbanization*, **15**(2), 87–114.

Camdessus, M. and J. Winpenny (2003), 'Financing water for all', Report of the World Panel on Financing Water Infrastructure, available at http://www.financingwaterforall.org/fileadmin/wwc/Library/Publications_and_reports/CamdessusReport.pdf (accessed 12 December 2011).

Dublin Statement on Water and Sustainable Development (1992), available at http://www.wmo.int/pages/prog/hwrp/documents/english/icwedece.html (accessed 20 March 2012).

Finger, M. and J. Allouche (2001), *Water Privatization: Trans-national Corporations and the Re-regulation of the Water Industry*, London: Spon Press.

Financial Times (1999), 'Making a big splash: with more than a century in the private sector, French water companies are showing the rest of the world how to manage their water utilities', *Financial Times*, 24 August.

Gentry, B.S. and A.T. Auyuan (2000), 'Global trends in urban water supply and waste water financing and management: changing roles for the public and private sectors', Paris: Organisation for Economic Co-operation and Development.

Global Water Partnership (GWP) (2000), *Towards Water Security: A Framework for Action*, Stockholm, Sweden and London: Global Water Partnership, available at http://www.gwptoolbox.org/images/stories/Docs/waterpercent20security_2000_doc_78_en.pdf (accessed 12 December 2011).

Gunatilake, H. and M.F.C. San Jose (2008), 'Privatization revisited: lessons from private sector participation in water supply and sanitation in developing countries', ERD Working Paper No. 115, Manila: Asian Development Bank.

Gunawansa, A. and S.F. Hoque (2012), 'Urban water infrastructure development in the third world: quo vadis PPP?', Lee Kuan Yew School of Public Policy

Research Paper No. LKYSPP12-02 IWP, available at http://papers.ssrn.com/sol3/papers.cfm?abstract_id=2004733 (accessed 1 April 2012).
Hachfeld, D., P. Terhorst and O. Hoedeman (eds) (2009), 'Progressive public water management in Europe: in search of exemplary cases', Reclaiming Public Water Discussion Paper, Amsterdam: Transnational Institute and Corporate Europe Observatory, available at http://www.waterjustice.org/uploads/attachments/Progressivepercent20publicpercent20waterpercent20managementpercent20inpercent20Europe.pdf (accessed 12 December 2011).
Hall, D. (2010), 'Public–public partnerships in water: an overview', Confluence of Ideas and Organisations: Centre of Excellence for Change, Chennai: Tamil Nadu.
Hall, D., E. Lobina, V. Corral, O. Hoedeman, P. Terhorst, M. Pigeon and S. Kishimoto (2009), 'Public–public partnerships (PUPs) in water', Public Services International Research Unit (PSIRU), Public Services international (PSI) and Transnational Institute (TNI), available at http://www.tni.org/sites/www.tni.org/files/download/pupinwater.pdf (accessed 12 December 2011).
Hanson, A.H. (1959), *Public Enterprise and Economic Development*, London: Routledge & Kegan Paul.
Harding, A. and A.S. Preker (2000), 'Understanding organisational reforms: the corporatization of public hospitals', Health, Nutrition and Population (HNP) Discussion Paper, World Bank Human Development Network.
Hearne, R.R. and G. Donoso (2005), 'Water institutional reforms in Chile', *Water Policy*, **7**, 53–69.
Hoque, S.F. (2012), 'Urban water sector reforms in India: financing infrastructure development through market-based financing and private–public partnerships', Lee Kuan Yew School of Public Policy Research Paper No. LKYSPP 12-05 IWP, available at http://ssrn.com/abstract=2004763 or http://dx.doi.org/10.2139/ssrn.2004763 (accessed 10 March 2012).
Jagannathan, N.V., A.S. Mohamed and A. Kremer (eds) (2009), *Water in the Arab World: Management Perspectives and Innovations*, Washington, DC: World Bank.
Johnstone, N. and L. Wood (eds) (2001), *Private Firms and Public Water: Realising Social and Environmental Objectives in Developing Countries*, Cheltenham, UK and Northampton, MA, USA: Edward Elgar Publishing.
Kauffmann, C. (2007), 'Private sector participation in water and sanitation infrastructure', Background Paper for the Regional Roundtable on Strengthening Investment Climate Assessment and Reform in NEPAD Countries, Lusaka, Zambia, OECD, Directorate for Financial and Enterprise Affairs, Investment Division, available at http://www.oecd.org/dataoecd/50/56/39663326.pdf (accessed 1 October 2011).
Kishimoto, S. (2009), 'Reforming public water services', Transnational Institute (TNI), available at http://www.tni.org/primer/reforming-public-water-services (accessed 1 October 2011).
Kuhlmann, S. and P. Fedele (2010), 'New public management in continental Europe: local government modernization in Germany, France and Italy from a comparative perspective', in Wollmann Hellmut and Gerard Marcou (eds), *The Provision of Public Services in Europe: Between State, Local Government and Market*, Cheltenham, UK and Northampton, MA, USA: Edward Elgar Publishing, pp. 49–74.
Le Strat, A. (2011), 'Paris defends re-municipalization record', *Global Water*

Intelligence, **12**(5), available at http://www.globalwaterintel.com/archive/12/5/general/paris-defends-re-municipalisation-record.html (accessed 12 May 2012).
Lenton, R., A.M. Wright and K. Lewis (2005), *Health, Dignity, and Development: What Will It Take?*, UN Millennium Project Task Force on Water and Sanitation, UK and USA: EarthScan.
Lewis, K. (ed.) (2004), *Water Governance for Poverty Reduction: Key Issues and the UNDP Response to the Millennium Development Goals*, New York: United Nations Development Programme.
Lovells Lee & Lee (2008), 'Privatisation and public–private partnerships in Singapore', Client Note, Singapore.
Maltz, H. (2005), 'Porto Alegre's Water: Public and for All', in Belén Balanyá, Brid Brennan, Olivier Hoedeman, Satoko Kishimoto and Philipp Terhorst (eds), *Reclaiming Public Water*, Amsterdam: Transnational Institute (TNI) & Corporate Europe Observatory (CEO), pp. 29–36.
Marin, P. (2009), *Public–private Partnerships for Urban Water Utilities: A Review of Experiences in Developing Countries*, Washington, DC: Public–private Infrastructure Advisory Facility (PPIAF), World Bank, available at http://www.ppiaf.org/ppiaf/sites/ppiaf.org/files/FINAL-PPPsforUrbanWaterUtilities-PhMarin.pdf (accessed 1 October 2011).
Marin, P., M. Fall and H. Ouibiga (2010), 'Corporatizing a water utility: a successful case using a performance-based service contract for ONEA in Burkina Faso', *GRIDLINES*, Note No. 53, March 2010, available at http://www.ppiaf.org/ppiaf/sites/ppiaf.org/files/publication/53-corporatizing-water-utility.pdf (accessed 1 October 2011).
Masten, S.E. (2004), 'Public utility ownership in 19th-century America: the "aberrant" case of water', *The Journal of Law, Economics, and Organization*, **27**(3), 604–54.
Ministry of Urban Development, Government of India (2008), 'Pooled Finance Development Scheme', available at http://urbanindia.nic.in/programme/ud/toolkit_pfds.pdf (accessed 10 October 2011).
Office of Water Regulation (OFWAT) (1993), 'Privatisation and history of the water industry', Information Note No. 18.
Pérard, E. (2012), 'Private sector participation in water infrastructure: review of the last 20 years and the way forward', a deliverable from the PPI Database to IFC, Washington, DC: Public Private Infrastructure Advisory Facility, World Bank.
Pigeon, M., D.A. MacDonald, O. Hoedman and S. Kishimoto (2012), *Remunicipalization: Putting Water Back into Public Hands*, Amsterdam: Transnational Institute.
Pinsent Masons LLP (2011), *Pinsent Masons Water Yearbook 2010–2011*, 12th edn, available at http://wateryearbook.pinsentmasons.com/historical_editions.aspx (accessed 13 March 2012).
Prasad, N. (2007), 'Privatisation of water: a historical perspective', *Law, Environment and Development Journal*, **3**(2), 217.
PSIRU (2010), 'ACP–EU water partnerships – briefing note', Public Services International Research Unit (PSIRU), University of Greenwich, available at http://www.waterjustice.org/uploads/attachments/ACPEUpercent20Waterpercent20Partnershipspercent20Briefingpercent20Notepercent20Augustpercent202010.pdf (accessed 1 October 2011).

Rogers, P. and A.W. Hall (2003), 'Effective water governance', Technical Committee (TEC) Background Papers No. 7, Global Water Partnership: Stockholm, available at http://www.tnmckc.org/upload/document/bdp/2/2.7/GWP/TEC-7.pdf (accessed 1 October 2011).

Roth, G. (1987), *The Private Provision of Public Services in Developing Countries*, Oxford: Oxford University Press.

Schwartz, K. (2008), 'The new public management: the future for reforms in the African water supply and sanitation sector?', *Utilities Policy*, **16**(1), 49–58.

Tropp, H. (2005), 'Developing water governance capacities', Feature Article, UNDP Water Governance Facility, Stockholm International Water Institute (SIWI).

United Nations (1948), *The Universal Declaration of Human Rights*, available at http://www.un.org/en/documents/udhr/ (accessed 1 October 2011).

United Nations (1992), 'Chapter 18 – Protection of the quality & supply of freshwater resources: application of integrated approaches to the development, management & use of water resources', in *Agenda 21* available at http://www.un.org/esa/dsd/agenda21/res_agenda21_18.shtml (accessed 30 October 2011).

United Nations (2006), 'Water: a shared responsibility', United Nations World Water Development Report 2, World Water Assessment Programme, available at http://unesdoc.unesco.org/images/0014/001444/144409e.pdf (accessed 1 October 2011).

United Nations Department of Economic and Social Affairs, United Nations Development Program and United Nations Economic Commission for Europe (UNDESA/UNDP/UNECE) (2003), 'Governing water wisely for sustainable development', in UN World Water Development Report, *Water for People, Water for Life*, Paris: UNESCO, pp. 369–84.

United Nations Human Settlement Programme (UN-HABITAT) (2008), *The UN-HABITAT Water and Sanitation Trust Fund Strategic Plan 2008–2012*, Nairobi: UN-HABITAT.

United Nations Children's (Emergency) Fund (UNICEF) (2010), 'Water, sanitation and hygiene', available at http://www.unicef.org/wash/ (accessed 29 May 2011).

Vaidya, C. and H. Vaidya (2008), 'Creative financing of urban infrastructure in India through market-based financing and public–private partnership options', Paper presented at the 9th Metropolitan Congress, Sydney, 22–26 October 2008, available at http://niua.org/present_series/sydney/sydney_paper.pdf (accessed 10 November 2011).

Veolia Water North America (2011), 'Winnipeg selects Veolia, bringing public and private expertise together in a new, innovative model', Press Release, 4 October, available at http://www.veoliawater.com/media/news/2011-10-04,winnipeg-collaborative-agreement.htm (accessed 27 April 2012).

Veolia Water North America (2012), 'Veolia Water signs partnership contract with New York City to optimize its public water and wastewater services', Press Release, 10 April, available at http://www.veoliawater.com/media/news/ (accessed 27 April 2012).

Water and Sanitation Partnership (WSP) (2007), 'How can reforming African water utilities tap local financial markets?', insights and recommendations from a Practitioners' Workshop in Pretoria, South Africa, July 2007 (revised 2009), available at http://www.wsp.org/wsp/sites/wsp.org/files/publications/African_water_utilities_local_financial_markets.pdf (accessed 1 October 2011).

Winpenny, J. (2003), 'Financing water for all', World Water Council, 3rd World Water Forum and Global Water Partnership.
Wolff, G.H. and M. Palaniappan (2004), 'Public or private water management? Cutting the Gordian Knot', *Journal of Water Resource Planning and Management*, **130**(1), 1–3.
World Bank (1997), *Toolkits for Private Participation in Water and Sanitation*, Washington, DC: World Bank.
World Bank (2007), *Making the Most of Scarcity: Accountability for Better Water Management Results in the Middle East and North Africa*, Washington, DC: World Bank, available at http://siteresources.worldbank.org/INTMENA/Resources/Water_Scarcity_Full.pdf (accessed 1 October 2011).
World Health Organization (WHO) and United Nations Children's Fund (UNICEF) (2010), 'Progress on sanitation and drinking water, 2010 update', available at http://whqlibdoc.who.int/publications/2010/9789241563956_eng_full_text.pdf (accessed 1 October 2011).
World Water Council (2000), 'Ministerial Declaration of the Second World Water Forum', available at http://www.idhc.org/esp/documents/Agua/Second_World_Water_Forum%5B1%5D.pdf (accessed 30 October 2011).
World Water Council (2003), 'Ministerial Declaration of the Third World Water Forum', available at http://www.mofa.go.jp/policy/environment/wwf/declaration.html (accessed 30 October 2011).
Wu, X. and N.A. Malaluan (2008), 'A tale of two concessionaires: a natural experiment of water privatization in Metro Manila', *Urban Studies*, **45**(1), 207–29.

2. Private sector participation in water supply: a global perspective

Edouard Pérard

2.1 INTRODUCTION

Access to clean water represents both a human and a financial challenge. To modernize obsolescent systems and to meet expanding demand in both high-income and low- and middle-income countries, it has been estimated that US$22.6 trillion should be invested in water infrastructure by 2030 (Doshi et al., 2007). Considering the tremendous investment needs for water infrastructure, the private sector has been playing an increasing role over the last 20 years. This chapter aims to put discussions on the issue of private sector participation in water supply in perspective, and to respond to the growing need for a neutral approach by providing a stocktaking of the development of private sector participation in water supply in the world.

This chapter provides a global perspective of the state of private sector participation in the water supply sector. Section 2.2 provides a statistical overview of the level of private sector engagement in high-, middle- and low-income countries, while section 2.3 addresses the question of the evolution of private sector participation in water over the recent years and the way forward.

2.2 SCOPE OF PRIVATE SECTOR PARTICIPATION IN WATER SUPPLY

What is commonly referred to as 'private sector participation' consists in fact of several very different types of contracts between a national or subnational entity and a private operator. The five major types of private sector involvement are management contract, lease contract, concession contract, greenfield project and divestiture. Differences between each type of contracts have been discussed in Chapter 1 in this volume.

The scope of private sector participation in water supply is particularly heterogeneous among countries regardless of their income level: from

Table 2.1 Private sector participation in water supply in high-income economies as of January 2012

Level of private sector participation in water supply (% of population served by the private sector)	Countries
Major (more than 70%)	France, United Kingdom
Important (between 40 and 70%)	Czech Republic, Greece, Italy, Spain
Moderate (between 10 and 40%)	Estonia, Germany, Hungary, Poland, Portugal, Slovak Republic, United States
Very limited (less than 10%)	Australia, Austria, Canada
Public water provision	Bahrain, Belgium, Brunei, Croatia, Cyprus, Denmark, Finland, Iceland, Ireland, Israel, Japan, Korea, Kuwait, Liechtenstein, Luxembourg, Malta, Netherlands, New Zealand, Norway, Oman, Qatar, Saudi Arabia, Singapore, Slovenia, Sweden, Switzerland, United Arab Emirates

Source: Author's estimate based on the review of institutional communication of major private operators, national and local governments, as well as literature review.

non-existent in Japan and Egypt, for example, to cases of divestiture in the United Kingdom and Chile. This situation is very different to what happens in other infrastructure sectors such as energy and telecommunications, which are predominantly privatized in high- and middle-income countries. This section consists of a major stocktaking exercise that aims to provide a global perspective on private sector participation in water supply by reviewing the introduction and the scope of private sector participation in water supply for each country in the world.

Overall, 84 countries in the world – 16 high-income economies and 68 low- and middle-income economies – have introduced private sector participation in water supply. The first section of this part reviews the situation in high-income economies, while the second part focuses on low- and middle-income economies.

2.2.1 Private Sector Participation in High-income Economies

In high-income economies, 16 countries have introduced some form of private sector participation in water supply. These countries can be classified into four groups, depending on the estimated percentage of population served by the private sector in each country (Table 2.1). The level of private sector participation in water supply is considerable in two countries (that is, more than 70 percent of the population is served by the private sector): the United Kingdom and France. In the Czech Republic, Greece, Italy and Spain, experience with private water supply is also important (that is, at least 40 percent of the population receives its water from a private operator). In Estonia, Germany, Hungary, Portugal, Poland, the Slovak Republic and the United States, the level of private sector participation is moderate (that is 10–40 percent of the population is supplied by the private sector). In Australia, Austria and Canada, private sector participation in water exists but is limited to a few municipalities. All other high-income countries, including Bahrain, Belgium, Brunei, Croatia, Cyprus, Denmark, Finland, Iceland, Ireland, Israel, Japan, Korea, Kuwait, Liechtenstein, Luxembourg, Malta, Netherlands, New Zealand, Norway, Oman, Qatar, Saudi Arabia, Singapore, Slovenia, Sweden, Switzerland and the United Arab Emirates, have opted for public water provision.

This section will present successively the case of each of the 12 countries with significant experience in private water provision: France, the United Kingdom, the Czech Republic, Greece, Hungary, Italy, Portugal, Spain, Estonia, Germany, the Slovak Republic and the United States.

Private water supply is predominant in only two high-income countries: the United Kingdom and France. In the United Kingdom, the situation in England and Wales differs very much from Scotland and Northern Ireland. While the water industry in England and Wales is fully divested, water companies are state-owned and operated in Scotland and Northern Ireland. In England and Wales, the water companies were divested in 1989 and are now all wholly owned by investors although they operate under strict economic and environmental regulation. The companies are obligated by license to provide the essential services within their areas. There are currently 34 companies: 10 regional companies provide both water and sewerage services, 11 regional companies provide water services only, 6 local companies provide either water or sewerage services or both, and 7 water supply licensees offer water and sewerage services to large use customers.

In France, water has been historically privately provided for more than two centuries; 36 800 municipalities are responsible for the management of drinking water supply and wastewater treatment, either grouped in syndicates or alone. Municipalities have a wide degree of flexibility in the selection of contractual arrangements for the provision of water supply and sewerage services. They can provide the services themselves by direct management or by delegation to a private company. The most common arrangements are lease contracts in which the private company is responsible for managing the system and the municipality for investments;

concession contracts, in which the private company also undertakes investments; and management contracts. Nowadays, it is estimated that private companies provide water supply services to about 75 percent of the population.

In addition, four other high-income economies have an important level of private water provision: the Czech Republic, Greece, Italy and Spain. In the Czech Republic, private water delivery has existed since 1993. Since then, 22 major water utilities, including Prague's network, have been delegated to private operators under management contract. The market share of private operators in terms of population served is estimated at between 60 and 70 percent.

In Greece, private sector participation was introduced in 2000 when the state partially divested the water supply and sanitation company of Athens. The following year, the water supply and sanitation company of Thessaloniki was also partially divested. These two partially private companies operate the water networks of both cities under a 20-year license. About 40 percent of the population is currently served by private operators.

In Italy, a significant process of restructuring the water sector has taken place since 1996, with the partial divestiture of the main municipally owned water companies. Since then, the water companies of Genoa, Rome, Como, Trieste, Brescia, Bologna, Modena and Naples have been partially divested or sold to private operators. In addition, several cities – including Florence, Perugia, Pisa and Sienna – have delegated water supply under concession contracts. It is estimated that more than 40 percent of the population is now served by a private operator.

In Spain, private involvement in the drinking water sector has been present since the nineteenth century. The largest private water company is Aguas de Barcelona, AGBAR, which has been engaged in the distribution and supply of potable water in Barcelona since 1911, and in 22 municipalities in the metropolitan area. AGBAR has the concession to supply Barcelona for an indefinite period. Nowadays, private participation reaches more than 40 percent of the supplied population, mostly in cities with populations higher than 20000 inhabitants.

In seven other countries – Estonia, Hungary, Portugal, Poland, the Slovak Republic, the United States and Germany – the level of private sector participation in water supply is moderate. Estonia introduced private sector participation in water supply in 2000, when the city of Tallinn divested 50.4 percent of AS Tallinna Vesi, the main water utility in the country. In 2005, the city of Tallinn proceeded to a second divestiture of 30 percent of AS Tallinna Vesi. The percentage of the country's population served by the private operator is close to 30 percent.

Hungary started delegating water supply in major cities in 1995. The government partially divested several water companies and awarded 15–25-year operation licenses. To date, six cities – Budapest, Debreen, Kapsovar, Pecs, Borsodviz and Erd – have been receiving water supply from private companies. About 20–30 percent of the population is now served by private operators. In Poland, four water supply companies were divested from 1999 to 2007 in Toszek, Glogow, Drobin and Bielsko. Four other networks are privately managed under concession and management contracts.

Portugal specifically outlawed private sector involvement in water supply in 1977. Twenty-six years later, in 1993, the law was changed and the state allowed for the establishment of concessions with private sector companies. Between 1995 and 1998, 12 long-term concessions were awarded to international water companies. It is estimated that private companies supply water to 20–30 percent of the population.

In the Slovak Republic, private involvement in water supply dates back to 1999, when the state divested 51 percent of the Trencin Water Company and awarded the newly privatized company a 20-year lease to operate the network. In 2006, two other 20-year lease contracts were signed with private operators for the water supply of the areas of Banska Bystrica and Poprad. About 10–15 percent of the population is served by a private operator.

In the United States, the legal entity responsible for water supply varies according to the region; it can be the responsibility of the county, the city, the borough, or the parish. As previously presented (see Chapter 1, section 1.3.1), water systems in the United States were historically privately owned and operated. However, during the nineteenth century, many water systems were municipalized. According to the Environment Protection Agency, about half of the waterworks are currently privately owned. The percentage of population served by the private sector ranges between 10 and 20 percent.

In Germany, water services are predominantly run by the public sector. However, a few cities, including Berlin, Gelsenkirchen, Leipzig and Kiel, have delegated water supply under concession contracts. About 10–15 percent of the population relies on private water supply.

In Australia, Austria and Canada, some water systems are privately operated under different forms of contracts. However, the number of private water supply systems is very small and not significant in comparison to the number of systems operated by the public sector in these three countries.

In other high-income economies – Bahrain, Belgium, Brunei, Croatia, Cyprus, Denmark, Finland, Iceland, Ireland, Israel, Japan, Korea, Kuwait,

Table 2.2 Private sector participation in water supply in East Asia and the Pacific

Country	Number of operational private water supply systems as of January 2012	Year of the first private water supply system	Forms of private sector participation that have been implemented
China	35	1998	Management/lease, concession, build–operate–transfer (BOT)/build–own–operate (BOO), divestiture
Indonesia	5	1995	Concession
Malaysia	5	1987	Management/lease, concession, BOT/BOO, divestiture
Philippines	4	1997	Management/lease, concession
Thailand	9	1997	Management/lease, concession, BOT/BOO, divestiture
Vanuatu	1	1994	Concession

Liechtenstein, Luxembourg, Malta, the Netherlands, New Zealand, Norway, Oman, Qatar, Saudi Arabia, Singapore, Slovenia, Sweden, Switzerland, the United Arab Emirates – water supply has remained public.

2.2.2 Private Sector Participation in Low- and Middle-income Economies[1]

Private sector participation in water supply is not a mode of service provision reserved for high-income countries. According to the World Bank/PPIAF Private Participation in Infrastructure Database (PPIAF, 2011), since 1984, 68 low- and middle-income countries have delegated 345 water supply systems to the private sector. This section summarizes, region by region, the experience of countries with private sector participation in water supply.

In East Asia and the Pacific, six countries – China, Indonesia, Malaysia, the Philippines, Thailand and Vanuatu – have introduced some form of private sector participation in water supply (Table 2.2).

China accounts for most of the private water supply infrastructure in the region. The first private water distribution project was implemented in Changli in the Hebei Province under a BOT agreement in 1998. A total of 35 private water systems are currently operational around the country. Twenty-two of them are concession contracts, four partial divestitures, two management contracts and seven BOT projects. Water supply is now under private operation in several large cities with more than 1 million inhabitants, including Shenyang, Fuzhou, Wuhu, Shenzhen, Tianjin, Lanzhou, Kuming, Changzhou, Wuhan, Wujiang and Linyi.

In Thailand, private sector participation in water infrastructure started in 1997 with the partial divestiture of East Water Company in charge of water distribution in the Chonburi and Rayong Province. Currently, nine Thai water supply networks are private: six of them are managed under concession contracts, one divestiture, one BOO and one management contract.

In Indonesia, private sector participation in water supply has existed since 1995. The state has handed over the operation of five water networks to the private sector, including the eastern and western network of Jakarta. All of these systems have been operating under concession contracts. Malaysia introduced private sector participation in water supply as early as 1987 with a BOT contract for the water supply of Sabah. Five private systems are currently operational: two under concession contracts, one divestiture, one BOT and one management contract. Water is privately provided in Selangor, Kuala Lumpur, Putrajaya, Pulau Langkawi city and Penang State.

In the Philippines, Manila was the first city to delegate the operation of water supply to private firms under two concession contracts in 1997. As of January 2012, four water supply networks are operated by private firms under concession contracts. Vanuatu also has some experience with private water provision. The water supply operations of Port Vila and Luganville were handed over to Suez under a 40-year concession contract in 1994.

In Europe and Central Asia, private water supply has been experienced in 15 countries: Albania, Armenia, Azerbaijan, Bulgaria, Georgia, Kazakhstan, Kosovo, the Kyrgyz Republic, Montenegro, Romania, the Russian Federation, Serbia, Turkey, Ukraine and Uzbekistan (Table 2.3). Poland[2] was the first country in the region to turn to private companies for water supply in 1993.

Management contracts have been the most used form of private sector participation. Indeed, as of January 2012, there were 32 water systems privately operated: 21 of them under management contracts, 7 under concession contracts and 4 were divested.

Russia, Poland, Georgia, Romania and Bulgaria have been the most active countries in delegating water supply. Russia accounts for most of

Table 2.3 Private sector participation in water supply in Europe and Central Asia

Country	Number of operational private water supply systems as of January 2012	Year of the first private water supply system	Forms of private sector participation that have been implemented
Albania	0	2002	Management/Lease, Concession
Armenia	2	2000	Management/Lease
Azerbaijan	1	2001	Management/Lease
Bulgaria	1	2000	Concession
Georgia	1	2007	Divestiture
Kazakhstan	1	1999	Divestiture
Kosovo	0	2002	Management/Lease
Kyrgyz Republic	1	2003	Management/Lease
Montenegro	0	2001	Management/Lease
Romania	4	2000	Management/Lease, Concession
Russian Federation	19	2001	Management/Lease, Concession, Divestiture
Serbia	0	2002	Management/Lease
Turkey	0	1996	Management/Lease
Ukraine	2	2003	Management/Lease, Concession
Uzbekistan	0	2004	Management/Lease

the private water networks in the region, but the responsibilities handed over to the private sector are very limited. Fourteen water systems are currently under lease contracts, one under management contract, two under concession contracts and two under divestitures. Water is privately managed in several cities with over 1 million inhabitants, including Perm, Omsk and Volgograd. Recently, Georgia divested the Tbilisi water system, which services 1.1 million inhabitants.

Romania has four private water networks: three concessions were signed in 2000 for the water systems of Ploiesti, Bucharest and Timisoara and one lease in 2007 for Otopeni. In Bulgaria, the water network of Sofia was also given in concession to a private operator in 2000.

The Latin America and Caribbean region has a strong experience with private water supply (Table 2.4). Seventeen countries have introduced private sector participation in water supply and 149 water distribution

Table 2.4 Private sector participation in water supply in Latin America and the Caribbean

Country	Number of operational private water supply systems as of January 2012	Year of the first private water supply system	Forms of private sector participation that have already been implemented
Argentina	10	1991	Management/Lease, Concession
Belize	0	2001	Divestiture
Bolivia	0	1997	Concession
Brazil	67	1994	Concession, Divestiture
Chile	17	1993	Concession, Divestiture
Colombia	45	1994	Management/Lease, Concession
Cuba	1	2000	Concession
Ecuador	3	2000	Management/Lease, Concession
Guatemala	1	2008	Concession
Guyana	0	2002	Management/Lease
Haiti	1	2009	Management/Lease
Honduras	1	2001	Concession
Mexico	4	1855	Management/Lease, Concession
Peru	1	2005	Concession
Uruguay	0	1993	Concession
Venezuela	0	1997	Management/Lease

systems are currently privately operated, most of them under long-term concession contracts (25–40 years). Indeed, 115 of them are under concession contracts, 22 under management and lease contracts, 11 of them were divested and one is a greenfield project.

Most of the private water systems are located in Colombia, Brazil, Chile, Argentina and Mexico. In Brazil, private sector participation was introduced in 1997. The number of private systems is very important. The two forms of private sector participation in water supply are concession and divestiture. Nationwide, 62 water networks are currently operated by private firms under concession contracts and one as a greenfield project. In addition, four states – Sao Paulo, Parana, Tocantins and Minas Gerais – divested their regional water companies in 1997, 1998, 1999 and 2006, respectively. Brazilian private water systems serve a very large population. For example, the privatized water company of Parana serves more than 7 million inhabitants.

In Colombia, private sector participation was introduced in 1994. As of January 2012, 45 distribution systems were privately operated: 25 under concession contracts and 20 management contracts. However in contrast with Brazil, most of these networks have a small capacity of only 50000 to 100000 inhabitants.

In Chile, 17 water systems are privately operated, 10 of them were handed out under 30-year concession contracts and 7 of them were divested. Private sector participation was introduced in two phases: between 1995 and 1999 the government divested the regional water companies, and then between 2001 and 2007 it delegated other water systems under 30-year concession contracts. As a result, all urban water companies in Chile, including Santiago, are now privately operated.

Since 1991, Argentina has delegated 18 water distribution networks under long-term concession contracts (mostly 30-year). Eight of them, including one for Buenos Aires province, were canceled after a couple of years of operation and returned to the state. Ten of them are still privately managed and serve large populated areas, such as Santiago del Estero, Mendoza and Corrientes Province, Posadas and Garupa in the Misiones Province and Cordoba.

Mexico has a long experience with private water provision. The first concession contracts for water supply of Pueble, Saltillo and Monterrey were awarded in 1855, 1899 and 1904, respectively. All private systems were returned to the state in the 1940s. In the 1990s, some Mexican cities have once again expressed their interest in private water provision. Since 1993, four 25–30-year concession contracts were awarded for the water supply of the cities of Nogales, Cancun, Aguascalientes and Saltillo.

Cuba also has experience with private sector participation. In 2000, Cuba signed a 25-year concession with Aguas de Barcelona to operate the water network for the 2.2 million inhabitants of Havana.

In the Middle East and North Africa, five countries and territories – Algeria, Jordan, Lebanon, Morocco and the West Bank and Gaza – have introduced some form of private sector participation in water supply (Table 2.5). Over the years, several major cities of the region have opted for private water supply management: Casablanca, Rabat, Tangiers and Tetouan, Amman, Algiers, Constantine, Oran, Annaba, Tripoli, Bethlehem and Hebron.

Morocco is the only country of the region to have granted concession contracts. The water distribution networks of Casablanca, Rabat, Tangiers and Tetouan have been run by the private sector since 1997, 1998, 2001 and 2001, respectively. As of 2011, about 20–30 percent of the population received drinking water from private operators.

In all other countries, management contracts were used to introduce

Table 2.5 Private sector participation in water supply in the Middle East and North Africa

Country	Number of operational private water supply systems as of January 2012	Year of the first private water supply system	Forms of private sector participation that have been implemented
Algeria	4	2005	Management/Lease
Jordan	1	1999	Management/Lease
Lebanon	0	2002	Management/Lease
Morocco	4	1997	Concession
West Bank and Gaza	0	1996	Management/Lease

private sector participation. In Jordan, private sector participation in water supply started in 1999, with the management contract for the water supply of Amman. Until the end of the contract in December 2006, Jordan had the largest share of private water supply in the region: about 40 percent of its population received drinking water from a private provider. The management contract for the water supply of Amman concluded in December 2006 and the network was handed back to a public operator. In June 2011, the Water Authority of Jordan signed a five-year management contract with Veolia to manage Yarmouk Water Company in the Northern Governorates.

Algeria signed for the first time a management contract for the water supply of Algiers in 2005. Since 2005, the Algerian government has awarded three more five-year management contracts for water supply in Oran, Constantine and Annaba and El Tarf. In 2011, the management contract for the water supply of Algiers was renewed. As of 2011, about 30 percent of the Algerian population was receiving water from private operators.

Lebanon also awarded a management contract for the water supply of Tripoli in 2002. The contract concluded in 2006 and the distribution system returned to a public operator. In the West Bank and Gaza, the water systems of Bethlehem and Hebron were operated by a private operator under a management contract between 1999 and 2002.

In South Asia, India and Nepal are the only two countries to have delegated the operation of some water supply systems (Table 2.6). In India, water distribution is privately operated under management contracts, concessions and BOT/BOO in eight cities: Latur, Kolkata, Jamshedpur, Mysore City, Tiruppur, Nagpur, Hubli, Dharwad and

Table 2.6 Private sector participation in water supply in South Asia

Country	Number of operational private water supply systems as of January 2012	Year of the first private water supply system	Forms of private sector participation that have been implemented
India	8	2000	Management/Lease, Concession, BOT/BOO
Nepal	1	2010	Management/Lease

Khandwa. However, these networks are usually small and private provision remains marginal for the moment. In Nepal, the first management contract for the water supply of Kathmandu was signed in 2010.

In sub-Saharan Africa (Table 2.7), 24 countries have delegated water supply to private operators. Over the years, a wide range of contracts has been awarded: management contract, lease, concession and divestiture. There are currently 12 privately operated water systems, in Cameroon, Côte d'Ivoire, Gabon, Ghana, Mali, Mozambique, Niger, Senegal and South Africa. However, the level of private sector involvement remains limited: only four of the 12 systems are operated under concession contracts; the others are under management and lease contracts.

Côte d'Ivoire was the first country to introduce private sector participation in water supply in 1960. The form of private sector involvement chosen was a 20-year lease and the contract has been renewed since then. The private operator is in charge of water supply nationwide. South Africa has only four privately run water systems in relatively small communities: two concession contracts for water networks in Nelspruit and on the Dolphin Coast, two management contracts in Queenstown and for the towns of Phuthaditjhaba, Harrismith, and Kestell. The other two water concessions are located in Gabon and Mali and are for combined water and electricity systems. Gabon delegated the water supply of Libreville, Franceville, Port-Gentil, Oyem and Tchibanga in 1997. Mali delegated the water distribution of Bamako in 2000. Other privately operated systems are located in Cameroon, Ghana, Mozambique, Niger and Senegal and are run under management and lease contracts.

2.3 CONCLUSION AND THE WAY FORWARD

While public operation of water supply remains the predominant mode of service delivery at the global level, the above review shows that experience

Table 2.7 Private sector participation in water supply in sub-Saharan Africa

Country	Number of operational private water supply systems as of January 2012	Year of the first private water supply system	Forms of private sector participation that have been implemented
Cameroon	1	2007	Management/Lease
Cape Verde	0	1999	Divestiture
Central African Republic	0	1991	Management/Lease
Chad	0	2000	Management/Lease
Comoros	0	1998	Concession
Congo, Rep.	0	2002	Management/Lease
Côte d'Ivoire	1	1960	Management/Lease
Gabon	1	1993	Management/Lease, Concession
Gambia, The	0	1993	Management/Lease
Ghana	1	2005	Management/Lease
Guinea	0	1989	Management/Lease
Guinea-Bissau	0	1991	Management/Lease
Kenya	0	1999	Management/Lease
Madagascar	0	2005	Management/Lease
Mali	1	1994	Management/Lease, Concession
Mozambique	1	1999	Management/Lease
Niger	1	2001	Management/Lease
Rwanda	0	2003	Management/Lease
São Tomé and Principe	0	1993	Management/Lease
Senegal	1	1996	Management/Lease
South Africa	4	1992	Management/Lease, Concession
Tanzania	0	2003	Management/Lease
Uganda	0	1998	Management/Lease
Zambia	0	2001	Management/Lease

with private sector participation has been extensive in both high-income and low- and middle-income countries. Thousands of projects with private participation have been implemented in more than 84 countries. As private sector participation in water supply has existed for several centuries, history shows that the level of private sector participation can fluctuate over time with upward and downward trends. Thus, one last question remains to complete this global review: what is the current trend in terms of the level of private sector participation? Is the model of private provision expanding or declining?

While data on high-income countries is relatively scarce, several publications (such as the *Pinsent Masons Water Yearbook 2011–2012*; Pinsent Masons LLP, 2012) suggest that despite some cases of conclusion or cancellation of contracts, private sector participation in water supply in high-income countries has been increasing over recent years in terms of number of contracts and population served.

Regarding low- and middle-income countries, some case studies on canceled water contracts suggest that the level of private sector participation has been declining. However, case studies do not provide a global picture. Only a comprehensive approach based on the review of the situation in all countries can provide a clear answer about the evolution of the level of private sector participation in water supply.

With the boom of desalination markets and the increasing need for water treatment, it is true that most new private activity in the water sector concerns water treatment rather than water supply. Over the period 2006–11, 78 percent of new water projects with private participation signed were for water treatment (PPI Database).

However, every year, in all regions, new water supply and sanitation networks are delegated to private operators. Over the period 2006–11, 64 water supply and sanitation networks were delegated in 19 low- and middle-income countries: 22 in East Asia and the Pacific, 8 in Europe and Central Asia, 21 in Latin America and the Caribbean, 3 in the Middle East and North Africa, 6 in South Asia and 4 in sub-Saharan Africa.

In fact, in 2011, the total number of water supply and sanitation networks operated by the private sector reached a record high of 271 utilities in 39 countries. A closer look reveals that the total number of water supply and sanitation networks operated by the private sector in low- and middle-income countries has actually never decreased over the last 20 years (Figure 2.1). The number of new and renewed contracts implemented across the years outweighs by far the number of contracts concluded or canceled. This trend is also verified at the regional level. Latin America and Caribbean is certainly the region with the highest number of water and sewerage utilities currently operated by the private sector (151 utilities in operation), followed by East Asia and the Pacific (59), Europe and Central Asia (32), sub-Saharan Africa (12), South Asia (nine) and the Middle East and North Africa (nine).

It therefore appears that despite the cancellation of several projects frequently quoted in case studies, overall the level of private sector

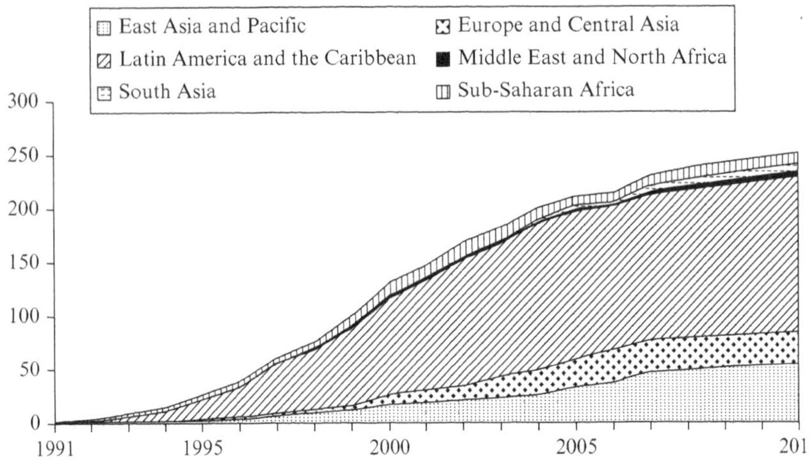

Source: Author, based on data from the World Bank/PPIAF PPI Database.

Figure 2.1 Number of water and sewerage utilities operated by the private sector in low- and middle-income countries

participation in water supply and sanitation in low- and middle-income countries has been increasing for the past 20 years. If the current trend follows the evolution of the last decade, the number of water supply and sanitation networks operated by the private sector in low- and middle-income countries should even exceed 300 by the end of 2017.

The aim of this chapter was to put discussions on the issue of private sector participation in water supply in perspective and to respond to the growing need for a neutral approach by providing a global review of the development of private sector participation in water supply in the world. Based on this review, several key characteristics (and possible misconceptions) have also been highlighted. First, private sector participation is neither a new phenomenon nor a new model for service delivery; it has been widely used for several centuries and has been experienced in 84 countries. The evolution of public and private sector engagement in the water sector, discussed in Chapter 1 in this volume, shows that the level of private sector participation in water supply fluctuates over long periods of time depending on economic factors. Increase and decrease in the level of private sector participation are not new phenomena either. Finally, the review of the evolution over the recent years shows clearly that the level of private sector participation has been expanding and not declining, despite the cancellation or conclusion of several contracts.

NOTES

1. The source of data used in this section is the World Bank/PPIAF Private Participation in Infrastructure Database (PPIAF, 2011), which tracks infrastructure projects with private participation in low- and middle-income economies.
2. Poland became a high-income economy according to the World Bank country classification released in July 2010, but is mentioned in this section for historical comparison.

REFERENCES

Doshi, V., G. Schulman and D. Gabaldon (2007), 'Lights! Water! Motion!', *Strategy and Business*, **46**, 1–16.

Pinsent Masons LLP (2012), *Pinsent Masons Water Yearbook 2011–2012*, 13th edn, available at http://wateryearbook.pinsentmasons.com/historical_editions.aspx (accessed 13 March 2012).

Public Private Infrastructure Advisory Facility (PPIAF) (2011), 'Private Participation in Infrastructure Database', available at http://ppi.worldbank.org/explore/Report.aspx (accessed 17 March 2012).

3. Good governance: the key to solving Asia's water supply challenges

Alex Wong

3.1 INTRODUCTION

3.1.1 Competing for a Finite Resource

More than a decade has passed since Ismail Serageldin, former Vice-President of the World Bank, stated in an interview to *Newsweek* in 1995 that 'Many of the wars this century were about oil, but those of the next century will be over water'. Still many parts of the developing world are no closer to solving this challenge of providing access to clean and reliable water supply to the growing population. Safe drinking water is one of the most basic human requirements and one of the Millennium Development Goals is to reduce by half the proportion of people unable to reach or afford safe drinking water by 2015. The proportion of people living in countries chronically short of water, which stood at 8 percent (500 million) at the turn of the twenty-first century, is set to rise to 45 percent (4 billion) by 2050 (Grimond, 2010).

Although about three-fourths of the Earth's surface is covered with water, availability of freshwater for human consumption is relatively very low. Shiklomanov (1999) estimated that the total volume of the earth's water is 1.4 billion km^3, of which freshwater accounts for only 35 million km^3, the rest being salt water covering the world's oceans. Moreover, of the 2.5 percent freshwater available, more than two-thirds is trapped as glaciers and permanent snow cover, leaving only less than one-third in the form of extractable groundwater and surface water. Thus, the total volume of usable freshwater supply for humans and ecosystems is 200 000 km^3, which is only 0.01 percent of all the water on earth (Gleick, 2003; Shiklomanov, 1999).

3.1.2 Asia's Plight

Asia faces the biggest challenge in the world in access to drinking water. Between 2002 and 2004, an additional 117 million people were served in the region, but progress has been insignificant and the gap is still considerable. The coverage of those in the region with access to drinking water moved up from 80.9 percent in 2002 to 81.1 percent in 2004. However, another 635 million people remain unserved (ADB, 2007). China and India, with over one-third of the world's population between them, have less than 10 percent of the world's fresh water supplies, which is being consumed at irreplaceable levels. Groundwater use in India has tripled since 1965 to over 230 km^3 each year, which amounts to a quarter of the world total. According to the World Bank, groundwater is overdrawn in over 75 percent of the blocks into which the state of Punjab is divided. Over half of the blocks of five other states – Gujarat, Haryana, Maharashtra, Rajasthan and Tamil Nadu – are judged to be in a critical or semi-critical condition, or are similarly overexploited (Grimond, 2010).

A part of developing Asia's plague of water scarcity arises from a lack of water preservation culture. It is, without doubt, difficult to price water using the usual cost–benefit analyses and to use water pricing to achieve a balance between supply and demand. But this failure to value water as a precious resource is at the forefront of the losing battle in bringing safe drinking water to large parts of developing Asia. For instance, in Punjab, the region that provides more than half of India's rice and wheat, agricultural use is responsible for the excess withdrawal of groundwater. Private tube-wells wastefully pump groundwater to the surface using cheap subsidized power in the absence of regulation and rationing. In Vietnam, high level of non-revenue water in the key urban centers is matched by Delhi's water supply system, where over 40 percent of the supply fails to reach customers and leads to severe water rationing. The Yamuna River, which is Delhi's main source of drinking water, receives sewage (95 percent of which is untreated) and industrial effluent, leaving the river clinically dead (Grimond, 2010).

However, the news is not all bad. Some parts of Asia demonstrate glittering examples of well-managed water resources and utilities amidst significant adversity in terms of supply, politics or otherwise. This chapter examines diverse and successful case studies of water delivery architectures from different parts of Asia. The relatively successful concession for East Manila granted to the Manila Water Corporation by the Metropolitan Waterworks and Sewerage System in the Philippines is compared to a similar concession granted in West Manila to Maynilad. The two Manila water concessions present one of the most obvious and closely observed contrasts between success and failure in municipal water delivery strategies using private sector participation. Moreover, the relative success of the public water utility, Phnom Penh Water Supply Authority (PPWSA), in Phnom Penh in Cambodia, presents a further contrast against similar public utilities in neighboring Ho Chi Minh City in Vietnam. There is no assumption that private sector participation or public–private partnerships (PPPs) are the only means of providing clean water supply. The generally well-documented success of PPWSA is often quoted as proof that a public water utility can be just as, if not more, successful than water delivery architectures that involve the private sector. Finally, the chapter discusses the case of Singapore, which involves a water strategy that revolves around both tight central public utility-type control and the leasing of water delivery infrastructure to the private sector through long-term PPPs. Unlike its neighbors, the challenge faced by Singapore is more in the administration of scarce water resources and less in the physical delivery of water.

Ultimately, what distinguishes a poorly organized water supply system from a well-run system? *Prima facie*, East Manila is hardly any less corrupt than West Manila and Phnom Penh is no less dysfunctional than Ho Chi Minh City. What explains the relative successes of East Manila and Phnom Penh in managing their water infrastructure when compared to its national and international neighbors? This chapter seeks to address the question of whether there is a single architecture or system of governance that will bring clean and reliable water supply to the parts of developing Asia that are currently afflicted by its absence.

3.2 PHNOM PENH: SUCCESSFUL MANAGEMENT BY PUBLIC SECTOR

In September 2010, Ek Sonn Chan, General Director of Cambodia's PPWSA, accepted the Stockholm Industry Water Award on behalf of PPWSA in recognition of PPWSA's performance in water supply and self-sufficiency. The International Award Jury noted that:

> PPWSA has successfully fought corruption and shown this can be achieved in a developing country on a large-scale basis using simple but effective management techniques that are based on well-accepted business principles and strategies. As a self-sufficient company, operating without subsidies from the state, the PPWSA today provides 24-hour service and 90 per cent coverage to a city of 1.3 million and fully recovers its costs as it continues to develop both its infrastructure and management. (SIWI, 2010)

The achievement marked a transformation in the fortunes of PPWSA and the water supply for the Cambodian capital which spanned 15 years. However, few would have predicted the drastic improvements that have occurred in a relatively short time span when Ek Sonn Chan was first appointed to the post of Director General of PPWSA in 1993. When Ek Sonn Chan first stepped into his post, only 20–25 percent of the Phnom Penh population were served by piped water, and that too for only ten hours a day. The workers were underpaid and corrupt, helping customers get illegal connections and avoid paying bills. This state of affairs, combined with very low tariffs and overstaffing, led to a situation where not even a fraction of operating costs were recovered from water users, making the utility effectively bankrupt (ADB, 2007). By the time the Director General stood up to receive the coveted Stockholm award, over 90 percent of the Phnom Penh population was connected (an increase of over 450 percent) and non-revenue water dropped from 72 percent to under 8 percent. PPWSA now collects 99.9 percent of its water bills (Duenas, 2006).

Public utilities as a whole struggle to make critical inward evaluations that help to foster change. Even when those evaluations are made, the steps needed for making and implementing recommendations are usually very difficult to undertake and can only be done gradually. PPWSA nonetheless managed to achieve this transformation, in strategic terms, in the blink of an eye. The following section explores PPWSA's journey from obscurity to championship.

3.2.1 The Era of Turmoil

Prior to Ek Sonn Chan's appointment to PPWSA in 1993, Cambodia had suffered almost two decades of social, political, economic and institutional turmoil at the hands of the Khmer Rouge. Phnom Penh and PPWSA did not escape this tumult as neglect in both infrastructure maintenance and investment left the water supply system in the Cambodian capital in disarray.

The turbulence of those two decades of Cambodian history also left its mark on the morale of PPWSA's staff where years of centralized, top–down management had contributed to a culture of bureaucracy, inefficiency and corruption. In particular, corruption was widespread, salaries and motivation were low and the working environment was poor. It was estimated that 80 percent of the staff worked for less than two hours a day (Biswas and Tortajada, 2010). This negligence had, by the mid-1990s, led to an epidemic of diarrhoeal diseases in Cambodia where 30 percent of all hospitalizations were caused by water-borne diseases, and the infant mortality rate was amongst the highest in the world (Biswas and Tortajada, 2010).

Cambodia is blessed with plentiful water resources. Groundwater is widely available both in terms of quality and quantity, and high annual precipitation (up to 3000 mm in some places) feeds the three main rivers (Mekong, Tonle Sap and Bassac) and their many tributaries. The challenge faced by the new Director General was thus not managing resources but managing infrastructure, people and, ultimately, finances.

3.2.2 The Road to Financial Self-sufficiency

One of the key strategic directions instituted for PPWSA in the early 1990s was that it would be run like a business (with the requirement of becoming financially self-sufficient) with the ultimate aim of providing the population of Phnom Penh with affordable and reliable water supply of good quality. Water was classified as an economic and social good such that any tariff charged had to be not only socially sensitive but also financially viable (Biswas and Tortajada, 2010). Given that, even today, many water utilities struggle to be financially viable as a result of the political pressure to keep tariffs at economically unsustainable levels, at the time the course that was set for PPWSA was progressive and, in hindsight, perceptive.

PPWSA made a three-pronged assault on the way towards becoming financially self-sufficient. The offensive was led first by the attack on non-revenue water closely followed by the introduction of a viable tariff structure and the initiation of a revised billing system.

Ek Sonn Chan admitted, when he took over the reins at PPWSA, that reducing the high level of non-revenue water (at over 70 percent) was a matter of 'do or die' (Duenas, 2006). With high operating expenses (partly caused by high power costs) and low water tariff, PPWSA's income barely covered half its operating expenses. A large part of non-revenue water came from physical losses, for example leakages caused by broken pipes. In 2006, 83 percent of non-revenue water was the result of pipes damaged in road construction although, by that time, the overall volume of non-revenue water had decreased dramatically from the previous decade. The decrease in non-revenue water has largely resulted from investment in a 1500 km long new distribution network and the division of this network into 41 zones. Each zone has its own non-revenue water control team which uses a pressure and flow rate data transmitter to provide online data for analyzing big leaks in the system. An incentive and fine system has been implemented for the teams that discover decrease or increase, respectively, in non-revenue water in their zones. Hotlines have also been established for reporting leaks and leak repair teams are constantly on standby to respond within two hours of reporting (Duenas, 2006).

The more highly publicized cause of non-revenue water has been commercial losses including those arising from illegal connections and lack of metering. As of 2006, PPWSA had managed to meter all 147 000 connections in the Cambodian capital (Duenas, 2006), and had upgraded its metering equipment to the more accurate volumetric water meter class C. The consumer base has been, and is constantly, updated using a computerized billing system. In many respects, this was the easy part: the difficult part was confronting the issue of illegal connections and collection problems. PPWSA confronted the issue head-on with its trademark carrot-and-stick approach by imposing heavy penalties on the holders of illegal connections and by ensuring that PPWSA staff associated with illegal connections were removed or punished, whilst providing incentives to those providing information on illegal connections. High-profile non-payers were also faced with the fact that their water supply would be cut off (Kea, 2007).

With measures firmly in place to achieve the objective of reducing non-revenue water, PPWSA next confronted the challenge of increasing revenues. Before 1994, the domestic water bill being charged by PPWSA was approximately Cambodian Riehl (KHR) 166/m^3 (US$0.04/$m^3$). However, given the low levels of service being offered by PPWSA, a number of consumers were seeking the services of private water vendors (most of which used the utility's resources and water for their own profit) and often paid significantly more than what was paid to PPWSA. By 1995, PPWSA had managed to make 24-hour uninterrupted service available in Phnom Penh and with the domestic water bill starting to creep towards KHR 250/m^3 (US$0.06/$m^3$) by that time, private water vendors were starting to be squeezed out of the market (Biswas and Tortajada, 2010). In 1997, increasing block tariffs were introduced and currently the rate for the lowest block of 0–7 m^3 is KHR 550/m^3 (US$0.13/$m^3$). Still the water tariff in Phnom Penh remains low by regional standards, with an average of US$0.22/$m^3$ since 1997 compared with East Manila at US$0.33/$m^3$ in 2008, Kuala Lumpur at US$0.45/$m^3$ in 2007 and Jakarta at US$0.70/$m^3$ in 2005 (Das et al., 2010). However, not all the residents of Phnom Penh can afford this tariff. PPWSA has made provisions to ensure that the poorest residents of Phnom Penh receive the benefits of the turnaround in PPWSA's fortunes. It established a revolving fund to finance domestic water connections to help more than 15 000 families in 123 urban poor communities with subsidized tariffs and connection fees which can be paid in instalments (ADB, 2007). Poor households are entitled to receive subsidies of 30–100 percent of the connection fee depending on their financial condition, which is jointly evaluated by a committee comprising representatives from the utility and local communities (Biswas and Tortajada, 2010).

3.2.3 A Culture of Change

Challenges in the twenty-first century are seldom solved without addressing the importance of human resources, people development and employee motivation. The challenges confronted by PPWSA in these areas were no less demanding than the tests it faced with respect to the physical degradation of the water transmission system and the ensuing fiscal difficulties, yet PPWSA set about its task in contending with this challenge with equal boldness.

The education and motivation of PPWSA staff was the starting point. The staff salaries were increased substantially during the reform process, with a staff member in a managerial position receiving ten times more in 2008 than in 1993. Profits are shared with the employees and the utility has created a retirement system, which is the first of its kind in Cambodia. Employees are evaluated four times a year with financial incentives being given for good performance, and disciplinary actions such as delays in pay rises or even pay cuts being implemented for poor performers. Substantial training is also provided by PPWSA (Das et al., 2010) as it professionalized a responsible, disciplined and self-motivated workforce. Ultimately, well-performing staff members are promoted and greater responsibility is placed on higher management. A culture of meritocracy is encouraged as PPWSA demonstrates that investing in staff yields radical results.

The challenge of human resource management extended to the Phnom Penh populace at large, who for years under the Khmer Rouge regime had been provided water for free. A campaign to educate PPWSA's customers was implemented where it was demonstrated that PPWSA would not be able to continue supplying water if water bills were unpaid.

3.2.4 From Third World to First

PPWSA was supported by international donors in its effort to achieve its objectives. The level of political support it received from the Cambodian government has also been steadfast. However, PPWSA managed to become entirely self-sustainable as it benchmarked itself against the best operators in the world and it is still looking to improve. At the Stockholm International Water Awards 2010, Ek Sonn Chan revealed that PPWSA intends to extend the water network into Phnom Penh's suburbs and surrounding regions, whilst reducing water losses from around 5.9 percent to 4 percent by 2020 (which will place PPWSA in the same league as Singapore and Tokyo).

3.3 MANILA WATER: EAST VERSUS WEST ZONE PRIVATIZATION

Metro Manila is an urban conglomerate of nearly 11 million inhabitants and covering an area of approximately 2400 km^2. The Metropolitan Waterworks and Sewerage System (MWSS), a public utility, has historically been tasked with providing water and sewerage services to Metro Manila.

With a national policy that prioritized the reduction and service of public debt, the MWSS (which itself owed huge liabilities to international financial institutions) had, to an extent, become a victim of underinvestment in its infrastructure. But that was only part of the story. By the mid-1990s, the MWSS was an inefficient organization, wasteful of resources and government money and unable to provide even basic standards of water and wastewater services. The MWSS supplied water to less than two-thirds of its coverage population and, even then, only for an average of 16 hours a day. Water losses from leakages and pilfering were at a staggering 63 percent and efficiency levels amongst employees were poor. According to a study by the Asian Development Bank, Manila ranked worst amongst all major cities in East Asia in almost all statistical categories of water delivery for utilities (Wolf, 2007).

The circumstances were made worse by the fact that it was often the poorest parts of the Metro Manila population that suffered the most severe consequences of the MWSS's inability to supply clean water. Without access to piped water, the poorest households often had to buy bottled water at significant price premiums of up to ten times the cost of piped water and, to add insult to injury, most of this bottled water was bottled with water taken from illegal connections (Wolf, 2007).

Drought ravaged the Philippines in the summer of 1995 and the then Philippines President Fidel Ramos was given emergency powers by the Congress through the Water Crisis Act 1995 to enter into contracts with private companies to assist the government in dealing with the drought. These powers served as the formal basis to privatize the MWSS in the hope that private sector participation would in some way shake the MWSS out of its lethargy and bring a new and brighter future to water supply in Metro Manila.

3.3.1 From Public to Private Hands

Although the private sector is commonly perceived to be more efficient and independent than a public utility like the MWSS, the privatization of basic public services such as water tends to attract partisan protest in almost any jurisdiction. It was thus surprising that opposition to the MWSS privatization from the Metro Manila populace was relatively muted. In fact, the *Manila Standard* (one of the main local Manila newspapers) reacted to the change with some positive optimism. It wrote that consumers 'had grown convinced that no other arrangement could be worse than the present situation . . . in which the poor bear the greatest economic burden of bad water service' (Wolf, 2007).

The International Finance Corporation (IFC), the private sector arm of the World Bank, was brought in to assist the MWSS with the privatization process and the privatization programme focused on water distribution and sewerage rather than on sourcing of raw water. Metro Manila was divided into two services areas – the East Zone and the West Zone. The IFC and the MWSS were of the opinion that this would best enable the MWSS to gain the benefits of scale and scope whilst promoting competition. Metro Manila was large enough to attract internationally experienced water companies and the scale of the service problems afflicting Metro Manila at that time required investment fixes that would overstretch any available public funds. It was decided that the MWSS would bid out the water services for each of the two zones to private sector concessionaires on a build–operate–transfer (BOT) basis where operation, maintenance and new investments are handed over to the private sector while the MWSS maintains ownership of the system assets and responsibility for raw water supply. Each concession is for 25 years, at the end of which the concessionaires will return the entire system (including upgrades) to the MWSS.

In January 1997, the Maynilad Water Services Inc. (Maynilad) and the Manila Water Company Inc. (Manila Water) emerged as the winning bidders for the MWSS's West Zone and East Zone, respectively. Both consortia were partnerships between renowned international and local corporations. The conclusion of the bidding process was a privatization undertaking hailed by the World Bank as the first large-scale water supply privatization in Asia (Freedom from Debt Coalition, 2005)

3.3.2 Achievements of the East Zone

In a relatively short period of eight years after Manila Water took over the East Zone, each category of key operational statistics of water service had improved dramatically. Non-revenue water had almost halved from 63 percent to 35 percent and tariff collections had almost doubled from 440 ml per day to 864 ml per day in billed volume. By 2009, 24-hour water supply, which was only available to 26 percent of the population when the concession began, was available to 99 percent of the East Zone population. Non-revenue water as of 2010 had dropped to 13.5 percent and Manila Water now serves 6.2 million customers. One key statistic, however, has been the subject of criticism. In January 2005, the tariff stood at Philippine peso (PHP) 8.92/m^3 and this had remained virtually unchanged from MWSS's tariff in 1997 at PHP 8.78/m^3. However, this was several times the tariff that Manila Water had originally submitted as part of its bid in 1999 and pointed to the conclusion that Manila Water had underestimated the tariff at the time of bid submission. This apparent failing should however be balanced against a strong counterargument that the service standards in the East Zone have shown a dramatic improvement over the eight years and the tariff has remained virtually unchanged in spite of the post-Asian financial crisis inflationary spiral that blighted the region and the Philippines during this period.

The management of the Filipino peso devaluation crisis that blighted the Philippines economy at the end of the last millennium should not be underestimated. The Filipino peso more than halved in value against the US dollar during the seven years from 1997 to 2004. This had a crushing impact on the economics and cash flow of the concessionaires including Manila Water where the underlying revenue was collected in pesos. As part of the concessionary structure, Manila Water had agreed to service (through concession fees) the MWSS's existing debt (including US dollar-denominated debt). Most of Manila Water's new credit lines (borrowed to pay for the East Zone infrastructure and service improvements) were also US dollar denominated. Admittedly, Manila Water had a smaller portion of the MWSS debt to service through the concession fees as compared to Maynilad, but on a plain reading of the statistics (that is, comparing the service standards and tariff levels of the two concessions measured against overall inflation/devaluation over time), it would be difficult to deny that the Manila Water consortium had largely succeeded in carrying out the tasks it had been set when the East Zone concession was awarded in 1997.

3.3.3 Management of Manila Water

Tony Aquino, the Chief Executive Officer (CEO) of Manila Water, faced many of the same issues that confronted Ek Sonn Chan in Phnom Penh when he first took over the reins at PPWSA: high levels of non-revenue water and a stagnant and entrenched culture of bureaucracy. Tony Aquino nonetheless attacked these challenges with great passion.

The old MWSS work culture was replaced with a new private sector mindset. Performance reviews, performance-linked pay, investment in employee training and a better work environment (from refurbished and modernized offices to fostering a sense of community and belonging) became part and parcel of everyday life at Manila Water. At the start of the concession, Manila Water had absorbed 2165 MWSS employees. By the end of 2004, Manila Water had 1541 employees (90 percent former MWSS staff members) with the others having left through natural attrition or the early retirement program. The human resource efficiencies also had a clear impact on labor productivity. Manila Water managed to reduce the measure of labor productivity from 8.4 employees per 1000 connections at the start of the concession to 2.8 employees per 1000 connections by the end of 2004 (Wolf, 2007).

The organizational structure of Manila Water was overhauled to reflect the new business culture and to ensure the amplification of a new customer-oriented focus. The East Zone was divided into seven business areas. Each of these business areas was subdivided into Demand Monitoring Zones (DMZs), each covering approximately 4000 connections. Each DMZ was further subdivided into District Metering Areas (DMAs) of about 500–1000 connections each. More often than not, the officers responsible for customer service in a DMA live in the same DMA as their customers and are in regular contact with them. The DMA officers form part of the wider DMZ teams empowered to oversee and address the overall needs of the DMZ relating to water supply and demand, water leakage and theft and any other customer concerns. DMZ feedback, concerns and ideas for providing better customer service are shared between DMZ teams and fed back to upper management, creating what Tony Aquino describes as a 'people-empowered learning organisation' (Wolf, 2007). Manila Water believes that this decentralized structure improves customer service by providing cheap, reliable and customer-focused water services and that Manila Water would be in a position to align the business interests of the company and its shareholders with the wider social and development interests of the communities that it serves (Wolf, 2007).

By 2005, more than half of the 2700 km of total pipeline covering the East Zone was newly laid by Manila Water. Due to underinvestment under the MWSS regime, the old pipeline network simply could not cope with the higher water pressures generated by the upgraded pumping stations. This was a step on the way to overcome the challenge of non-revenue water, but it was not the only step. Although in comparison to other countries in the region, for example Singapore, non-revenue water in the East Zone remained relatively high at 35 percent in 2005, it was still well below the target of 45 percent agreed with the MWSS and almost half of 63 percent when the concession started. The MWSS does not charge Manila Water for raw water. Therefore, reduction in non-revenue water does not result in financial savings for Manila Water but it increases revenues.

The establishment of DMZs and DMAs was instrumental in reducing non-revenue water in the East Zone. Flow measurement devices were installed between the different areas such that territory teams could monitor inflows and outflows of water and their billed volume. Areas with high non-revenue water were thus identified and isolated so that its cause could be eliminated; whether illegal connections, broken pipes or otherwise. Community water connections were established in the Barangays (the poorest areas of the zone where individual inhabitants and households often could not afford their own connections), where the people were desperate for a reliable and cheap alternative to bottled water. The communities were then encouraged (with support from Manila Water) to form local water cooperatives and take responsibility for distribution within the Barangay. This strategy not only helped to achieve Manila Water's aim of reducing non-revenue water (in tackling illegal connections within the Barangays) but also brought considerable social and health benefits to the poorest parts of the East Zone (Wolf, 2007).

The MWSS privatization is often quoted as an example of privatization gone wrong but the West Zone's struggles (see section 2.5.2) are often not distinguished from the successes of Manila Water in the East Zone. As of 2007, Manila Water served at least 2.5 million customers who in 1997 still had to rely on vended water at a cost several times greater than the piped water they received (Wolf, 2007). No amount of hindsight could have predicted the condition of Metro Manila's water supply network had the MWSS not been privatized; but given the MWSS's variable history of achievement in providing clean and reliable water supply to Metro Manila, and the external and internal economic shocks that the Philippines as a whole had to endure in the post-1997 era, the counterfactual argument that the East Zone is most likely better off under Manila Water's stewardship than if the MWSS continued as the guardian of Metro Manila's water supply system is correct.

3.4 PUBLIC UTILITIES BOARD SINGAPORE: SAVING EVERY DROP

With over 5 million people living on approximately 700 km^2 of land, Singapore is one of the most densely populated countries in the world. High population density places immediate pressure on water resource organization, with catchment area management and water storage being key issues. Singapore faces the additional challenge of having no natural aquifers or groundwater. Out of a list of 190 nations, Singapore is ranked as low as 170 in fresh water availability (Tan, 2011).

Up until the last few years, Singapore relied on Malaysia for its raw water imports through two water treaties; one that expired in 2011 and the second that will expire in 2061. One of the key events that forced Singapore's surrender by the British in 1942 was the blowing up of water pipelines from Malaysia by the invading Japanese army. Therefore, upon gaining independence in August 1965, the island republic felt vulnerable enough to register the water agreements with Malaysia with the United Nations Charter Secretariat Office in June 1966 (Seneviratne, 2005). In 1970, when referring to the water agreements with Malaysia, former Singapore Prime Minister, Lee Kuan Yew, said, 'If these chaps do not observe the agreements, it will be a very serious matter for us . . . it is a matter of life and death . . . it can lead to war' (Tan, 2011).

The huge challenge of striving to achieve self-sufficiency in water was thus placed in the hands of Singapore's Public Utilities Board (PUB). PUB, as its name suggests, is a public utility run independently as a statutory corporation with oversight and, where needed, funding from the government. Over the last 15 years, with the onset of new water technologies, PUB has accelerated its programs to achieve self-sufficiency in water, resulting in PUB receiving the Stockholm Industry Water Award 2007 and, in a moment of significant symbolism, allowing the earlier of the two water agreements with Malaysia to expire in 2011.

3.4.1 The Four Taps

In spite of the well-documented shortcomings (limited catchment areas and groundwater availability), Singapore has been blessed with two resources that helped it to achieve self-sufficiency in water. The first is that it is surrounded by water (albeit saltwater) and the second is that it receives more than twice the global average rainfall (2400 mm annually). Moreover, there is a sense of seriousness in undertaking this task that reaches up to the highest levels of government and politics.

Armed with these resources, PUB has implemented its strategy to create a greater number of 'taps' to address Singapore's water crisis, and the 'tap count' currently stands at four. The first tap remains the import of water from Malaysia under the agreement that expires in 2061. The Singapore government does not appear to have made a firm decision on whether it intends to let this water import agreement expire in the same way that the first agreement was allowed to expire in 2011, although in a seemingly pointed speech to Parliament in 2002 the then Minister for the Environment, Lim Swee Say, stated that Singapore could be completely self-sufficient in water by 2061 (Seneviratne, 2005). The other three taps, however, required more innovation, determination and organization. These are desalination, water recycling and better management of catchment and storage resources.

If cost was not an issue, the most obvious and immediate solution to solving a water resource crisis on an island would be to opt for desalination. As desalination is a power-intensive operation, PUB recognized that desalination can only be a part of the solution to manage water resources and not the only solution. What is of greater interest regarding Singapore's first foray into the desalination industry was the manner in which it procured expertise. Until 2001, when the tender for Singapore's first desalination plant was launched, public procurement in Singapore was carried out mostly in the traditional top–down fashion. The tender for Singapore's first desalination plant in Tuas broke new ground in that the government opted to involve the private sector to develop the project through a BOT model. It may not have appeared apparent when the tender was launched, but the appointment of Hyflux as the winning bidder for the project sparked the beginning of a water industry revolution in the city-state. By the time Hyflux won the tender for Singapore's second desalination plant in 2010 (with double the capacity of the first plant), Hyflux had become a global name in water technologies and Singapore had become a regional and global centre for excellence for the water industry, with the annual Singapore International Water Week attracting immense interest worldwide. Together with recycled 'NEWater' (see below), desalination makes up approximately 40 percent of the current daily consumption of water in Singapore. By 2061, Singapore expects this number to rise to 80 percent (Tan, 2011).

'NEWater' is a uniquely Singaporean term. In summary, NEWater is high-grade reclaimed water produced from treated used water that is further purified using advanced membrane technologies (microfiltration, reverse osmosis and ultraviolet disinfection), making the water ultra-clean and safe to drink (Guan, 2007). Originally supplied only for non-domestic industrial uses, NEWater is ultimately intended to form a large part of both domestic and non-domestic consumption in Singapore. Therefore, an international panel of scientists and specialists was assembled to provide independent advice on the suitability of NEWater for potable use. The panel ultimately concluded that NEWater is of a consistently high quality and is well within the US Environment Protection Agency and the World Health Organization's standards for drinking water (Guan, 2007). As of 2011, Singapore had five NEWater plants, three of which have been procured using the BOT model and allowing Singaporean conglomerates such as Keppel and Sembcorp to apply their skills in the field of NEWater technology. A deep tunnel sewerage system has been built on the east side of the island to maximize the used water management potential, and this system flows directly to the large Changi Water Reclamation Plant, adjacent to a large NEWater facility. A similar system is in the works on the west side of Singapore.

The final tap has a longer history. From the time of independence in 1965, Singapore started a general cleaning-up process that was at least partially aimed at making it easier to capture and store rainwater. Industry and commerce were shifted into estates, and pig and duck farms were closed. By the time the Marina Barrage was completed in 2008 (with the aim of forming a fresh-water reservoir within the commercial and financial heart of the city), Singapore's water catchment area amounted to two-thirds of its land area. Currently, there are 17 fresh-water reservoirs compared to only three in 1965. As much as possible, rainwater is fastidiously collected in streets and ponds, even on skyscrapers and bridges, before being taken by a system of drains, rivers and canals into reservoirs. Singapore is one of the first cities in the world to tap an urbanized catchment area (Guan, 2007).

3.4.2 Every Drop Counts

Like PPWSA, PUB also recognizes the critical role of education and encouraging a culture of change in managing and valuing scarce water resources. Subjected to constant water-consciousness campaigns, Singaporeans are obliged to install low-use taps and toilets, and expected to be equally thrifty with their showers and washing machines. The first 'Water is Precious' campaign in 1971 reduced water consumption by 5 percent (Tan, 2011). Four decades later, the public education drive continues in schools, factories and the media, whether it is exemplifying 'water-efficient homes' with toilets that use cistern water-saving bags, or mandating self-shutting delayed-action taps in buildings. A water conservation tax has also been introduced, the current rate for which is 30 percent for the first 40 m^3 per month, rising to 45 percent beyond that. Domestic water use per person has fallen from 165 litres a day in 2003 to 155 litres a day in 2010. PUB's aim is to reduce per capita water consumption to 140 litres per day by 2030 (Tan, 2011).

3.4.3 Turning Vulnerability to Strength

The Singapore Prime Minister, Lee Hsien Loong, believes that the city-state's prior water dependency on mainland Malaysia has been a blessing in disguise, but that Singapore has now turned the tables and transformed a 'vulnerability into strength' (Seneviratne, 2005). In 1963, Singapore had to endure ten months of water rationing as a result of a drought in the Tebrau River area of Johor, Malaysia. Less than half a century later, Singapore leads the world in water resources management and water recycling innovation. PUB expects the gross domestic product (GDP) contribution from the water sector to reach Singapore dollar (SGD) 1.7 billion by 2015, and Singapore now looks upon water scarcity as a matter that is more likely to evoke hope and opportunity than anxiety and despair (Tan, 2011).

3.5 MANAGEMENT AND MISMANAGEMENT

3.5.1 The Art of the Possible

In recent years, the words 'water' and 'crisis' have become synonymous with one another. Water, it is said, is the new oil: a long-squandered resource that is growing increasingly expensive and is soon to be overwhelmed by insatiable demand. But as the trio of case studies above has shown, bringing relatively cheap, clean and reliable water to the most challenging parts of urban Asia is conceivable; it is an art – but the art of the possible (Grimond, 2010).

Certain key factors however have to be available to turn these possibilities into reality. Funds are one important factor and these were available in Phnom Penh (through multilateral help), Manila (through a privatization structure) and Singapore (with PUB's deep balance sheet and government support). Political will and strong policy support are not always forthcoming, but they were also at hand in the difficult political environments in Phnom Penh (where the government supported Ek Sonn Chan and PPWSA in spite of their challenging some important elements of the establishment on water bills) and Manila (where politics is notorious for meddling with regulation).

3.5.2 The Factors that Count

The art of success in each of these examples is in managing the varied resources and commitments into a complete product. In Singapore, it is a system of management that breeds success with the professionally run PUB using highly motivated (and paid) professionals who are not afraid to bring in private sector partners to do what they believe needs doing, so that money is invested in everything from dams to drains to membranes and bioreactors. Arguably, faceless technocrats can be thrown into a well-managed and highly structured system and the end product will remain constant. Where there is no such systemic quality, strength of leadership becomes the determining factor, and Phnom Penh and the East Zone of Manila are fortunate to have found Ek Sonn Chan and Tony Aquino.

In Phnom Penh, it became clear to PPWSA that one of the most significant causes of non-revenue water was road construction. It was apparent that the transport department (and its contractors) in the Cambodian capital had not been particularly stringent with checks for underground services when carrying out their work. Rather than attacking them head on and risking a bureaucratic war that was likely to make little headway in resolving the concern, Ek Sonn Chan instead strengthened PPWSA's own leak repair teams so that leaks could be spotted and addressed as efficiently as possible.

The ability and knowledge to address strategic challenges and practical matters distinguishes PPWSA from its neighbors. At about the same time that Ek Sonn Chan was appointed to PPWSA, Ho Chi Minh City in Vietnam also received a loan of US$65 million to improve its water supply and sanitation network (ADB, 2004). In February 2011, Ho Chi Minh City received another loan of US$138 million from the Asian Development Bank (ADB) as part of the Multi-tranche Financing Facility of the Water Sector Investment Program for Vietnam as a whole (ADB, 2011). In spite of this, as of 2008, Ho Chi Minh City's levels of non-revenue water still remained relatively high at 40 percent and its coverage stood at around 75 percent of the city's population (ADB, 2008). Having said that, Saigon Water Corporation (SAWACO), the water utility in Ho Chi Minh City, runs a relatively commendable service in that it maintains an operating profit and had a respectable five workers per 1000 connections as of the end of 2006 (ADB, 2008). Although the relatively unfavorable comparisons that SAWACO receives when compared to PPWSA can be partly explained by the relatively different sizes of the cities (Ho Chi Minh City with a population of about 8 million, and Phnom Penh at about 2.5 million), PPWSA is, undoubtedly, the winner in the race for water supply excellence in South-East Asia.

A similar comparison can be drawn between Manila Water in the East Zone of Metro Manila and the failures of its neighbor, Maynilad in the West Zone. In fact, the statistical comparisons are so strikingly poor for Maynilad that it is often cited as an example of why privatization of a public service water utility should not be undertaken (Freedom from Debt Coalition, 2005). Non-revenue water actually increased from 63 percent in 1997 to almost 70 percent by the end of 2005, and even though the West Zone serves a larger part of the overall Metro Manila population, by 2005 Manila Water was actually billing some 28 percent more than Maynilad in the West Zone. In contrast to Manila Water, Maynilad was unable to stem the losses that had been inflicted on it by the Asian financial crisis and by 2001, Maynilad had stopped paying the concession fee to the MWSS. By 2003, it was decided that Maynilad's position as the West Zone concessionaire was no longer tenable and the MWSS terminated the concession and stepped into the company's shoes (Wolf, 2007). A new contract was awarded to a consortium of DM Consunji Inc., Philippines (42 percent), First Pacific, Hong Kong (42 percent) and SUEZ (16 percent) in 2006, with significantly smaller investment commitments (US$504 million) than the first concession.

Maynilad's failure was not all of its own doing, as a significant part of its financial pressures came from the uneven distribution of the MWSS debts amongst the East and West Zone concessionaires (with the West Zone taking up 90 percent of the MWSS's previous debts). Yet some poor management decisions certainly contributed to Maynilad's downfall. Contrary to what seemed to be the conventional wisdom of aggressively attacking non-revenue water, Maynilad failed to undertake that task effectively. Unlike its across town neighbor, there was no clear or innovative vision to create a decentralized and cross-functional organizational structure to increase accountability and performance and, ultimately, revenue. Manila Water put together the right 'soft touches' as part of its strategy, such as a whole regulatory planning team to understand and ultimately liaise with the MWSS as the regulator. This proved crucial especially during and immediately following the years of peso devaluation where tariff adjustments eventually made or broke each concession.

3.5.3 Moving Beyond the Manager

In 2005, Abdullah Al Hussayen, Saudi Arabia's Minister for Water and Electricity, said that he believed the secret behind Singapore's water success story is in 'keeping unaccounted-for-water low, pricing water to reflect its scarcity value and encouraging everyone, be it households, communities or the industries, to use and manage water wisely' (Seneviratne, 2005). Without missing a beat, this quote can almost be replicated to describe the successful water supply stories for Phnom Penh and the East Zone in Manila. The stories of PPWSA, PUB and Manila Water demonstrate that good policies, direction and funding can only take one so far – good management is needed to take that additional step towards becoming a leader in the service of water supply. However, there is one more step in the process for PPWSA and Manila Water. They must prove that they can continue to succeed without their talismanic leaders. In that regard, Singapore has in many ways commoditized the management of PUB. Ek Sonn Chan and Tony Aquino will need to do likewise with PPWSA and Manila Water so that they can achieve their high management expectations through systemic excellence rather than through force of character.

3.6 CONCLUSION

More than a decade into the twenty-first century, there have not been any true water wars, although with a commodity as scarce as water it would be complacent to think that can never happen. The Six-Day War in 1967 was arguably partly prompted by Jordan's proposal to divert the Jordan River to Israel's dismay and, to this day, Israel extracts some 65 percent of the upper Jordan River to the chagrin of many in the West Bank. Even relatively benign neighbors like Singapore and Malaysia flexed their muscles over a water disagreement in the late 1990s (although that has largely subsided, given Singapore's water self-reliance policies). Of greater concern is the surprisingly durable Indus Water Treaty of 1960 between India and Pakistan. Having endured and survived no less than three wars between the two nuclear neighbors, it is coming under increasing pressure as a result of India's economic boom (leading to new hydropower projects in Kashmir and a water diversion scheme from the Kishanganga River in Kashmir) and unhappiness from Pakistan that the spirit (but not the letter) of the agreement has been infringed (Grimond, 2010).

Any such conflicts would, however, be an indictment of our inability to solve an eminently solvable issue. There is no logical or physical reason why access to safe drinking water should be so difficult to accomplish. Many of the obstacles are political, some are cultural, and these obstacles are also hampered by water's astonishing ability to defy economic analysis (Grimond, 2010). But the strategic remedies are no secret – the examples of Singapore, Phnom Penh and Manila illustrate that setting the right policies and strategies coupled with changes in behavior (in terms of valuing water both physically and fiscally) and financial resources to implement these goals are the tripartite keys to success – but the final piece that is always needed to complete the puzzle is, undoubtedly, good management.

REFERENCES

Asian Development Bank (ADB) (2004), 'Project completion report – Ho Chi Minh water supply and sanitation project', available at http://www.adb.org/Documents/PCRs/VIE/pcr-vie-25095.pdf (accessed 20 October 2011).

Asian Development Bank (ADB) (2007), 'Country water action: Cambodia Phnom Penh Water Supply Authority: an exemplary water utility in Asia', available at http://www.adb.org/water/actions/cam/PPWSA.asp (accessed 20 October 2011).

Asian Development Bank (ADB) (2008), 'Socialist Republic of Viet Nam: preparing the Ho Chi Minh water supply project', Technical Assistance Report, available at http://www.adb.org/Documents/TARs/VIE/41070-VIE-TAR.pdf (accessed 20 October 2011).

Asian Development Bank (ADB) (2011), 'Proposed multitranche financing facility Socialist Republic of Viet Nam: water sector investment program', available at http://www2.adb.org/Documents/RRPs/VIE/41456/41456-01-vie-rrp.pdf (accessed 28 October 2011).

Biswas, Asit and Cecilia Tortajada (2010), 'Water supply of Phnom Penh: an example of good governance', *International Journal of Water Resources Development*, **26**(2), 157–72.

Das, Binayak, Ek Sonn Chan, Chea Visoth, Ganesh Pangare and Robin Simpson (2010), 'Sharing the reform process: learning from the Phnom Penh water supply authority', available at http://data.iucn.org/dbtw-wpd/edocs/2010-046.pdf (accessed 20 October 2011).

Duenas, Christina (2006), 'Water champion: Ek Sonn Chan, pulling the plug on nonrevenue water', Asian Development Bank, available at http://www.adb.org/water/champions/chan.asp (accessed 20 October 2011).

Freedom from Debt Coalition (2005), 'ADB and the privatization of Metro Manila's water distribution system: corporate greed takes over public welfare', paper prepared for the Regional Conference on the Asian Development Bank Water Policy, Quezon City, Philippines, available at http://www.adb.org/water/policy/consultations/NGO-FORUM-PHI.pdf (accessed 20 October 2011).

Gleick, P. (2003), 'Soft path's solution to 21st-century water needs', *Science*, **320**, 1524–8.

Grimond, John (2010), 'For want of a drink', *The Economist*, 20 May, available at http://www.economist.com/node/16136302 (accessed 20 October 2011).

Guan, Yap Kheng (2007), 'Sustainable water management – the Singapore water story', available at http://www.wcponline.com/pdf/0709Guan.pdf (accessed 20 October 2011).

Kea, Puy (2007), 'Assassination threats couldn't stop Cambodian official in providing safe drinking water', MaximsNews Network, available at http://www.amic.org.sg/new/adb/Cambodia.pdf (accessed 20 October 2011).

Seneviratne, Kalinga (2005), 'Technology turns water weakness into strength', Asian Development Bank, available at http://www.adb.org/water/actions/SIN/Technology.asp (accessed 20 October 2011).

Shiklomanov, Igor (1999), State Hydrological Institute (SHL, St Petersburg) and United Nations Educational, Scientific and Cultural Organization (UNESCO, Paris).

Stockholm International Water Institute (SIWI) (2010), 'Water – everybody's business', available at http://www.siwi.org/documents/WWW_PDF/Media/2010/SIWA_Water_Everybodys_Business_20100908.pdf (accessed 3 September 2011).
Tan, Elgin (2011), 'Quenching Singapore's thirst', *Straits Times*, 5 September, available at http://www.greenbusinesstimes.com/2011/09/05/quenching-singapores-thirst-news/ (accessed 3 October 2011).
Wolf, Christian (2007), 'Manila Water – Privatising Regulated Services in a Developing Country', available at http://papers.ssrn.com/sol3/papers.cfm?abstract_id=1133437 (accessed 3 September 2011).

4. The UK water industry: infrastructure, governance and procurement

Akintola Akintoye and Suresh Renukappa

4.1 INTRODUCTION

In the early part of the twenty-first century, there is an intense debate over the water industry in the United Kingdom (UK) as well as in other countries. The water industry plays a key role in meeting economic, environmental and social objectives at both national and international levels (Water UK, 2010). There is a general view that the companies and authorities that constitute the UK water industry provide reliable and safe water supply and sewerage services to UK citizens and businesses.

The water industry is large and unique. Goldman Sachs (2008) estimates the size of the global water market at US$425 billion with an annual growth rate of 4–6 percent. Maxwell (2005) estimates that worldwide revenues from the water industry range from US$350 billion to US$650 billion. However, according to Abbott and Cohen (2009), the industry structures in the water sector vary across the world in terms of the range of activities that individual businesses undertake, the geographical size, the number and nature of customers they service, the extent of private sector involvement, the scope of competition, the nature and extent of regulation, and the responsible bodies for oversight and/or implementation of regulation.

The OECD (2008) projects that 67 percent of the world's population is expected to lack a connection to public sewerage in 2030. It also estimates that 44 percent of the world is currently living under water stress and an additional 1 billion people will be living in areas with severe water stress by 2030. Developing and managing water and wastewater infrastructure constitutes an urgent and major challenge because it is suggested that by 2025 2 billion people will be living in areas with water shortages (Black and King, 2009).

The need for maintaining ageing physical water infrastructure and building infrastructure has grown dramatically around the world in the last decade (Weber and Alfen, 2010). The key challenges of the water and wastewater sector include the need to expand access to water and wastewater services, invest in building new infrastructure and replacing and maintaining ageing infrastructure, and address security and environmental concerns (OECD, 2007). The massive pressure for new water-related infrastructure stems from a conflation of interrelated factors, including population increase, migration flows towards cities, deterioration and obsolescence of existing assets, and the globalization of supply chains (WEF, 2011).

Addressing the above challenges will require large capital investment for new infrastructure, ongoing investment in maintenance, repair, upgrading and operation of existing facilities, and integration of new ways of organizing, funding and producing revenues and reducing costs within utilities (Palaniappan et al., 2007). There is a range of estimates of required annual investment in the water and wastewater infrastructure sector. According to the World Bank (2009), it is expected that the world will invest approximately US$23 trillion in water infrastructure by 2030. Ashley and Cashman (2006) noted that in high-income countries, 0.35 to 1.2 percent of gross domestic product (GDP) will be required to finance water-related infrastructure. In middle-income countries, 0.54 to 2.60 percent of GDP is needed and in low-income countries, an estimated 0.70 to 6.30 percent of GDP will be needed. According to the OECD (2009), halving the proportion of people without access to drinking water and sanitation by 2015 requires an investment of approximately US$72 billion per year. However, the water sector is experiencing renewed and significant interest from investors due to increasing global scarcity and very large end-markets.

The UK water industry today faces its greatest challenges for a generation. The enormity of the task of ensuring security of supply in the twenty-first century while simultaneously addressing environmental, social and economic sustainability issues is fast pushing the UK water industry towards its greatest transformation in decades (Marketforce and KPMG, 2010). Hence, increasingly, questions are being raised about the appropriateness of the existing water industry business models.

This chapter presents the status of the UK water industry and the different business models that are used for water and wastewater services in the UK. It draws upon the literature in the areas of water industry and infrastructure procurement. Sections 4.2 and 4.3 discuss the status of the UK water industry and distinctive characteristics of the water infrastructure. Sections 4.4 and 4.5 are devoted to privatization and business models and governance of the UK water industry. In section 4.6, water-related infrastructure investment gaps and opportunities are presented. Section 4.7 looks at water-related infrastructure procurement options. In section 4.8, the main challenges for the UK water industry are discussed. Finally, a conclusion on the future of the UK water industry is presented.

4.2 OUTLOOK OF THE UK WATER INDUSTRY

According to Water UK (2010), the UK water industry is generally known for its diversity, complexity, universality, reliability and high quality of service. In addition, the UK water industry is generally known for its excellence and success, and it is an international benchmark for some best practices in this sector worldwide (Grigg, 2010). However, the water industry has a low profile compared to other infrastructure sectors such as telecommunications, energy and transportation. Perhaps this is because of the fragmented nature of the sector. However, in recent years, its profile has been rising because of highly visible social and environmental problems.

The UK water industry employs more than 35000 people and the total revenue of the regulated water industry was at £11.35 billion in 2009/2010, which represents an increase of 2.3 percent over the previous year (Water UK, 2010; BIS, 2011). The water industry in England and Wales accounts for approximately 90 percent of the total UK water industry. The UK water industry collects, treats and then supplies over 16 billion litres of water per day to domestic and commercial customers. It then collects and treats over 10 billion litres of the resulting wastewater before returning it to the environment. To do this, the industry abstracts water from over 1500 boreholes, 650 reservoirs and at 600 river abstraction points, taking two-thirds of its water from surface sources and one-third from groundwater. It operates approximately 2500 water treatment works and some 9000 sewage treatment works and discharges the treated effluent back to the environment. In addition, the UK water industry uses over 350000 km of water mains, a further 350000 km of sewers to collect sewerage, and uses over 25000 intermittent discharges from its wastewater systems (Water UK, 2008).

According to OFWAT (2010), the UK water industry has invested approximately £90 billion (at 2010 prices) over 21 years to improve services, increase environmental and drinking water quality, and maintain and build new critical infrastructure. Because of this huge investment, it is claimed that drinking water quality has improved, leakage levels are 35 percent lower than they were in the mid-1990s, the companies' compliance with strict environmental standards is higher, and customer bills are 30 percent lower than they would otherwise have been without regulation. The Consumer Council for Water (2009)'s annual tracking survey shows that 95 percent of customers are satisfied with the reliability of their water supply and 84 percent with their sewerage service. This suggests that customers are highly satisfied with the services of the UK water industry.

Although the UK water industry has been considered a success, it is undergoing profound and far-reaching changes. The unique drivers of these changes include water scarcity, European Union (EU) directives, UK regulations, climate change, demand for high-quality water, sustainability issues, ageing infrastructure, higher water rates, water security concerns and smart technologies (for example, meter reading, billing) (Means et al., 2005).

Ashley and Cashman (2006) consider climate change and water pollution to have the greatest impact on increasing the cost of water and wastewater services. While the UK water industry's greenhouse gas emissions are less than 1 percent of the UK's annual greenhouse gas emissions, there is still a compelling case for action. There is a real risk that this figure will continue to rise as water demand is likely to increase with population growth, new developments and lifestyle changes.

According to Cohen (2010), the UK water companies have already plucked the low-hanging fruits of potential efficiency improvements and mitigating price increases, and now more structural changes are required for the industry to progress. A McKinsey Report (2009) estimates that over the next 40 years an investment of US$66.4 billion per year on average in water demand management will be needed to improve water efficiency.

Deloitte (2010) has identified the significant challenges faced by the UK water industry in respect of the ongoing heavy investment requirements that are needed to fund the EU Water Framework Directive, to tackle water scarcity and to ensure resilience. This will require an investment of another £22 billion in the five years to 2015, and rising to £25 billion in the next Price Control Review of the water industry regulator, that is, the Office of Water Services (OFWAT). These challenges have prompted varying responses in terms of management, governance and institutional reform by policymakers, businesses and civil society at local, regional, national and international scales.

4.3 DISTINCTIVE CHARACTERISTICS OF THE WATER SECTOR

Water infrastructure systems utilize large-scale fixed assets, provide essential public services, involve both public and private sectors, and are regulated by the government for quality, health, safety and performance (Grigg, 2010). Grigg (2010) presented figuratively a distinctive characteristic of water infrastructure in relation to regulatory agencies, water and wastewater providers and customers. From these characteristics of the industry, it is reasonable to assume that the scope of the water sector is complex, owing to its diversity, interactions and synergies with other industrial, commercial and financial sectors, and its international nature (Palaniappan et al., 2007). The water-related infrastructure comprises vast physical assets that support everyday life and allow a country to operate efficiently. It has certain characteristics that make it unique and play a key role in meeting economic, social and environmental objectives at both national and international levels (WEF, 2011).

The specific characteristics of the water infrastructure sector can be summarized as follows (Seidenstat et al., 2002; OECD, 2009; Weber and Alfen, 2010; WEF, 2011):

1. High fixed costs coupled with long-term irreversible investments and relatively inelastic demand, which tends to make it a monopolistic sector in which competition is difficult to introduce and regulation plays a central role.
2. The water infrastructure and service market is fragmented and accommodates a large variety of different agents: international investors, local and regional actors, private sector firms whose core activity is not water but they are an important user of water, joint ventures between public and private companies as well as public companies operating abroad as private participants in water projects. Therefore, decision-making in the water infrastructure sector is difficult because of the fragmented nature of the industry.
3. The size and complexity of financing required, and the risks peculiar to the water industry, imply not only the need for both private and public sector finance, but also the need to invest on both the demand and supply sides.
4. The responsibility for water infrastructure development and service provision often rests with local authorities. Given the importance of the externalities in the sector and the need to take into account the full water cycle and to optimize economies of scale, this requires an integrated approach to development and management of water infrastructure and service provision.
5. Long-term relationships, limited competition and irreversibility of infrastructure and technology may expose the sector to risks, particularly of vested interests. Investors in the water infrastructure sector are, therefore, faced with commercial risk, contractual risk, foreign

exchange risk, sub-sovereign risk, arbitrary political interferences, and complex pricing policies with multiple objectives, such as cost recovery, economic efficiency, environmental objectives, equity and affordability.
6. Given the high level of capital investment and the history of government-subsidized services, full-cost pricing of water service has yet to take hold fully in many countries. Furthermore, the public perception in many countries is that water is 'free', which means that it remains broadly underpriced, which discourages private investment, or forces investors to rely on subsidies to obtain sufficiently attractive rates of return on invested capital.

As an investment sector, the water industry can look appealing because demand for water is certain to grow with population (WEF, 2011). As a mostly government-run sector, the water industry needs several reforms that promote privatization and innovative financial strategies. Although these possibilities may seem attractive, the heavily regulated and politicized water sector has disappointed many would-be investors and reformers (Jones, 2010).

4.4 PRIVATIZATION OF WATER AND WASTEWATER SERVICES

According to Prasad (2006), privatization of the water industry is one of the most controversial topics in the area of business research and development given that the modern water industry operates sophisticated and costly water and wastewater networks. The water industry is generally a natural monopoly. As a result it is usually run as a public service by a public utility which is owned by a local or national government. However, during the late 1980s, with the encouragement of the World Bank, the International Monetary Fund and others, water privatization was proposed or implemented in many countries including Chile, New Zealand, France, the UK and the Czech Republic (Marin, 2009; Jones, 2010).

Privatization of water services tends to be associated with neoliberal reform strategies (Prasad, 2006). According to Robison and Hewison (2005), neoliberal reform strategies emphasize the importance of the market, fiscal discipline, trade, investment and financial liberalization, deregulation, decentralization, privatization and a reduced role for the state.

The principal argument in favor of privatization is that private companies can go to the stock market and banks for funding and overcome the problems of underinvestment that had led to ageing and inefficient water infrastructure (Harris, 2003). World Bank (2004) argues that privatization of the water industry is expected to improve efficiency, extend the service, bring in more investment and relieve governments of budget deficits. However, it is estimated that over 90 percent of the world's population is currently supplied by the public sector (Jones, 2010). The funding comes generally from taxation, borrowings and user fees.

Shaoul (1997) conducted a financial analysis of the UK water industry pre- and post-privatization and found higher costs, prices and profits but little improvement in the level of net investment or service quality. Saal and Parker (2001) conducted a productivity analysis of the privatized industry and found that while labor productivity improved after privatization, total factor productivity declined. The authors concluded that privatization resulted in higher profits but few efficiency gains.

There is a general belief that compared with the public sector, the private sector can afford to hire managers with strong career experience as professional managers rather than engineers with less experience of the marketplace. Consquently, it is argued that a private operator would operate more efficiently because of its profit motive and the fact that its contract contains clear and consistent objectives rather than the multiple and often conflicting goals assigned to public sector-owned water utilities. Thomas and Ford (2005) noted that the impact of private sector competition should be good for customers and result in price reductions. Table 4.1 summarizes the general arguments for and against water industry privatization.

Holder (1998) argued that privatization is merely a change in the ownership of a firm and that if both public sector and private sector owners had direct control over the behavior of managers, then there might be no reason to assume that a publicly owned firm would be more or less efficient than a privately owned firm. The two firms may have different underlying objectives, in particular the privately owned firm will be profit-seeking rather than welfare-maximizing, but most of these differences can be addressed through an appropriate regulatory framework.

Creating a blueprint for water industry privatization is difficult because water rights vary from nation to nation. Some countries treat water and water utilities as public goods under government ownership, while other nations leave water dispersion to the private sector. In the UK, England and Wales fully privatized their water industry in 1989, resulting in direct competition for water utilities. Some countries however have used the private sector to provide specific services as part of the overall water and wastewater systems. In Scotland, for example, a public–private partnership (PPP) provides wastewater treatment services to Scottish Water (a public sector utility company). Similarly, in Australia, public sector water

Table 4.1 Arguments for and against water industry privatization

Criterion	For privatization	Against privatization
Ideology	Water is an economic good and ought to be managed by the private sector; smaller government is better.	Water is a public good and ought to be managed by the government; government is needed to resist excesses of private sector.
Efficiency	The private sector is more competent than the public sector. Cost savings occur in construction, procurement, hiring and training and tax obligations.	The public sector is efficient; cost savings in privatization are fictional; tax benefits are a shell game; negative aspects of long-term contracts; potential for rate increases.
Social impacts	Government is not reliable, and people do not always trust it.	Loss of political control; loss of jobs; loss of public benefits of water.
Risk reduction	Private firm will guarantee performance.	Effective management by government minimizes risk.
Capital generation	Access to capital in private markets; government debt limits not imposed.	The public sector should generate capital for infrastructure.

Source: Grigg (2010).

utilities receive water treatment services through PPPs. Due to this discrepancy of state or privately owned water, the terms of privatization vary by transaction, with key differences including the division of responsibility of asset ownership, capital investment, operation, user fee collection, and so on (OECD, 2010).

Even though privatization of publicly owned water systems has received considerable attention, interest has waxed and waned (Grigg, 2010). It seems that many failures and controversial deals put the brakes on privatization of water industry in many countries. Nickson and Vargas (2002) and Kohl (2004) showed how vested interests, combined with politics, lack of proper communication, street protests, and poor understanding of the social and political realities resulted in the cancellation of the water concession project in Cochabamba, Bolivia.

Oher examples where controvesies have been levied against water industry privatization include a civil-society-initiated referendum banning water privatization in Ghana and Uruguay in October 2004, and a law banning privatization of public water supply in the Netherlands in September 2004 with broad cross-party support. The 2010 Italian campaign against water privatization collected over 1 million signatures in two months in support of a demand for a referendum to prevent the privatization of water services. The campaign aims to block the proposed new law of the Berlusconi government which would force public authorities to use private water companies. According to Clarke (2010) under the new law, from 2012 water supplies must be managed exclusively by private companies or by mixed public–private enterprises where the private investor holds at least 40 percent and contracts must be awarded by tender. In addition local authorities with a stake in utilities listed on the stock exchange are asked to reduce their shareholding gradually to a maximum of 30 percent by the end of 2015.

4.5 UK WATER INDUSTRY: BUSINESS MODELS AND GOVERNANCE

For water and wastewater services, several business models involving different relationships between private and public actors have emerged and become predominant in many developed and developing countries. There are many types of organizations which manage water throughout the world, as follows (Agthe et al., 2003):

1. Water and wastewater infrastructure is owned by a commercial company which is also responsible for its operation.
2. Commercial firms build water and wastewater facilities at their own expense, operate those facilities for a specified length of time and then transfer the responsibilities to a government unit. These schemes are known as build–operate–transfer (BOT).
3. The local government maintains ownership of the water supply and wastewater treatment infrastructure and a private company operates the infrastructure, sometimes being responsible for investment and construction to increase its capacity.
4. A private company is responsible for the operation and maintenance of the infrastructure but the assets are still owned by the local government, which also carries out any increases to capacity and bills the customers.
5. Larger companies provide support to the managers of a water and wastewater organization on a partnership or consultancy basis.

The UK water industry (Scotland, Northern Ireland, and England and Wales) is made up of 39 main water and wastewater services companies (Box 4.1). With this large number of companies, there are different water and sewerage services models in the UK. This ranges from the model of

BOX 4.1 THE UK WATER AND WASTEWATER SERVICE PROVIDERS

ENGLAND AND WALES

Regional water and sewerage service companies

1. Anglian Water Services Ltd
2. Dwr Cymru Cyfyngedig (Welsh Water)
3. Northumbrian Water Ltd
4. Severn Trent Water Ltd
5. South West Water Ltd
6. Southern Water Services Ltd
7. Thames Water Utilities Ltd
8. United Utilities Water plc
9. Wessex Water Services Ltd
10. Yorkshire Water Services Ltd

Regional water-only companies

11. Bournemouth and West Hampshire
12. Bristol Water
13. Cambridge Water
14. Cholderton and District Water
15. Dee Valley Water
16. Essex and Suffolk Water
17. Anglian Water Services
18. Portsmouth Water plc
19. South East Water
20. South Staffordshire Water
21. Sutton and East Surrey Water
22. Veolia Water Three Valleys
23. Veolia Water East
24. Veolia Water Southeast

Local water companies

25. Veolia Water Projects
26. Albion Water Ltd, Deeside
27. Independent Water Networks Ltd
28. SSE Water Ltd
29. Peel Water Networks
30. Albion Water Ltd, Fareham

Local water supply licensees

31. Avon Valley Water Limited
32. Osprey Water Services Limited
33. Satec Limited
34. Scottish Water Business Stream Limited
35. Severn Trent Select Limited
36. United Utilities Water Sales Ltd
37. YorWater Limited

SCOTLAND

38. Scottish Water

NORTHERN IRELAND

39. Northern Ireland Water Service

purely public sector in Scotland and Northern Ireland to the full divestiture (ownership, management and juridical-formal aspects) of water services in England and Wales. Currently, there are ten major regional private companies in England and Wales which supply both water and sewerage services; these are the former water authorities. There are also 14 private regional companies which are responsible for the provision of water only and have no sewerage functions. In addition to this, there are six local companies which supply either water or sewerage services or both, and seven water supply licensees offering water and sewerage services to large-use customers. Scotland and Northern Ireland, with devolved parliamentary power, each have a single water and sewerage service provider. These are in public ownership but rely upon private companies for delivery of services.

The UK water companies differ in size, turnover, operating profit and population served. For instance, during the financial year 2010–11, Thames Water had a turnover of £1600 million, an operating profit of £510.7 million, and it served a population of 8.8 million, compared with Veolia Water East, which had a turnover of £14.4 million and an operating profit of £4 million, with a population served of 154000 (OFWAT, 2011). Unlike gas and electricity infrastructure provision, there is no national grid for the UK water industry which has resulted in a monopolistic market in the regions where these companies operate.

4.5.1 England and Wales' Water Sector Model

In 1989, the Thatcher government introduced privatization to bring in private sector investment and improved, more dynamic management to solve the public sector challenges and shortcomings (OECD, 2009). As a result, in England and Wales, municipal water companies were combined into a small number of regional water and sewerage companies, and then fully privatized. This allowed public sector water infrastructure and assets to be placed into private shareholder ownership. All the overground and underground assets, buildings, associated land, and operation of the water and sewerage infrastructure were sold to the private sector.

Cowan (1998) commented that the privatization of water and sewerage utilities in England and Wales was the largest effort to sell off public water agencies. In essence, water and wastewater provision in England and Wales is through private operators which, apart from being responsible for managing the system, also own the assets. It is highlighted by Jones (2010) that privatization of the water industry provided considerable advantages of enhanced efficiency and access to finance to fund major water infrastructure development programs aimed at dramatically improving service quality.

Sheshinski and Lopez-Calva (2003) identified four major objectives of privatization: to achieve higher allocative and productive efficiency; to strengthen the role of the private sector in the economy; to improve the public sector's financial position; and to free resources for allocation to other important sectors such as social policy. At the time of the privatization, the UK government argued that the change in ownership was necessary to make the industry more efficient by opening it to competition and exposing it to the profit motive. Privatization was also intended as a means to inject new money into the industry in order to counter decades of underinvestment and to enable it to comply with European environmental legislation (Bakker, 2003).

Thomas and Ford (2005) highlight the motivations for the UK water industry privatization. These include anecdotal experience, better management, best practices management, competition beliefs, economic growth, efficiency from new technology, fiscal issues, global context, governance and accountability issues, ideology embedded in policy arguments, management techniques, power issues and strategic planning. Furthermore, they noted that market forces and competition and associated private sector ethos was better than the nationalized industry legacy and culture. An argument proffered by Thomas and Ford (2005) was that the drive for profitability was believed to incentivize employees and managers beyond the survival ethic. Associated benefits derived could be job security, diminishing takeover threats, and higher potential salaries and pensions available in the private sector. This argument leads to the general belief that privatization was inevitable and preferable to public sector operation.

Marin (2009) commented on the major role played by the water industry privatization in England and Wales in convincing policymakers that private financing of water utilities could be viable. The privatization was a momentous event for the water industry worldwide, raising massive amounts of private money from international financial markets with the attendant new regulatory scheme that was put in place. Marin (2009) noted that this promising new approach could be replicated on a large scale in developing countries, where a considerable amount of money was needed to fund the investment backlog.

Figure 4.1 presents an overview of the water and wastewater services model in England and Wales. With this model, public interest is guaranteed and supervised by the state through an independent regulatory agency, OFWAT.

Many of the opponents of privatization argue that the transfer of assets and decision-making control to a private corporation will change water services from a public good to a source of private profit (OECD, 2009). Dore et al. (2004) noted that water prices charged by private companies

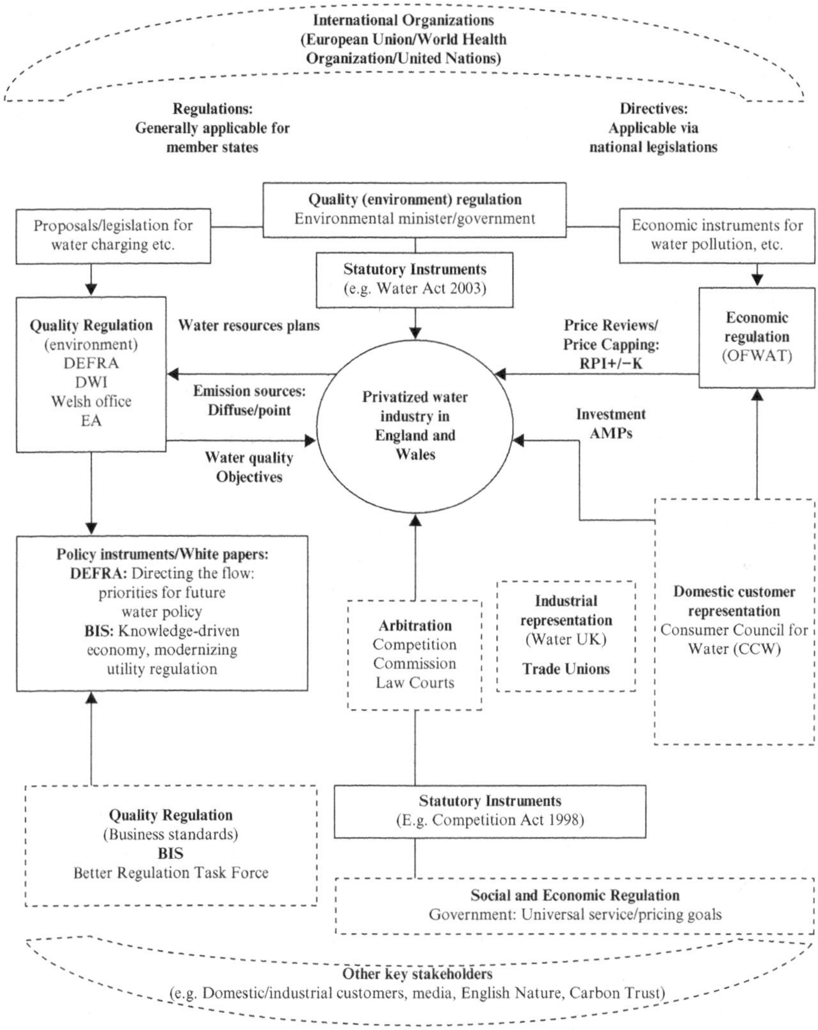

Note: BIS: Department for Business, Innovation and Skills; AMPs: asset management plans.

Source: Thomas and Ford (2005).

Figure 4.1 An overview of England and Wales water and wastewater services model

to customers in England and Wales are relatively high compared with those charged by public companies. In addition, the rate of return and the profits of the private companies have been extremely high. This position is supported by the Walker Review (2009), which notes that since privatization, the water prices in England and Wales have risen about 42 percent on average in real terms. This suggests that privatization of the water industry in England and Wales may not have led to the expected efficiency gains or welfare gains. However, water prices remain relatively cheap for most households when compared to energy bills (Walker Review, 2009).

Grigg (2011) commented that in many countries, water governance is in a state of confusion with a lack of water institutions or fragmentation of authorities, and decision-making structures. Therefore, privatization of the UK water industry and the accompanying regulation led to the emergence of new institutional arrangements and changes in the structure and operation of the industry. With the transfer of responsibilities to the private sector, there has been an accompanying growth of economic, environmental, resource and quality oversight that constitutes part of the governance and regulation of the water industry. The increasing institutionalization and formalization of regulation have led to the emergence of new forms of governance and accountability which applies as much to the regulatory agencies as the water companies. The goverance also encompasses the inclusion of other stakeholders representing a much wider polity than under state-centered service provision (Page, 2001).

Ogden and Watson (2007) argued that privatization implied not merely ownership changes but also radical changes in corporate governance and the risks and responsibilities of management. As Stoker (1998) noted, 'the essence of governance is its focus on governing mechanisms which do not rest on recourse to the authority and sanctions of government'. Water sector governance is both context dependent and subject to ongoing change; it provides an organizing framework for the identification of relationships and issues concerning the development of regulations together with regulatory and stakeholder relationships, the changing nature, roles and legitimacy of stakeholders and reporting. This mode change in the UK water industry brings in the increasing involvement of un-elected bodies such as private firms and their trade organizations (for example, Water UK), quangos (for example, OFWAT), civil society bodies (for example, Consumers Association and trade unions) and other lobby groups (for example, RSPB) (Kaïka and Page, 2002).

Table 4.2 presents the main bodies that are responsible for the water industry in the UK. OFWAT is the economic regulator of the water sector in England and Wales. It is responsible for protecting consumers' interests; developing long-term approaches to environmental planning to

Table 4.2 Main bodies responsible for the water industry in the UK

Country	Water resources management	Water and wastewater services	Environmental regulation	Water quality	Economic regulation
England and Wales	DEFRA	10 water and wastewater companies and 12 water-only companies	EA	DWI	OFWAT
Scotland	SEPA	Scottish Water	SEPA	DWQ	WICS
Northern Ireland	DRD	Northern Ireland Water	NIEA	DWI	NIAUR

Notes: DEFRA: Department for Environment, Food and Rural Affairs; SEPA: Scottish Environment Protection Agency; NIEA: Northern Ireland Environment Agency; DRD: Department for Regional Development, EA: Environment Agency; DWI: Drinking Water Inspectorate; DWQ: Drinking Water Quality; OFWAT: Office of Water Services; WICS: Water Industry Commission for Scotland; NIAUR: Northern Ireland Authority for Utility Regulation.

ensure sustainable water and sewerage services now and for the future; and providing the regulatory stability necessary for efficient water companies to attract investment from competitive financial markets (OFWAT, 2010). OFWAT makes its decisions independent of the UK government but works closely with the Secretary of State for Environment, Food and Rural Affairs and the Welsh Assembly Government.

The success of the water industry model in England and Wales is closely associated with the leading role played by regulation and its effectiveness (Marques, 2010). In England and Wales, the Consumer Council for Water is an independent organization that represents customers' interests and deals with customers' complaints. The Drinking Water Inspectorate sets standards for the quality of drinking water; the Environment Agency regulates and enforces water abstraction consents and quality standards in inland, estuarial and coastal waters; and Natural England and the Countryside Council for Wales are concerned with environmental issues.

4.5.2 Scottish Water Sector Model

The approach to water provision in Scotland is different from Engand and Wales. Scotland was not prepared to go along the privatization route because the Scottish were concerned about the commercialization of a natural monopoly. Therefore, Scottish Water was created in response to the Water Industry (Scotland) Act 2002 using a public sector model (Marques, 2010). Scottish Water, although managed and operated as a company, is in the public sector with annual revenue and borrowing caps set by Scottish ministers. Private sector investment and engagement by Scottish Water is in practice through granting long-term operating contracts to the private sector, while retaining assets in public ownership. This was deemed more acceptable to Scottish society. Since devolution, the new Scottish Parliament has discussed the water industry but privatization has been ruled out except for some build, operate and transfer (BOT)-financed wastewater treatment plants.

Birdsall and Nellis (2002) noted that water industry privatization in Scotland remains widely and increasingly unpopular, largely because of the perception that it is fundamentally unfair, in both conception and execution. For instance, the Scottish have a strong sense of ownership of their land and, by implication, its water resources. This was illustrated by the rejection of the option of privatization of the water industry in the referendum held in Strathclyde in 1997.

The Water Industry Act 2002 created Scottish Water by merging the three existing public operators, namely the North of Scotland Water, East of Scotland Water and West of Scotland Water, in 2002 (Pinsent Masons, 2010). Scotland is the only part of Great Britain where the water industry is still in the public sector. Scottish Water has an annual turnover of around £1 billion and provides clean, safe and high-quality drinking water to 2.4 million households and 123 000 business premises across Scotland. Scottish Water has invested over £3.2 billion to improve water industry infrastructure in Scotland since it was created.

Scottish Water's Annual Report (2010) revealed that the average combined water supply and wastewater collection bill for households across Scotland in 2010–11 is £324, which is £15 lower than the average household bill in England and Wales. The publicly owned body claimed water quality was at an all-time high with 99.83 percent of samples meeting standards. It also cut leaks by nearly 100 million litres a day despite a rise in burst pipes over the winter. During the period 2009–10, Scottish Water had invested over £648.3 million to improve drinking water. The environment and customer service of Scottish Water operates 1800 water treatment works and 1127 service reservoirs linked by nearly 29 391 miles of water mains and delivers about 844 million litres of safe drinking water to homes across Scotland every day. Its 280 wastewater treatment works connect to over 31 122 miles of sewer pipes (Scottish Water, 2010). However, critics have called on the Scotland government to consider changing the ownership structure of Scottish Water to give it more financial freedom.

The Scottish Parliament holds Scottish Water responsible for provision of water and the Scottish Ministers set the objectives. Scottish Water is responsible for providing water and wastewater services to household customers and wholesale Licensed Providers across Scotland. The Water Industry Commission for Scotland (Scotland's equivalent of OFWAT) is the economic regulator and it is responsible for setting charges and reporting on costs and performance whereas Waterwatch Scotland is responsible for representing the interests of Scottish Water customers. The Drinking Water Quality (DWQ) regulator is responsible for protecting public health by ensuring compliance with drinking water quality regulations whereas the Scottish Environment Protection Agency (SEPA) is responsible for protection and improvement of the environment.

4.5.3 Northern Ireland's Water Sector Model

The Northern Ireland government has chosen a different water services model due to its special political environment. In Northern Ireland, water and wastewater services are provided by a government-owned company, Northern Ireland Water, similar to Scottish Water in Scotland. Northern Ireland Water is a publicly owned company and it is subject to public sector borrowing requirements. The key reasons for public ownership in Northern Ireland are strong public resistance to privatization, unfavorable economic conditions, and uncertain political environment.

In 2003, the Northern Ireland government started considering the reform of water services, that is, its water and sewerage services. The Northern Ireland Water Service was created as an agency of the former Department of the Environment for Northern Ireland in 1996. Investment in water and sewerage services in Northern Ireland had been significantly lower than Scotland and England and Wales over a number of decades (NIW, 2007); this is not surprising considering its population and land coverage. Northern Ireland Water delivers drinking water to approximately 798 700 households and businesses across Northern Ireland against a backdrop of historic underinvestment and ageing water and sewerage infrastructure. It supplies 622 million litres of drinking water every day to customers through more than 26 435 km of water mains. In addition, it collects 338 million litres of wastewater per day from 654 000 businesses and households that are connected to the sewerage system, and transports sewage through approximately 14 465 km of sewers to works where it is treated and disposed of.

4.6 WATER-RELATED INFRASTRUCTURE INVESTMENT GAPS AND OPPORTUNITIES

The perennial challenge for water-related infrastructure development is raising project finance (WEF, 2011). The lack of available finance is widely regarded as the main constraint on the development of water-related infrastructure and is typically linked to risks specific to the sector: commercial (tariffs, cash flow and credit risks), political (expropriation, political interfaces and devaluation), legal, regulatory, contractual, water resources-related (scarcity, flooding and pollution), and reputational risks. The OECD (2009) reported that for water infrastructure investment around the world, there is a large gap between what is financed and what is built.

Deutsche Bank (2010) reports a huge need for investment in the global water sector, though the focus varies from region to region. According to the OECD (2007) report, the spending required to update infrastructure in developed countries and to develop appropriate infrastructure in emerging economies is estimated at a staggering US$53 trillion between 2007 and 2030. The OECD (2007) report also exhorts developed nations to invest at least 2.5 percent of their GDP in infrastructure. The OECD (2006) estimates that the European Union will need to invest over US$178 billion annually on water and wastewater services by 2025. These needs have triggered an increased involvement of private sector capital in infrastructure renewal, development and operations.

The global demand for infrastructure expansion is large as many countries seek to upgrade systems to serve growing populations and rising standards of living. Dzierwa (2009) summarized the global situation and concluded that substantial opportunities were emerging. His evidence was a series of international trips and reports by consultants, policy groups and investment bankers. One report was from Britain's Foresight Programme, which has focused attention on infrastructure and recently completed studies on intelligent infrastructure and coastal defence against floods. Dzierwa reported that global infrastructure needs are at levels of US$32 trillion to US$41 trillion for the next 20 to 25 years. The drivers for this are identified to include demographics, globalization, government policies and a desire of nations to upgrade infrastructure for entry to the European Union (for example, in the case of Poland).

However, an increasing number of analysts view the high future demand for water-related infrastructure in both developed and developing countries as an attractive investment opportunity (UBS, 2009). Pension funds also embrace water stocks because they involve secure and multi-year contracts, while private bankers see water as a safe choice, offering steady, low-volatility returns for their demanding clients. This may be due to the fact that water is a commodity with no alternative and no substitute, which all human activities require and for which analysts suggest that we will face a 40 percent global shortfall between demand and supply by 2030 (WEF, 2011). Therefore, many large-scale investors are interested to invest in water-related infrastructure because of dwindling supply, soaring demand, no alternative or substitute goods for water, limited competition and a relative lack of price volatility on revenues due to public sector regulations on price making, makes it easy to model the returns and identify key risks.

The water industry uses an extensive infrastructure to provide services to customers. This has to be maintained and extended to meet the needs of existing and new population centres. The level of investment in different parts of the industry's infrastructure – water treatment, water distribution, wastewater treatment and sewage treatment – provides the context for its performance in these areas of activity (Weber and Alfen, 2010).

Helm et al. (2010)'s report concluded that the importance of infrastructure investment in the UK cannot be underestimated. According to the report, all infrastructure-related sectors are faced with the need both to renew ageing existing infrastructure assets and to invest to deal with anticipated future pressures deriving from social, economic, political and environmental changes. After carrying out a detailed analysis of the state of the infrastructure in the UK, Helm et al. (2010) concluded that the UK needed a total of £434 billion of new investment into transport (£120 billion), energy (£264 billion), water (£45 billion) and telecommunications (£5 billion) related infrastructure over the period. However, the report warns that the actual figure could be closer to £500 billion or £50 billion a year over the next ten years. This suggests that other infrastructure developments are competing for limited available resources. The water-related infrastructure sector developments are needed in respect of water and wastewater infrastructure (£37 billion) and flood and coastal defence systems (£8 billion), a total new investment of up to £45 billion, or £4.5 billion per annum over the next ten years.

The current performance of the UK's infrastructure as well as its preparedness to meet present and future challenges varies across sectors (HM Treasury, 2010). For example, water infrastructure requires ongoing maintenance and renewal to maintain service levels, meet environmental standards and ensure sustainability, whereas the energy sector requires increased investment: electricity generation capacity that is going offline needs to be replaced with new low-carbon generating stations (which are more capital-intensive), and declining indigenous gas production increases demand for new gas import and storage facilities.

The level of underinvestment in the UK water industry was noticeable in December 2010, when nearly 60 000 householders in Northern Ireland were unable to access their water supply, which soon became an international embarrassment for the UK. Northern Ireland Water, on its website, admits there has been chronic underinvestment and that there is a big gap in performance compared to water companies on the mainland (Hipwell, 2011).

There are major water infrastructure-related projects in the pipeline in the UK to address these underinvestments (HM Treasury, 2010). This includes OFWAT's final price limits for 2010–15, which allows for £22 billion to be invested by the UK water companies; and Thames Tideway, a water infrastructure project intended to improve the capacity of London's wastewater system and prevent wastewater overflows, which will require approximately £6 billion. Other projects are by Anglian Water (£3 billion), Severn Trent Water (£3 billion) and Yorkshire Water (£3 billion), a total investment of around £9 billion in the next ten years (Helm et al., 2010).

4.7 WATER-RELATED INFRASTRUCTURE PROCUREMENT OPTIONS

The different types of water governance architectures have been discussed in Chapter 1 of this volume. This section further elaborates on the topic with particular reference to the UK and the EU. Table 4.3 provides an outline of the procurement models in the water sector.

4.7.1 Pure Public Sector

The key argument for the pure public sector infrastructure procurement model is that water is a common good, the essence of life itself, and access to it is a human right. Therefore, water should never be treated as a commodity based on market principles and it is the government's responsibility to provide such a vital resource to everyone and it should not be in the hands of the private sector. Furthermore, developing and maintaining water-related infrastructure facilities, treating raw water, making it available to the public, and manageing wastewater are expensive processes.

KPMG (2008) reported that the average capital investment in water projects runs between US$200 million and US$300 million. Given the level of capital expenditure required, traditionally the government has been fully responsible for water supply and the development and management of water-related infrastructure facilities. To fund, build and manage the

Table 4.3 Participation of the private sector in water and wastewater provision

Features	Pure public sector models	Public–private partnership models						Pure private sector models
	Fully public	Service contract	Management contract	Affermage/ Lease	Concession	BOT	Joint Venture	Full Divestiture
Scope		Multiple contracts for a variety of support services such as meter reading, billing, etc.	Management of entire operation or a major component	Responsibility for management, operations and specific renewals	Responsibility for all operations and for financing and execution of specific investment	Investment in and operation of a specific major component, such as a treatment plant		
Fixed assets ownership	Public	Public	Public	Public	Public/ Private	Public/ Private	Public/ Private	Private
Capital investment	Public	Public	Public	Public	Private	Private	Shared	Private
Commercial risk	Public	Public	Public	Shared	Private	Private	Shared	Private
Overall level of risk assumed by private sector	NA	Minimal	Minimal/ Moderate	Moderate	High	High	Shared	High

Table 4.3 (continued)

Features	Pure public sector models	Public–private partnership models						Pure private sector models
	Fully public	Service contract	Management contract	Affermage/ Lease	Concession	BOT	Joint Venture	Full Divestiture
Operation and maintenance responsibility	Public	Public/ Private	Private	Private	Private	Private	Shared	Private
Contract duration	Without duration	Short (1–2 years)	Short or medium (3–5 years)	Medium (8–15 years)	Long (25–30 years)	Long (20–30 years)	Long or without duration	Without duration
Source of remuneration of operator	Operator collects user fees	Munici-pality	Municipality: fee is fixed or based on performance	Operator collects user fees; Lease: fee paid by municipality Affermage: revenue shared	Users	Municipality	Users	Users

Problems and challenges		Requires ability to administer multiple contracts and strong enforcement of contract laws	Management may not have adequate control over key elements, such as budgetary resources, staff policy, etc.	Potential conflicts between public body which is responsible for investments and the private operator	How to compensate investments and ensure good maintenance during last 5–10 years of contract	Does not necessarily improve efficiency of ongoing operation; may require guarantees		
Examples	Scotland	Mexico City, Chennai and Mysore (India)	Johannesburg (South Africa), Amman (UAE)	France, Senegal	Jakarta, Manila	China, India, Malaysia, Mexico, Morocco	Netherlands, France	England and Wales, Chile
Occurrence 1991–2009 (World Bank PPI Database)	Not part of scope	Not part of scope	Together: 111 of 715 projects		278 of 715 projects	294 of 715 projects	Not a separate category	32 of 715 projects

Sources: Marques (2010), World Bank (2010), Skilling and Booth (2007).

cost of water infrastructure assets, governments have long used available funding sources from tax revenues, user fees, foreign aid, international lending agencies, issuance of municipal bonds or pension funds. In many cases, even under the traditional procurement method, design, construction, operation and maintenance were carried out by the private sector under different contracts (Tan, 2007).

4.7.2 Public–Private Partnership

An alternative is the public–private partnership model. Jones (2010) has argued that the conventional procurement model does not tap the full range of expertise and funds from the private sector. In addition, due to shortages of capital funding and rising maintenance costs of ageing water infrastructure, sustainability issues and more stringent environmental standards, many governments are in the position of needing to undertake substantial capital replacement, refurbishment or expansion of their water and wastewater facilities without sufficient available capital (KPMG, 2009).

Faced with the challenge of meeting the EU Water Framework Directive, and in particular the inability to fund the massive investments required to comply with the EU Urban Waste Water Treatment Directive and its 2000 and 2005 compliance deadlines, countries across Europe have adopted the public–private partnership (PPP) mode of procurement (Mandri-Perrott, 2009).

Akintoye (2009) defined PPP as a contractual agreement of shared ownership between a public agency and a private company, whereby they pool resources together and share risks and rewards, to create efficiency in the production and provision of public or private goods. Jones (2010) noted that profit is the prime motive for private company involvement; that PPPs frequently do not offer the best of both worlds. In addition, private sector involvement in water-related infrastructure development has made it harder for governments to uphold public sector values.

Table 4.4 presents the global public–private partnership deals in the water sector by region between 1985 and 2010, which suggests that this model is popular across many developed and developing countries for water and wastewater provision.

There is an array of options for private sector involvement, characterized by a variety of contractual forms: service contract, management contract, affermage/lease, concession, build–operate–transfer (BOT) and joint venture. These contractual forms differ in the degree of responsibility and risk transferred from the public sector to the private sector. Some

Table 4.4 Global public–private partnership deals in the water sector by region between 1985 and 2010

		Number of projects	Cost (in US$ million)
United States	Total planned and funded since 1985	194	19094
	Funded as of October 2010	141	13678
Canada	Total planned and funded since 1985	31	2333
	Funded as of October 2010	17	1263
Mexico, Latin America and the Caribbean	Total planned and funded since 1985	161	19003
	Funded as of October 2010	85	11655
Europe	Total planned and funded since 1985	223	33838
	Funded as of October 2010	171	23263
Africa and Middle East	Total planned and funded since 1985	113	45971
	Funded as of October 2010	55	29290
Asia and Australia	Total planned and funded since 1985	184	51170
	Funded as of October 2010	121	37677
Worldwide	Total planned and funded since 1985	906	171408
	Funded as of October 2010	590	116825

Source: Public Works Financing (2010).

governments have embraced concessions and other forms of PPPs to help turn a significant short-term financial cost into a long-term financial proposition for sponsors (KPMG, 2009).

Under PPP, for example, water project delivery and finance methods such as design–build–operate and build–operate–transfer enable governments to take advantage of the capital and expertise of the private sector while securing public benefits of water infrastructure systems. The long-term contractual relationships between public and private entities involve aligning a significant investment of private capital, transferring some risk to the private sector, and increasing the public benefit from water infrastructure. However, critics contend that PPPs are an expensive and inefficient method of delivering infrastructure, and can undermine the capability of governments to plan for public interests (Shaoul et al., 2006).

4.7.3 Pure Private Sector

At the other extreme, some governments have opted for an outright sale of the water assets and wastewater services. The principal argument in favor of the pure private sector mode of procurement is that private companies can go to the stock market and banks for funding and overcome the problems of underinvestment that had led to ageing and inefficient water infrastructure (Harris, 2003).

However, critics argue that the prices charged for water and wastewater services provided by purely private operators tend to be higher than for services provided by government-operated bodies (Grigg, 2011). Furthermore, Hall (2004) argues that privatization of water services has failed in many parts of the world and is falling apart in many cities and countries, including Buenos Aires, Atlanta, Manila, Cochabamba, Jakarta, Nelspruit, Kelantan, Mozambique, Nkokebde, Conakry, Gambia, Parana, Trinidad and Tobago, Belize, La Paz and Dar es Salaam. Some of the key reasons for the failure of water industry privatization include projects confronted with controversies relating to price increases, and problems of non-payment by consumers; lack of understanding of the local context in which reform is taking place; flawed theoretical foundation of privatization of water industry; the economic and financial crises; and natural disasters like droughts and floods (Harris et al., 2003; Prasad, 2006).

World Bank (2004) argues that privatization of the water industry may be expected to improve efficiency, extend the service, bring in more investment and relieve governments of budget deficits. The most notable example of this is the sale of water companies in England and Wales. Because of increasing controversy that is associated with purely private provision of water, there is a trend towards repackaging privatization through different means such as PPP initiatives, arguing that PPP offers the same benefits as privatization (IMF, 2004).

In the UK, the PPP mode of procurement is being used to help fund the non-privatized water services in Scotland and Northern Ireland. The key wastewater projects that have been procured through the PPP scheme in Scotland and Northern Ireland, their value and the consortium involved are listed in Table 4.5. Some members of the consortium (Thames Water, United Utilities, etc.) involved in these projects are water companies operating in England and Wales.

According to the Pinsent Masons (2010) report, there are a total of 23 PPP sewerage schemes in Scotland, with an original value of £587 million.

Table 4.5 The key wastewater projects that have been procured through PPP schemes in Scotland and Northern Ireland

Project name	Value (£ million)	Consortium
Almond Valley and Seafield	100	Thames Water, MJ Gleeson and Montgomery Watson
Esk Valley	20	Thames Water, MJ Gleeson and Montgomery Watson
Levenmouth	47	Northumbrian and Degrémont
Ayrshire	50	Northumbrian, Degrémont and AMEC
Inverness and Fort William	45	United Utilities
Tay	90	Bechtel, United Utilities, Henderson, Anglia Water Group Project Investments Limited (AWGPIL) and Infrastructure Investors Ltd
Moray Coast	42	Bechtel, United Utilities and Anglia Water Group Project Investments Limited (AWGPIL)
Aberdeen and Peterhead	80	Yorkshire Water, BICC and Earth Tech
Daldowie and Shieldhall	57	Caledonian
Dalmuir	50	Taylor Woodrow, SAUR, Stereau, Barr and Halcrow
Inverclyde	50	Northumbrian and Degrémont (Suez) and AMEC
Kinnegar	12.4	Laing and Lagan Holdings
Alpha*	110	Northstone, Dalriada Wastewater, Earth Tech and Kelda
Omega	122	Thames Water and Laing O'Rourke

Note: * Drinking Water project.

Source: PartnershipsUK, and Pinsent Masons (2010).

4.8 MAIN CHALLENGES FOR THE UK WATER INDUSTRY

The UK water industry faces a range of challenges (Environment Agency, 2009). For example, energy use in the water utilities is considered a major problem due to the rising cost of energy for treatment and distribution, as well as pressures for a low-carbon economy (Ainger et al., 2009). HM

Treasury (2010) listed the major challenges facing the UK water industry. These include rising demand and increased pressure on supply in the South and East due to climate change causing drier summers, decreased river flows and less predictable supply; maintenance and renewal of ageing network infrastructure combined with the need to meet EU environmental standards; and the increasing need for expenditure on flood defence infrastructure. It is estimated that investment in England will need to double in real terms between 2010 and 2035 to maintain risk at current levels; and water companies are expected to adopt a twin-track strategy of demand management and investment in new infrastructure.

Similarly, the OFWAT (2010) report noted that the UK water and wastewater companies now face a set of very different challenges, including a changing and unpredictable climate; population growth, particularly in the South-East of England, where water is already scarce; economic uncertainty and the consequent affordability issues; rising environmental standards, including implementing the EU Water Framework Directive; and rising consumer expectations from an increasingly sophisticated customer base.

4.8.1 Climate Change Issues

Stern (2009) noted that the two defining challenges of the twenty-first century are overcoming poverty and avoiding dangerous climate change. Schwartz and Randall (2003) emphasize that it is not the threat from terrorism, but the vulnerability of countries and societies to natural catastrophes and climate change impacts that offers the greatest future challenges to water security. According to Levina and Adams (2006), the potential climate change impacts on water resources include variations in snowmelt and water supply; increase in floods, flash floods and drought; insufficient water; and saltwater intrusion. In the UK, drought is predicted to have a major impact on water security, with decreasing runoff in the South-East and growing populations leading to greater water scarcity (Table 4.6). Winter flooding is also anticipated to be an expensive problem.

Recently, the UK government has passed legislation which introduces the world's first long-term legally binding framework to tackle the dangers of climate change. The UK Climate Change Act 2008 sets legally binding emissions reduction targets for 2020 (a reduction of 34 percent in greenhouse gas emissions from 1990 levels) and for 2050 (a reduction of at least 80 percent in greenhouse gas emissions from 1990 levels), and introduces five-yearly carbon budgets to help ensure those targets are met. Furthermore, in April 2010, the UK government introduced a new mandatory Carbon Reduction Commitment (CRC) energy efficiency scheme

Table 4.6 Climate change impacts related to water in the UK

Countries	Climate change impact			
	Floods	Drought	Insufficient water	Saltwater intrusion
England and Wales	X	X	X	X
Scotland	X			
Northern Ireland	X			

Source: Levina and Adams (2006).

aimed at improving energy efficiency and cutting carbon emissions in large public and private sector organizations. These organizations are responsible for around 10 percent of the UK's emissions. This is vital to achieve the UK's overall target of reducing greenhouse gas emissions by at least 80 percent by 2050 compared to the 1990 baseline.

Defra (2008) presented that the UK water industry used almost 7900 GWh of energy in its total operations during 2006–07, and emitted over 5 million tonnes of greenhouse gases, as carbon dioxide equivalents. Even though the UK water industry's greenhouse gas emissions are less than 1 percent of the UK's annual greenhouse gas emissions, there is still a compelling case for action.

The water sector is a major energy user; 3 percent of the total UK electricity demand is accounted for by the water industry. The water industry is the fourth most energy-intensive sector in the UK and only about 10 percent of the water industry's total energy use is from renewable sources. Therefore, water companies are now placing greater emphasis on renewable energy through wind turbines and sludge conversion. Regulators are working with suppliers to help promote water efficiencies and to develop improvements on both the supply and demand sides, including innovative tariffs. The environmental driver of climate change will make it more challenging to operate and maintain built assets, and may require increasing levels of investment.

4.8.2 Population Growth, Distribution and Changing Lifestyles and Consumer Expectations

Apart from climate change, the Cave review report (2009) predicted that the UK water demand will be severely influenced by population growth. Wohland et al. (2010) noted that the UK population is expected to rise to 78 million in 2051 (up from 60 million in 2008), leading to higher average household use and intensifying the strain on current water and wastewater services. Much of this population growth is likely to be in the water-scarce South-East England.

Furthermore, the number of single-occupancy households, which use more water per person, is increasing. According to Waterwise (2010), each person in the UK uses 150 litres of water a day, which has been rising by 1 percent a year since 1930. Against the backdrop of mounting environmental challenges and huge population growth, the Environment Agency states that 20 million extra people will potentially be living and working in England and Wales by the 2050s and concludes that this level of water consumption is not sustainable in the long term. Given this problem, the McKinsey Report (2009) has called for a greater need for water demand management. In addition, it is necessary to develop new water industry-related strategies based on the future demographic shifts, social issues, and changing lifestyle and consumer expectations within the UK.

4.8.3 European Union Water Framework Directive

According to HM Treasury (2010), a major challenge facing the UK water industry is meeting the European Union Water Framework Directive (WFD), which entered into force in December 2000. This introduced key elements intending to shift EU water governance towards the application of the principle of Integrated Water Resources Management. The WFD has a number of objectives (EU, 2000). These include: (1) expanding the scope of water protection to all waters, surface waters and groundwater; (2) achieving 'good status' for all waters by a set deadline; (3) water management based on river basins; (4) 'combined approach' of emission limit values and quality standards; (5) getting the prices right; (6) getting the citizens involved more closely; and (7) streamlining legislation. This implies the consideration of economic, environmental, social and ethical issues in water management at the river basin level, which presents major challenges to the sector. For example, the Cave review report (2009) noted that in England and Wales, on average, only 10 percent (approximately 60000 megalitres per day in 2006–07) of freshwater resources are used for abstraction; however, in South-East and Eastern England this rises to over 22 percent of available water abstracted in 2007. There are significant challenges from current levels of abstraction contributing to the risk of those water bodies failing to meet the 2015 objectives of the EU Water Framework Directive. This is an issue for a number of lakes, rivers and estuaries and in particular groundwater in some areas.

4.8.4 Economic Environment

The current economic pressures place greater emphasis on the need for the UK's businesses to maintain and improve their competitiveness. The ability of businesses to choose and control input prices, including for water and sewerage services, is becoming even more important. Water supply will become an increasing problem in the twenty-first century, according to Underhill (2010). Improving water delivery, minimizing costs and increasing effectiveness of water recycling necessitates an increase in investment in water infrastructure. However, spending requirements to maintain ageing and build new physical water infrastructure are too large for the public sector to absorb. This presents an economic challenge for the sector. It is not surprising therefore that there is an increasing call for private sector participation in developing and delivering water infrastructure.

4.8.5 Ageing Infrastructure

The HM Treasury (2010) report on national infrastructure in the UK identified significant vulnerabilities, capacity limitations and a number of national infrastructure components nearing the end of their useful life. The report also highlighted serious fragmentation in the arrangements for infrastructure provision in the UK. An inventory of the UK's water-related infrastructure assets and their condition suggests that this is a major challege for the sector. There is a growing need to analyse the condition of infrastructure assets to help assess risks of water infrastructure failure. The analysed data and information should greatly help to improve decision-making about investment needs for maintaining, upgrading and expanding water-related infrastructure.

4.8.6 Water-related Infrastructure Financing

Financing is critical for ongoing operation and maintenance as well as responding to needs for new infrastructure. The UK water-related infrastructure sector including water, wastewater, flood and coastal defence systems requires new investment (Helm et al., 2010). Given current budgetary challenges and competing pressures for government spending cuts in the UK, it is unlikely that a step change in official development assistance will fill the gap. Water and wastewater services are more capital-intensive than other utilities, requiring twice the capital compared to electricity utilities with the same annual operating expenses. With growing financial needs, along with a decline in public investments in water and the lack of private investment being directed to this sector, new strategies are required to make the necessary investments.

In order to attract the private sector to provide much needed investment in the sector, a combination of good national policy, forward-thinking international finance institutions which can work to improve the risk–return ratio that private investors face in water infrastructure development, and a scaled deal flow of investment-worthy project opportunities will be the key ingredients required to fill the financing gap for water infrastructure. WEF (2011) has called for a need to explore alternative sources and transformational change in the approach to financing water infrastructure given that future financing of the water industry cannot be based on past practices and historical measures.

4.9 CONCLUSION

The unique nature of the water industry, combined with high levels of regulation, sustainability issues, ageing infrastructure, demographic changes, increasing energy costs, increased cost of capital and decreased capital availability, makes the UK water industry a very dynamic sector. It is unique because it deals with personal matters as well as social issues, and it is heavily regulated. The UK water industry plays a key role in meeting the economic, environmental and social objectives at both national and international levels. As a business, water is complex to define and has distinct parts. At the same time, the water industry has a low profile compared to other infrastructure sectors such as telecommunications, energy and transportation.

Since 1989, the market structure of the water industry differs amongst the four nations that comprise the UK. Water and wastewater services in England and Wales are fully privatized compared with water and wastewater services in Scotland and Northern Ireland, which are provided by public operators, albeit controlled by private law.

Governance of the water industry is generally regarded as poor, and until recently only a few countries have recognized water to be a scarce resource. Transfer of assets and decision-making control in the water sector to private sector providers has been proffered as a solution to the water governance problems as well as to increase access to water. In addition, it is seen as providing some relief to the public sector expenditure demands created by water stress, and in some cases lowering the price payable by the customers. The privatization of the UK water industry and the accompanying regulation that saw the emergence of new institutional arrangements and changes in the structure and operation of the industry is seen as providing a more sustainable footing for the sector.

Water governance is generally regarded to be in a state of flux in many countries given its fragmentation of authorities and decision-making structures. However, the England and Wales water business model is considered an interesting option, not so much because of the total privatization of the water sector, but rather for its governance framework. Difficulties in accessing knowledge and lack of awareness on water governance concepts are hampering the growth of the water industry, particularly in many developing countries. It is therefore essential to give more attention to the institutional framework arrangements when it comes to water resources management. The framework should cover a comprehensive water governance system, and should be dynamic and adaptable to the changing needs of business, society and people at large. A collaborative partnership within the water industry and industry-wide awareness-raising program(s) on the concept of water governance need to be developed and implemented on an ongoing basis if the industry is expected to meet the inceasing demand for water and wastewater provision.

Although the OECD (2009) reported that intelligently targeted infrastructure investment can facilitate significant improvements in productivity, unlock areas of competitive advantage, distribute economic activity more evenly around the UK, reduce business costs and create jobs, the level of investment in the water-related infrastructure in the UK is low compared to other infrastructure sectors. The general view is that the UK's approach to infrastructure investment is generally timid, uncoordinated, incremental, wasteful in its procurement and insufficiently targeted to support balanced and sustainable growth, both economically and environmentally (ICE, 2011). There is a need for a substantial level of water-related infrastructure investment in the UK in order to deal with anticipated future pressures deriving from social, economic, political and environmental changes.

Given its diversity and mix of public–private partnerships, it is a challenge to find the right formula to finance the water industry. Three different procurement models are put forward for water-related infrastructure development, that is: pure public sector model, public–private partnership model and pure private sector model. These models however vary in terms of level of decentralization, who owns assets, who finances investments, and who defines the price and level of service. The UK water industry needs to step forward and develop solutions with the public sector and create new physical infrastructure procurement models that will underpin mutual goals. It will then be possible to evolve new procurement models for the delivery of broader long-term sustainability agenda, particularly in relation to climate change issues, and to attract long-term, large-scale infrastructure investments.

The UK water industry faces a diverse set of major challenges. Many of these are of an unprecedented nature in respect of climate, demographics, population and economics. These challenges, no doubt, put pressure on the long-term sustainability of the industry in terms of scale, complexity and level of uncertainty. The UK water industry cannot continue to rely on doing things in the same way as before; hence new water industry business models must evolve to deliver products and/or services in an economically viable, socially responsible and environmentally sustainable manner.

REFERENCES

Abbott, M. and B. Cohen (2009), 'Productivity and efficiency in the water industry', *Utilities Policy*, **17**, 233–44.

Agthe, D.E., R.B. Billings and N. Buras (2003), *Managing Urban Water Supply*, Netherlands: Kluwer Academic Publications.

Ainger, C., D. Butler, I. Caffor, D. Crawford-Brown, D. Helm and T. Stephenson (2009), 'A low carbon water industry in 2050', Bristol: Environment Agency.

Akintoye, A. (2009), 'PPPs for physical infrastructure in developing countries', in Akintola Akintoye and M. Beck (eds), *Policy, Finance and Management for Public–Private Partnerships*, Oxford: Blackwell Publishing, pp. 123–44.

Ashley, R. and A. Cashman (2006), 'The impacts of change on the long-term future demand for water sector infrastructure', *Infrastructure to 2030: Telecom, Land Transport, Water and Electricity*, Paris: OECD.

Bakker, K. (2003), *An Uncooperative Commodity: Privatizing Water in England and Wales*, Oxford: Oxford University Press.

Birdsall, N. and J. Nellis (2002), 'Winners and losers: assessing the distributional impact of privatization', CDG Working Paper 6, Washington, DC: Center for Global Development.

BIS (2011), 'Infrastructure supply chains: barriers and opportunities', London: Department for Business Innovation and Skills.

Black, M. and J. King (2009), *The Atlas of World Water: Mapping the World's Most Critical Resource*, London: Earthscan Publications.

Cave, M. (2009), 'Independent review of competition and innovation in water markets: final report', available at www.defra.gov.uk/environment/water/industry/cavereview (accessed 8 June 2011).

Clarke, L. (2010), 'Public services: water and water everywhere', available at http://www.wantedinrome.com/articles/complete_articles.php?id_art=1041 (accessed 20 September 2011).

Cohen, R. (2010), *Meeting Future Infrastructure Needs*, London: Deloitte.

Consumer Council for Water (2009), 'Annual Tracking Survey 2009', London: FDS International.

Cowan, S. (1998), 'The water industry', in D. Helm and T. Jenkinson (eds), *Competition in Regulated Industries*, Oxford: Oxford University Press, pp. 150–75.

Defra (2008), 'Future water: the government's water strategy for England',

London: Department for Environment, Food and Rural Affairs, HM Government.
Deloitte (2010), 'Integrated assets for the water industry', London: Deloitte.
Deutsche Bank (2010), 'Infrastructure research report', Frankfurt: Deutsche Bank.
Dore, M., J. Kushner and K. Zumer (2004), 'Privatization of water in the UK and France: what can we learn?', *Utilities Policy*, **12**, 41–50.
Dzierwa, J. (2009), 'Infrastructure: a global opportunity for investors', *US Global Investors*, available at www.usfunds.com (accessed 14 January 2011).
Environment Agency (2009), 'Transforming wastewater treatment to reduce carbon emissions', Bristol: Environment Agency.
European Union (EU) (2000), 'Directive 2000/60/EC of the European Parliament and of the Council of 23 October, 2000, establishing a framework for Community action in the field of water policy', *Official Journal of the European Communities*, **327**, 1–72.
Goldman Sachs (2008), *The Essentials of Investing in the Water Sector*, Goldman Sachs, USA.
Grigg, N.S. (2010), *Infrastructure Finance – The Business of Infrastructure for a Sustainable Future*, New York: John Wiley & Sons.
Grigg, N.S. (2011), *Water Finance: Public Responsibilities and Private Opportunities*, New York: John Wiley & Sons.
Hall, D. (2004), *Reclaiming Public Water: Achievements, Struggles and Visions from Around the World,* Amsterdam: Transnational Institute.
Harris, C. (2003), 'Private participation in infrastructure in developing countries: trends, impacts, and policy lessons', Working Paper, Washington, DC: World Bank.
Harris, C., J. Hodges, M. Schur and P. Shukla (2003), 'Infrastructure projects: a review of cancelled private projects', Public Policy for the Private Sector 252, Washington, DC: World Bank.
Helm, D., J. Wardlaw and B. Caldecott (2010), 'Delivering a twenty-first century infrastructure for Britain', London: Policy Exchange.
Hipwell, P. (2011), 'The Northern Ireland Water – biting the bullet', available at http://thewaterplaceltd.blogspot.com (accessed 10 January 2011).
HM Treasury (2010), *Strategy for National Infrastructure*, London: HM Treasury Office.
Holder, S. (1998), 'Privatization and competition: the evidence from utility and infrastructure privatization in the UK', Paris: Organisation for Economic Co-operation and Development.
ICE (2011), 'Submission: National Infrastructure Plan 2011 (NIP2)', London: Institution of Civil Engineers.
IMF (2004), 'Public–private partnerships', a paper prepared by Fiscal Affairs Department of International Monetary Fund in consultation with World Bank and IADB.
Jones, J.A.A. (2010), *Water Sustainability: A Global Perspective*, London: Hodder Education.
Kaïka, M. and B. Page (2002), 'The making of the EU Water Framework Directive: shifting choreographies of governance and the effectiveness of environmental lobbying', Oxford University School of Geography and the Environment, Oxford, UK.
Kohl, B. (2004), 'Privatization Bolivian style: a cautionary tale', *International Journal of Urban and Regional Research*, **28**(4), 893–908.

KPMG (2008), 'Delivering water infrastructure using private finance', London: KPMG International.
KPMG (2009), 'Bridging the global infrastructure gap: views from the executive suite', London: KPMG International.
Levina, E. and H. Adams (2006), 'Domestic policy frameworks for adaptation to climate change in the water sector', Paris: Organisation for Economic Co-operation and Development.
Mandri-Perrott, C. (2009), *Public and Private Participation in the Water and Wastewater Sector: Developing Sustainable Legal Mechanisms (Water Law and Policy)*, London: IWA Publishing.
Marin, P. (2009), *Public–Private Partnerships for Urban Water Utilities: A Review of Experiences in Developing Countries*, Washington, DC: World Bank.
Marketforce and KPMG (2010), 'The future of the UK's Energy and Water Industries', Marketforce and KPMG report, London.
Marques, R.C. (2010), *Regulation of Water and Wastewater Services: An International Comparison*, London: IWA Publishing.
Maxwell, S. (2005), 'How big is the "water business"?', *Journal of the American Water Works Association*, **97**(1), 1–10.
McKinsey Report (2009), 'Charting our water future: economic frameworks to inform decision making', Washington, DC: McKinsey & Company.
Means, E.G. III, L. Ospina and R. Patrick (2005), 'Ten top trends and their implications for water utilities', *Journal of AWWA*, **97**(7), 64–75.
Nickson, A. and C. Vargas (2002), 'The limitations of water regulation: the failure of the Cochabamba concession in Bolivia', *Bulletin of Latin American Research*, **21**(1), 99–120.
NIW (2007), 'Drinking water quality annual report 2007', Northern Ireland Water, UK.
OECD (2006), 'Infrastructure to 2030: telecom, land transport, water and electricity', Paris: Organisation for Economic Co-operation and Development.
OECD (2007), 'Infrastructure to 2030: mapping policy for electricity, water and transport', Volume 2, Paris: Organisation for Economic Co-operation and Development.
OECD (2008), 'Environment outlook to 2030', Paris: Organisation for Economic Co-operation and Development.
OECD (2009), 'OECD reviews of regulatory reform: China 2009: defining the boundary between the market and the state', Paris: Organisation for Economic Co-operation and Development.
OECD (2010), 'Going for growth, economic policy reforms', Paris: Organisation for Economic Co-operation and Development.
OFWAT (2010), 'The form of the price control for monopoly water and sewerage services in England and Wales', Discussion Paper, Birmingham: The Office of Water Services.
OFWAT (2011), 'Water companies in England and Wales: financial summary 2010–11', Birmingham: The Office of Water Services.
Ogden, S. and R. Watson (2007), 'The influence of comparative pay, customer service measures and accounting profits upon CEO pay in the UK privatised water industry', *Accounting and Business Research*, **37**(3), 199–215.
Page, B. (2001), 'Achieving sustainable and innovative policies through participatory governance in a multi-level context: results of the case study on the

water supply in London and the UK', School of Geography and Environment Working Paper, University of Oxford, Oxford.
Palaniappan, M., H. Cooley, P.H. Gleick and G. Wolff (2007), 'Water infrastructure and water-related services: trends and challenges affecting future development', in *Infrastructure to 2030: Mapping Policy for Electricity, Water and Transport*, Volume 2, Paris: Organisation for Economic Co-operation and Development.
Pinsent Masons (2010), *Water Yearbook 2010–2011*, London: Pinsent Masons LLP.
Prasad, N. (2006), 'Privatization results: private sector participation in water services after 15 years', *Development Policy Review*, **24**(6), 669–92.
Public Works Financing (2010), 2010 Projects Database, *Public Works Financing Newsletter*, 253, October 2010.
Robison, R. and K. Hewison (2005), 'Introduction: East Asia and the trials of neoliberalism', *Journal of Development Studies*, **41**(2), 183–96.
Saal, D. and D. Parker (2001), 'Productivity and price performance in the privatised water and sewerage companies of England and Wales', *Journal of Regulatory Economics*, **20**(1), 61–90.
Schwartz, P. and D. Randall (2003), 'An abrupt climate change scenario and its implications for United States national security', available at http://www edf.org/documents/3566_AbruptClimate Change.pdf (accessed 10 April 2011).
Scottish Water (2010), 'The Scottish Water 2009–10 Annual Report and Accounts', Dunfermline.
Seidenstat, P., D. Haarmeyer and S. Hakim (2002), *Reinventing Water and Wastewater Systems: Global Lessons for Improving Water Management*, New York: John Wiley & Sons.
Shaoul, J. (1997), 'A critical financial analysis of the performance of privatized industries: the case of the water industry in England and Wales', *Critical Perspectives on Accounting*, **8**(5), 479–505.
Shaoul, J., A. Stafford and P. Stapleton (2006), 'Highway robbery? A financial analysis of design, build, finance and operate (DBFO) in UK roads', *Transport Reviews*, **26**(3), 257–74.
Sheshinski, E. and L.F. Lopez-Calva (2003), 'Privatization and its benefits: theory and evidence', *Economic Studies*, **49**, 429–59.
Skilling, H. and K. Booth (2007), 'Public–Private Partnership Handbook', Manila: Asian Development Bank.
Stern, N. (2009), 'Deciding our future in Copenhagen: will the world rise to the challenge of climate change?', Policy Brief, London School of Economics and Political Science, London.
Stoker, G. (1998), 'Governance as theory: five propositions', *International Social Science Journal*, **50**(155), 17–28.
Tan, W. (2007), *Principles of Project and Infrastructure Finance*, London: Taylor & Francis Publications.
Thomas, D.A. and R. Ford (2005), *The Crisis of Innovation in Water and Wastewater*, Cheltenham, UK and Northampton, MA, USA: Edward Elgar Publishing.
UBS (2009), 'Outlook for global infrastructure', London: UBS Global Asset Management.
Underhill, M.D. (2010), *The Handbook of Infrastructure Investing*, New York: John Wiley & Sons.

Walker Review (2009), 'The independent review of charging for household water and sewerage services', London: Department for Environment, Food and Rural Affairs.

Water UK (2008), 'Sustainable water state of the water sector report', London: Water UK.

Water UK (2010), 'Meeting future challenges a blueprint for policy action', London: Water UK.

Waterwise (2010), 'Mainstreaming water efficiency in the UK: helping to meet the challenges of climate change through wasting less water', London: Waterwise.

Weber, B. and H.W. Alfen (2010), *Infrastructure as an Asset Class: Investment Strategies, Project Finance and PPP*, New York: John Wiley & Sons.

WEF (2011), *Water Security: The Water–Food–Energy–Climate Nexus*, Washington, DC: World Economic Forum Publications.

Wohland, P., P. Rees, P. Norman, P. Boden and M. Jasinska (2010), 'Ethnic population projections for the UK and local areas, 2001–2051', Working Paper, University of Leeds, Leeds, UK.

World Bank (2004), 'Water resources sector strategy, strategic directions for World Bank engagement', Washington, DC: World Bank.

World Bank (2009), 'Summary of key INFRA projects by region', available at http://siteresources.worldbank.org/INTSDNETWORK/Resources/SummaryofKeyINFRAProjectsbyRegion_FinalCR.pdf (accessed 23 September 2010).

World Bank (2010), 'Global economic prospects: crisis, finance, and growth', Washington, DC: World Bank.

5. Performance of urban water governance in France

Eshien Chong

5.1 INTRODUCTION

In the past two decades, the way by which public services are being produced and provided has undergone a major transformation. Indeed, public services, such as urban water supply, public transportation, electricity production and distribution and so on, exhibit general public interest attributes and, sometimes, natural monopoly characteristics. Such characteristics prevent their provision from being left entirely to private operators through full privatization. Traditionally, the provision of such services is generally organized through public firms. In recent years, however, while the adequate provision of these public services still remains the prerogative of the state, there has been a change in the paradigm on how the state can achieve this result. In particular, while direct public provision or in-house provision of a public service is still possible and continues to be widely relied on to provide public services, legislation in various countries generally provides new contractual tools to allow some form of private sector participation in the provision of public services. Public authorities these days can use concession contracts, lease contracts, private finance initiatives (in the UK) and so on to delegate the production and provision of a public service to the private sector. Such contracts allow public authorities to involve the private sector in providing public services, while at the same time avoiding outright privatization. The development of these contractual tools, which will be referred to collectively as public–private partnerships (PPPs)[1] for the purpose of this chapter, therefore give greater flexibility to public authorities to decide on how to organize the provision of a public service.[2]

A positive consequence of the development of these contractual forms, arguably, is that they allow public authorities to choose among various organizational alternatives to ensure the provision of public services. In other words, because PPPs are made available to public authorities, this increases the flexibility with which public authorities can ensure the

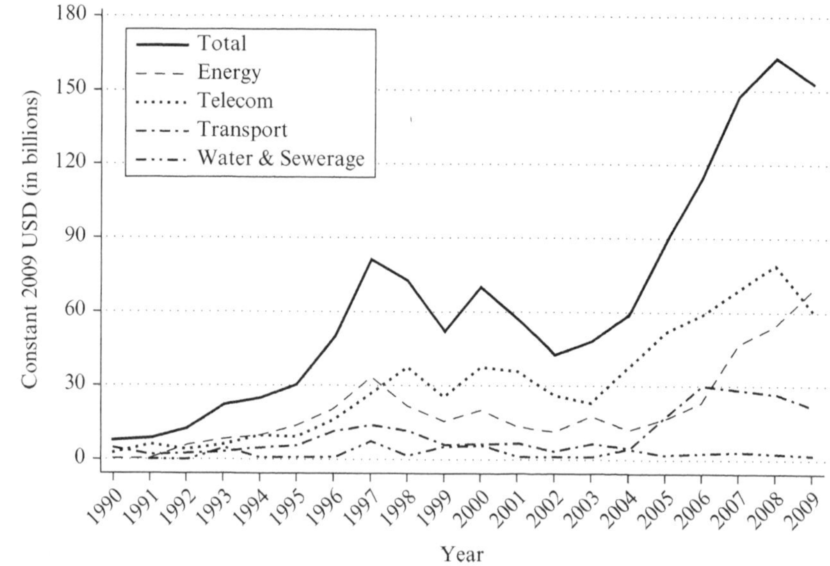

Source: World Bank and PPIAF, PPI Project Database.

Figure 5.1 Private sector investment commitments in infrastructure projects in developing countries (1990–2009)

provision of a public service. Notwithstanding the issue of public finance, alternative ways of organizing the provision of public services may have their own strengths and weaknesses, and may be more or less adapted to specific situations and public policy objectives with which public authorities are confronted. The extension of public authorities' choice set, through the introduction of PPPs, *inter alia*, can potentially contribute to improve the efficiency with which public services are provided. More specifically, this can potentially lead to better use of public funds, and better and/or cheaper service for users.

For those services that require heavy infrastructure investments, investment costs can be assumed by the private sector instead of the public sector under certain types of PPPs. In a context where public funds are becoming scarce, PPPs may seem appealing. As Figure 5.1 shows, there is an increasing trend to use PPPs for infrastructure projects in developing countries, where infrastructure needs are high and resources are limited.[3]

Figure 5.1 also shows that the use of PPPs in fact concentrates on some types of infrastructure projects. In particular, PPP seems to be relatively absent in the water and sanitation sector, while being fairly important in the telecommunication sector. A natural question then arises: why do public authorities seem to be reluctant to appeal to private sector participation in the water industry, when legislation allows them the possibility to do so? There are several potential explanations: first of all, political and ideological reasons may make public authorities more reluctant to use PPP for water and sewerage services. After all, water is quite unlike other goods and services in the economy. This may also be due to inertia in the policymaking process. Second, PPPs may not be the most efficient way to organize water and sewerage services. Consequently, public authorities may be making the right decision to avoid using PPPs to ensure the provision of water services. Obviously, the answer to this question has important consequences for policymaking.

This chapter intends to investigate the performance of PPPs and in-house production in delivering water services to end users. In particular, it intends to undertake an empirical and statistical assessment of the performance of PPPs and in-house provision for delivering water services. To this end, this chapter looks into the organization of water supply services in France.[4] The French case is particularly relevant since there is a long experience with private sector participation in the production and provision of public services in France. Moreover, under French legislation, water services are organized at the municipal level, and municipalities can freely choose among various organizational choices to ensure that water services are provided to end users.

As a consequence, a large variety of organizational structures for water services can be observed in France, ranging from in-house provision or direct public management to concession contracts. This provides a fertile empirical basis to examine the outcome of various organizational choices. The empirical analysis relies on a representative panel data of French municipalities, collected by the French Institute of Environment. The data comes from a survey conducted in over 4000 French municipalities, in 1998, 2001, 2004 and 2008. The database contains a rich set of information on water services in France, ranging from prices paid by consumers to information on the water distribution network, water production technology and organizational choices made by municipalities. The analysis relies on information from about 3600 municipalities.[5] Using this database, the analysis shows that, on average, users pay a higher price when their water services are organized through a PPP. There was no evidence that this price premium is related to higher investment undertaken by a PPP operator, even though water networks managed under a PPP seem to perform better than those run in-house.

Economic literature has explored the relationship between governance structures and their performances in delivering water services to consumers (Chadwick, 1859; Demsetz, 1968; Goldberg, 1976; Williamson, 1976; Hart et al., 1997; Hart, 2003; Bajari et al., 2009; Levin and Tadelis, 2010). Among those that are most closely related to the present study, Ménard and Saussier (2002) are among the first to investigate this issue. They show that municipalities that have chosen to run water services in-house tend to have a higher probability to deliver water that conforms to legal norms than outsourced services. However, they do not investigate issues related to pricing. Carpentier et al. (2006) and Chong et al. (2006) investigated the performance of PPPs in the French water sector using prices paid by consumers as a performance measure. Their studies rely on the same source of information as the present study, but they only have information for the year 1998. Similar to the present study, they show that prices are higher for water services organized under PPPs. Ruester and Zschille (2010) investigated this issue for Germany, and found similar results as Chong et al. (2006).[6] This study differs from the previous literature in that panel data is exploited to shed light on the issue at hand. Arguably, the panel dimension of the present data allows a better consideration of unobservable heterogeneity that could motivate a municipality to choose a particular organizational choice and to have a long-term perspective on the consequences of organizational choices.

Section 5.2 of this chapter gives an overview of the French water industry, while section 5.3 presents the data and provides results of the empirical analysis, followed by the conclusion.

5.2 FRENCH WATER SECTOR: AN OVERVIEW

In France, as in most European countries, local public authorities[7] are in charge of the organization of local public services. More precisely, they are responsible for creation and operation of these services. The reason is that these activities have general public interest attributes. Because of these general public interest attributes, provision of these services cannot be left entirely to the private sector in the sense that any private firm(s) may only be granted monopoly rights over these services for a limited period of time at best. To fulfill their responsibilities, local public authorities must define the general principles governing the service.[8] There is generally no national regulator for these services. The following section will briefly discuss how local public authorities may ensure the provision of public services and the type of legal constraints faced by them.

5.2.1 Wide Range of Contractual Arrangements

There are a wide range of instruments available to local public authorities to help them organize the provision of a given local public service. Some of these arrangements allow local public authorities to benefit from the expertise of the private sector in providing such services. In France, for instance, local public authorities may decide to transfer some of their decision rights and revenue rights to an external operator (Desrieux, 2006). In other words, a local public authority can choose to either manage the service itself (direct public management or in-house provision) or initiate a PPP that leaves a private operator in charge of managing the service. In the latter case, there are a variety of contractual instruments at a local public authority's disposal to serve as a basis for forming such partnerships.

Briefly, these contracts differ with respect to two major dimensions: first of all, the degree of a firm's involvement in the service (who makes the investment); and secondly, the risks that the private operator bears (how is the operator paid). In this sense, a *gérance* contract is the closest to direct public management: the operator is only responsible for the management of the service and is paid a fixed amount by the public authority. The operator is free of all investment obligations. The 'intermediary management' contract has similar contractual arrangements, with an exception concerning the private operator's compensation for managing the service: instead of a fixed compensation, part of the operator's revenues will depend on his performance.

A public authority can also choose to use a lease contract or a concession contract. These contractual forms leave more responsibility to the private operator in terms of investment. Under a lease contract, the private operator is in charge of some investment, even if the most important investment remains public. In contrast, a concession contract leaves a substantial part of the investment to the private operator. Under a lease or concession contract, the operator is paid through receipts from end users rather than by the public authority itself, which increases the proportion of risk they bear as compared to *gérance* and 'intermediary management'. In terms of risks, concessions are even riskier than lease contracts because of the higher proportion of investment the operator must realize. Investment made by the private operator under both types of contracts automatically reverts back to the public authority once the contract terminates. Generally, operators do not receive any supplementary financial compensation when this happens.

In lease and concession contracts, water prices in France are adjusted over time by weighting the initial price in the contract with a coefficient

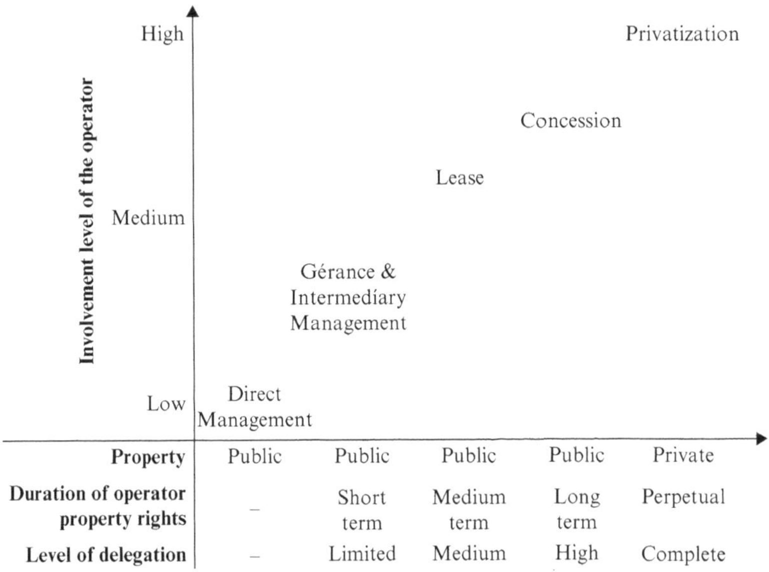

Source: Huet and Saussier (2003).

Figure 5.2 Contractual options for local public services in France

that accounts for variations in the costs of production factors over time. It should be noted that cost differences are measured at the national level, and therefore do not account for local conditions.

Lastly, if a local public authority wishes to organize the service in-house, the French legislation requires that production costs are entirely covered by receipts from the service. In other words, in France, local public authorities are not allowed, legally at least, to subsidize the costs of running a particular service in-house.

The various differences between the types of contracts that may be used by a local public authority to organize the provision of a local public service are summarized in Figure 5.2. Huet and Saussier (2003) provide a detailed description of these organizational modes.

Figure 5.3 shows the distribution of various organizational choices in the sample. As can be easily observed from Figure 5.3, a majority of French municipalities organize their water services using either direct in-house provision or lease contracts. On average, about 54 percent of water services in France are organized through lease contracts, whereas 37 percent of the municipalities prefer to provide water to consumers through in-house provision. It is also interesting to note that concession

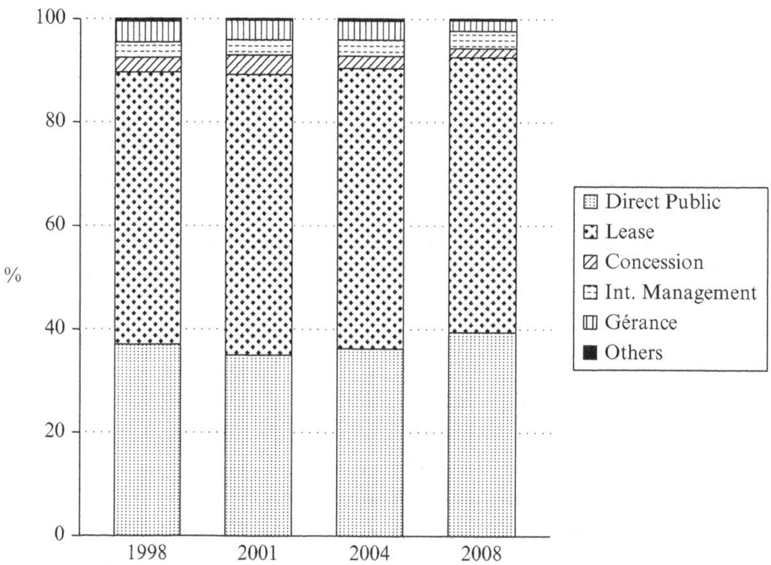

Figure 5.3 Organization choices for water supply

contracts account for a weak share of how water is supplied to the French population: this contract is only chosen by 2.4 percent of the French municipalities.

5.2.2 Specific Institutional Framework

Local public authorities in France enjoy a great degree of flexibility in organizing the provision of local public services. The institutional framework in which this freedom of choice is embedded amplifies the discretionary power of local public authorities. Indeed, the *intuitu personae* principle[9] governs how local public authorities organize a service. Furthermore, the fact that such contracts are considered to be 'administrative contracts' also gives great power to the public contracting party, as explained below.

Negotiations and competition for the field: the *Intuitu Personae* Principle

If the public authority chooses a lease or a concession contract, the selection mechanism follows a two-step procedure:

1. The public authority chooses a number of potential candidates using the competitive tendering process.

2. There is a phase of negotiation between the public authority and the potential providers. At the end of the negotiation, the public authority chooses its final partner for the duration of the contract.

The municipality is not obliged to select its partner by complying with any objective criteria defined by law, as would be the case in a strict competitive tendering process. The use of such a two-step procedure gives the public authority considerable latitude to select its partner freely, using not only objective but also subjective criteria that are not necessarily specified by law. Furthermore, the local public authority is not required to justify its decision to organize the service through a PPP or through direct public management.[10]

PPP and the rules on administrative contracts

In France, contracts signed between local authorities and private operators are considered as 'administrative contracts'. As such, a local authority may unilaterally change the terms of the contract after it is signed. Hence, theoretically, a local public authority may be in a more favorable position. Of course, such changes must be justified (for example, through reasons of general public interest) and the private operator may claim fair compensation for the implementation of modifications required by the local authority.[11] Nevertheless, in case of conflict, the private operator must first implement those changes required by the local authority, even in case of disagreement over the changes. In practice, local authorities use this power sparingly. Nevertheless, it provides a credible threat that helps to prevent the private operator from acting opportunistically because it generates fear of the contract being terminated prematurely or changed unilaterally.

Evolution towards anti-corruption laws

As of 1993, under the Anti-Corruption Law (Loi Sapin), local authorities are required to advertise their willingness to contract out the provision of local public services and to organize an auction for interested candidates. Nevertheless, the *intuitu personae* principle implies that local authorities are not obliged to choose the best offer. Thus prices might be lower for contracts signed after 1993 compared to those signed before, to the extent that anti-corruption laws can lead to increased *ex ante* competition by reducing the discretion of local authorities insomuch as they are politically constrained by the publication of the auction's results.

The benefits of using a PPP also depend on the number of potential operators that are capable of running the service. It is well known that the higher the number of bidders, the more efficient an auction mechanism can be (Demsetz, 1968; McAfee and MacMillan, 1987). Therefore, the anti-corruption law may actually help to foster competition and its benefits.

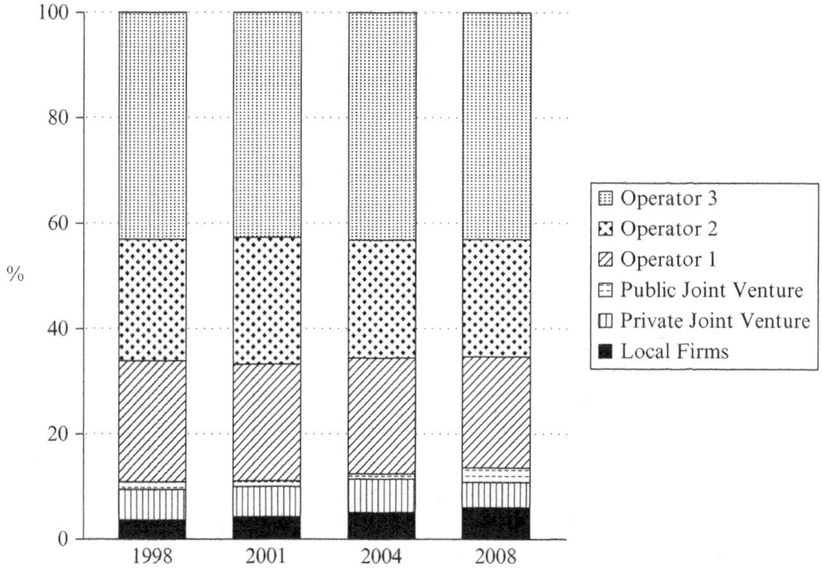

Figure 5.4 Market shares of private operators

5.2.3 Market Structure of the French Water Sector

The water industry in France is highly concentrated and it is dominated by three major players who form an oligopoly. Together, they provide water services to about 75 percent of the French population.[12] According to the French Competition Authority (Conseil de la Concurrence), in 2000, as much as 98 percent of the outsourced contracts in the water sector were awarded to one of these three operators. More specifically, the French Competition Authority also observed in 2001 that these three operators and their joint ventures controlled 56 percent, 29 percent and 13 percent of market shares in the sector, respectively. The remaining 2 percent is controlled by independent and small local water companies.[13] This leads to a Herfindahl index of 4250,[14] indicating that the French water sector has a particularly concentrated structure. Furthermore, the High Council for Public Service (Haut Conseil du Service Public) observed in a report in 1999 that only about 5 percent of the contracts changed hands after expiry (Haut Conseil du Service Public, 1999). This indicates that market shares of these operators are relatively stable over time.[15]

Figure 5.4 shows the structure of the water market in France computed based on the sample and on the number of services that each operator operates. The figure shows that the concentration of the operators is relatively high for water supply contracts in France. It is also worthwhile to note that the market share of local and independent water producers has increased slightly within the observation period.

5.3 GOVERNANCE STRUCTURE AND PERFORMANCE IN WATER SERVICES IN FRANCE: AN EMPIRICAL ANALYSIS

Do organizational choices made by municipalities have any impact on the outcome of a water service? This section will attempt to answer this question. To this end, the relationship between various outcome variables for water services and how they are organized is examined. The performance measure used is the price paid by consumers for water services. Arguably this is one of the important variables to measure performance. Indeed one could consider that lower prices are beneficial to society, so long as the prices do actually cover the cost of producing the service. In the following section, a statistical analysis is presented and then the econometric results are discussed.

5.3.1 Performance of PPPs at First Glance

A first and preliminary answer to the question can be found in Figure 5.5. This figure shows the average price paid by consumers per 120 m^3 of water

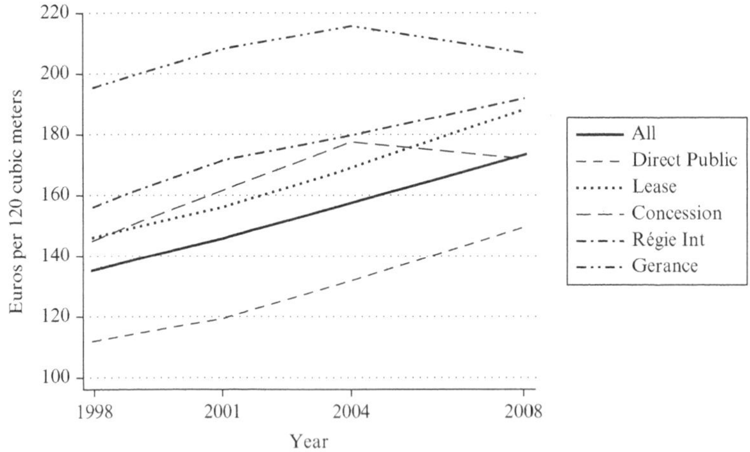

Figure 5.5 Average prices paid by consumers for 120 m^3 of water under various governance structures

per year.[16] Strikingly Figure 5.5 shows that, on average, consumers who pay the lowest price for their water services are those whose services are run through direct public management. Water services in which there is an involvement of the private sector tend to be associated with higher prices, on average. Of course, this observation may also reflect subsidy issues under direct public management. However, the author is fairly confident that this is not the case, since cross-subsidies are theoretically not allowed by law. Indeed, municipalities cannot legally choose to finance an in-house water service beyond receipts from the water service.

Nevertheless, a large variety of factors may explain higher prices under a PPP: water services run under a PPP may be more complex, hence leading to higher observed prices paid by consumers on average. One may also argue that if investments are higher under a PPP, it is not surprising that consumers will be charged higher prices when their water service is run under a PPP. Therefore, to go a step further in the analysis, various factors that may influence prices paid by consumers for their water services must be considered, and it must be confirmed whether prices paid by consumers still remain higher under a PPP, *ceteris paribus*. This is the goal of the econometric exercise in the following subsection.

5.3.2 Simple Econometric Analysis of Performance and Organizational Choices

In order to rigorously compare whether different organizational choices lead to different prices paid by consumers, *ceteris paribus*, a simple econometric analysis is conducted. For simplicity, the performance of in-house provision of a water service is compared to the performance of an outsourcing strategy without taking into account the contract type used in an outsourcing strategy. In other words, for simplicity, the performance of PPP as a whole is compared to the performance of in-house provision for water services.

To this end, the author exploited the panel dimension of the data to estimate the following equation:

$$y_{it} = \omega O_{it} + x'_{it}\beta + u_i + \varepsilon_{it} \tag{5.1}$$

where y_{it} is the deflated price paid by consumers in municipality i at time t, O_{it} is the organizational choice observed for municipality i at time t, x'_{it} is a vector containing factors that may influence the price paid by consumers, u_i is a term that captures unobservable heterogeneity for a municipality and ε_{it} is the usual error term. More specifically, let $O_{it} = 1$ if the water service in municipality i is outsourced at time t, and $O_{it} = 0$ otherwise. ω is therefore the average impact of organizational choice on the price of water paid by consumers, that is, ω measures the *ceteris paribus* effect of PPPs on water prices.

Differences in service characteristics are accounted for by including as explanatory variables the type of raw water treatments (*Treat1*, *Treat2*),[17] the origins of raw water (*Surface*, *MixOrigin*), the density of the water distribution network (*N/w density*), the length of network replaced (*N/w replace*), whether there is an investment program for water services (*InvstProg*), the amount of water produced and its square, and the total volume of water distributed over the sum of volume distributed and volume of water imported (*IndepRatio*) as a measure of the size of a municipality's water resources. Characteristics of a municipality are also taken into account with variables such as whether a municipality is classified as a tourist area (*Tourist*), whether some water restriction measures have been imposed (*WaterRestr*), and whether a municipality belongs to an association of municipalities for the organization of water (or other) public services (*Intermuni*). Additional control variables include information on the contracts that are signed after the Sapin Law of 1993 (*SapinLaw*), and after the Barnier Law of 1995 (*BarnierLaw*). Finally, year (fixed) effects are included in the regression in order to account for unobservable heterogeneity in different years. A description of these variables can be found in Table 5.1.

The estimation results for equation (5.1) are presented in Table 5.2. The results for three different estimation procedures are shown: the first column shows results from a simple Ordinary Least Squares (OLS) estimator, and the second and third columns exploit the panel dimension of the data and use random-effects estimator and fixed-effects (within) estimator respectively. The difference between the latter two estimators resides in the assumption about the correlation between control variables (O_{it} and x_{it}) and the unobservable individual heterogeneity term (u_{it}). The random effects estimator assumes that there is no correlation between u_{it} and the explanatory variables while the fixed effects estimator assumes that there is some correlation between the two types of variables, giving rise to an endogeneity problem. Hence, if the explanatory variables are correlated with the error terms (u_{it} or ε_{it}), the fixed-effects estimator is unbiased, whereas OLS and random-effects estimators are biased.

The estimation results show that consumers pay a positive price premium when they live in a municipality whose water services are organized through a PPP, *ceteris paribus*. This means that controlling for various factors that could have an impact on water prices, consumers still pay more when a service is outsourced through a PPP. The size of this premium ranges from an average of about €31 to about €15. In relative

Table 5.1 Definition of variables and descriptive statistics

Variable	Definition	N	Mean	Std. Dev.	Min.	Max.
Price	Deflated price per 120 m^3 of water	12053	143.74	45.866	0	681.13
Outsourced	Takes value 1 if the local authority externalizes water services	12053	0.699	0.459	0	1
SapinLaw	Takes value 1 if the PPP contract is signed after the Sapin Law comes into effect	12053	0.33	0.47	0	1
BarnierLaw	Takes value 1 if the PPP contract is signed after the Barnier Law comes into effect	12053	0.286	0.452	0	1
Treat1	Takes value 1 when raw water only needs a light disinfection treatment	12053	0.53	0.499	0	1
Treat2	Takes value 1 when raw water needs a medium disinfection treatment	12053	0.151	0.358	0	1
Surface	Takes value 1 when origin of the water is surface water	12053	0.185	0.389	0	1
MixOrigin	Takes value 1 when origin of the water is mixed (underground and surface)	12053	0.167	0.373	0	1
Production	Volume of water produced	12053	860.6	5131.2	0	250058
Density	Density of the water distribution network (length in km/population size)	12053	0.0232	0.033	0	1.44
Replacement	Length of water distribution network replaced (in km)	12053	0.576	1.537	0	53.2
Intermuni	Takes value 1 if the local authority is organizing water services in cooperation with other local authorities	12053	0.699	0.459	0	1
InvstProg	Takes value 1 when the contract specifies an investment program	12053	0.651	0.477	0	1
WaterRestr	Takes value 1 if consumed water volume is constrained by regulations in the year	12053	0.0651	0.247	0	1
Tourist	Takes value 1 if the municipality is a tourist area	12053	0.128	0.334	0	1
IndepRatio	Total volume distributed / (total volume distributed + imported volume)	12053	0.894	0.219	0	1

Table 5.2 Organizational choices and prices for water

Estimator	(1) OLS	(2) Random Effects (GLS)	(3) Fixed Effects (Within)
Dep. Variable	Price	Price	Price
Outsourced	30.990***	29.625***	14.963***
	(0.917)	(1.235)	(2.885)
SapinLaw	1.413	−3.528+	−6.361**
	(1.936)	(2.126)	(2.417)
BarnierLaw	−5.780**	−3.074	−0.998
	(1.995)	(2.029)	(2.252)
Treat1	−4.521***	−3.554***	−1.511
	(1.085)	(1.062)	(1.247)
Treat2	3.625**	2.850*	0.075
	(1.222)	(1.182)	(1.372)
Surface	19.196***	11.088***	−2.266
	(1.268)	(1.470)	(2.103)
MixOrigin	8.518***	6.960***	−1.470
	(1.253)	(1.385)	(1.938)
Production	−0.001***	−0.001*	0.001
	(0.000)	(0.000)	(0.001)
Production2	0.000***	0.000*	0.000
	(0.000)	(0.000)	(0.000)
Density	226.779***	53.511**	−21.485+
	(40.590)	(20.596)	(12.058)
Replacement	0.479	−0.242	−0.117
	(0.273)	(0.265)	(0.322)
Intermuni	21.283***	16.497***	6.543***
	(0.959)	(1.172)	(1.793)
InvstProg	1.805*	0.687	0.194
	(0.888)	(0.592)	(0.615)
WaterRestr	−0.742	−2.000+	−2.435*
	(1.466)	(1.032)	(1.089)
Tourist	5.118***	4.953***	5.518**
	(1.186)	(1.373)	(1.905)
IndepRatio	−17.455***	−11.844***	−4.283+
	(1.765)	(1.900)	(2.445)
Const	113.144***	117.471***	131.514***
	(2.220)	(2.365)	(3.423)
Year Fixed Effects	Yes	Yes	Yes
Muni. Fixed Effects	No	No	Yes
R^2	0.2810	0.2602	0.0353[a]
Obs	12053	12053	12053

Notes: Significance stars: + $p < 0.10$, * $p < 0.05$, ** $p < 0.01$, *** $p < 0.001$. Standard errors within parentheses. [a] Corresponds to within R^2.

terms, this premium is about 21.5 percent to 10.4 percent of the mean deflated price of about €144 per 120 m^3 paid by consumers in the sample used for the regressions.

Obviously, one of the shortcomings of the estimation results is that organizational choices may be endogenous. In particular, if there are some explanatory variables that have not been controlled for in the regression, this may lead to higher prices for consumers and municipalities may prefer to organize these water services through a PPP, and then the estimations will be biased because of this endogeneity problem. This bias occurs because of a correlation between existing explanatory variables and the error term in the regression. However, the fixed effects estimator takes into account the possibility of this correlation. As such, to the extent that the organizational choices made by municipalities depend on some unobservable and time-invariant characteristics of municipalities, estimates in column (3) have adequately dealt with this problem of endogeneity. This leads to focus the analysis more particularly on the estimation results of column (3) in Table 5.2.

Interestingly, according to the results in column (3) of Table 5.2, consumers also seem to pay a lower price for their water services when the corresponding PPP is awarded after the Sapin Law came into effect. This effect is significant and shows that consumers pay about €6 less for their water services, *ceteris paribus*, than their counterparts whose water services are organized through a PPP contract signed before the law. This observation suggests that a compulsory call for tender that a municipality has to organize when it decides to outsource its water services under the law has generated positive effects for consumers. This result is consistent with the idea of an increase in competition faced by PPP operators. Nevertheless, one should note that even with this positive effect of the Sapin Law, consumers whose services are organized through a PPP still pay a significant and positive premium.[18]

In a nutshell, both simple statistical analysis and econometric analysis point to negative performance of PPPs in water services in France, when price is considered as a measure of performance. The statistical analysis shows that consumers pay on average a higher price when water services are organized as a PPP in France, and controlling for specific characteristics of a municipality's water services and its characteristics, the higher price premium associated with services run under a PPP still persists. A conservative estimate situates this effect at about €15 per 120 m^3 of water paid by consumers. The estimation results also show that the Sapin Law, by making call for tenders compulsory when a municipality seeks to outsource its water service through a PPP, has a moderating effect on the size of this premium.

This can potentially be explained by competitive pressures that accompany the process of a call for tenders, allowing municipalities to receive better terms, at least for the initial price charged to consumers, under a PPP contract.

5.4 CONCLUSION

In recent years, there has been an increasing political demand for private sector involvement in the operation and provision of services that are traditionally undertaken directly by public entities. One of the reasons that underpin this trend is greater productive efficiency: private operators are widely perceived to be more efficient than their public counterparts. This trend can also be observed for local public services such as water services, garbage collection and urban transportation. For such services, a way to benefit from private expertise is to use a PPP to form a partnership with a private operator. Indeed, full privatization of these services is often politically hard to achieve due to their inherent general public interest attributes.

This chapter investigates whether PPPs are adapted for water services. To this end, it considers the French experience in the organization of water services. The French case is particularly relevant for this exercise given the long tradition of using PPPs for various public services in France. To examine the performance of various organizational forms to deliver water services, a representative data set of water services organized by the French municipalities is used, with information observed in 1998, 2001, 2004 and 2008. Using the price paid by consumers for their water services in various municipalities as a performance measure, this study shows that water services provided to consumers through a PPP lead, on average, to the payment of a higher price by consumers.

This observation is corroborated when the specific characteristics of water services in various municipalities are accounted for using econometric techniques. In this analysis, it is shown that the average price premium associated with a PPP for water services is about €15 per 120 m^3, accounting for various factors that may influence water prices.

The results suggest that PPPs are not adapted for water services, at least in France. Nevertheless this interpretation of the estimation results should be subject to some caution. Indeed the estimation corresponds to an average price premium associated with PPP. This does not mean that PPPs may not fare better in some situations even though, on average, PPPs lead to higher prices. Hence, one should not consider that PPPs are always inefficient based on these results: they may be more adapted in some situations when compared to public in-house provision, while being maladapted for other situations. An important question is to identify the conditions under which PPPs may outperform other organizational choices, and when they will not. This would be an interesting direction for future work.

Although attempts have been made to control for factors that could have an impact on water prices, an important set of variables relating to the quality of the service being provided to consumers is absent from the explanatory variables. One may argue that services provided through PPPs are better in quality and this is reflected in higher prices paid by consumers. This is another caveat of this study. Unfortunately, such information is not currently available in order to estimate the price premium of PPPs that is adjusted for quality of the service. If quality of the service provided indeed differs between PPPs and public in-house provision, the price premium in this study may be overestimated. This may be another direction for future work.

Nevertheless, to conclude briefly, this analysis provides some insights into the impact of organizational choices to ensure the delivery of a public service, and more specifically water services. In particular, PPPs may not always be the best solution to organize the provision of a public service and greater transparency in how a PPP is awarded may be beneficial for consumers. Several useful directions that could contribute to a better understanding of PPPs as a means of ensuring the provision of a public service have also been pointed out.

NOTES

1. Our broad definition of PPPs follows Grout and Stevens (2003). PPPs are sometimes defined in a narrower way, encompassing only those contractual arrangements in which a public authority delegates both investment in infrastructure and operation of the infrastructure to a single operator.
2. Note that this flexibility extends not only to allow the participation of the private sector in the provision of a public service. Contractual tools also provide for cooperation between public authorities and non-governmental organization (NGOs) or associations to ensure the provision of a public service. Such tools are often referred to as public–public partnerships, to contrast with public–private partnerships.
3. There is also an increasing trend towards the use of PPPs in industrialized countries for the provision of public services. See, for instance, European PPP Expertise Center (2010, 2011). Note, however, that the use of PPPs for infrastructure investments remains small in comparison with public investments (Kappeler and Nemoz, 2010).
4. One of the first experiences with PPP is the concession contract used to construct the Midi Canal in France, as early as 1663.
5. This chapter restricts attention to municipalities which organize both the production and distribution of water using the same organizational choices. This is the case in a majority of French municipalities.

6. Readers who are interested in an assessment of water PPPs in developing countries can also refer to Marin (2009).
7. The authorities are essentially municipalities or associations formed by several municipalities.
8. An exception to this is the UK case, where water services are economically regulated by the Office of Water Services (OFWAT).
9. A contract concluded *intuitu personae* is a contract involving a customized relationship between the buyer and the seller. Employment contracts and mandates are examples of *intuitu personae* contracts. For instance, an employer is not obliged to engage the candidate with the highest qualifications who is willing to accept a job at the lowest wage, but may also consider intrinsic characteristics of the candidates themselves (dynamism, kindness, ability to communicate with colleagues and/or customers, etc.) before deciding whether or not to employ a candidate. These features cannot be incorporated into a contract and are not verifiable by a court. The same reasoning applies to a local public authority choosing a firm within the framework of an intermediary management, a lease or a concession contract.
10. Flexibility is not necessarily less efficient. It may help to overcome many of the identified problems concerning the choice of the operator in a more rigid franchise bidding process (Bajari et al., 2009; Athias and Saussier, 2010).
11. The legal right of a local public authority to unilaterally change the terms of a contract is conferred by French jurisprudence. This right is affirmed in a series of judgments by the French 'Council of the State' (Conseil d'État): l'arrêt CE 10 janvier 1902 Gaz de Deville–les–Rouen, l'arrêt CE 21 mars 1910 Compagnie générale des tramways, and l'arrêt CE 2 février 1983 Unions des transports publics. In the second case, it was decided that the local public authority in question had the right to oblige the private operator to increase the number of streetcar services it ran (with respect to the number agreed upon in the initial contract). This judgment was motivated by the fact that such an increase is necessary to ensure that the streetcar service would 'function normally' (in the original text of l'arrêt CE 21 mars 1910 Compagnie générale des tramways, page 216, 'pour assurer, dans l'intérêt du public, la marche normale du service').
12. The remaining population is served through direct public management.
13. See Bulletin officiel de la Concurrence, de la Consommation et de la Répression des fraudes, http://www.minefi.gouv.fr/dgccrf/boccrf/01_01/a0010006.htm.
14. The Herfindahl index is a measure of market concentration. It is calculated as the sum of the squared market shares of firms, and can take a maximum value of 100000. An industry with an index value higher than 2500 is usually considered to be concentrated.
15. Note that this does not necessarily imply that competition is nonexistent. Sometimes a threat of competition is sufficient to induce an operator to adopt competitive behavior.
16. This is a standard measure of water price for statistical studies in France.
17. These two treatments are simple treatments for raw water. Because of collinearity, the third type of treatment, more complicated, is used as the reference variable.
18. The Fisher statistic corresponding to the null hypothesis that *Outsourced – SapinLaw* = 0 is 28.5. This test statistic leads to a rejection of the null hypothesis.

REFERENCES

Athias, L. and S. Saussier (2010), 'Contractual flexibility or rigidity for public private partnerships? Theory and evidence from infrastructure concession contracts', Chaire EPPP Discussion Paper series No. 2010–3.

Bajari, P., S.R. McMillan and S. Tadelis (2009), 'Auctions versus negotiations in procurement: an empirical analysis', *Journal of Law, Economics, and Organization*, **25**(2), 372–99.

Carpentier, A., C. Nauges, A. Reynaud and A. Thomas (2006), 'Effets de la délégation sur le prix de l'eau potable en France: une analyse à partir de la littérature sur les "effets de traitement"', *Economie and Prévision*, **174**(3), 1–19.

Chadwick, E. (1859), 'Results of different principles of legislation in Europe', *Journal of the Royal Statistical Society Series A*, **22**, 381–420.

Chong, E., F. Huet, S. Saussier and F. Steiner (2006), 'Public–private partnerships and prices: evidences from water distribution in France', *Review of Industrial Organization*, **29**(1), 149–69.

Demsetz, H. (1968), 'Why regulate utilities?', *Journal of Law and Economics*, **11**, 55–66.

Desrieux, C. (2006), 'Le rôle de l'autorité publique dans la gestion déléguée des services publics locaux: Une approche par la théorie des contrats incomplets', *Revue Economique*, **57**(3), 529–39.

European PPP Expertise Center (2010), 'Market update: review of the European PPP market in 2010', European PPP Expertise Center Library.

European PPP Expertise Center (2011), 'Market update: review of the European PPP market, first semester of 2011', European PPP Expertise Center Library.

Goldberg, Victor P. (1976), 'Regulation and administered contracts', *Bell Journal of Economics*, **7**(2), 426–48.

Grout, P.A. and M. Stevens (2003), 'The assessment: financing and managing public services', *Oxford Review of Economic Policy*, **19**(2), 215–34.

Hart, O. (2003), 'Incomplete contracts and public ownership: remarks, and an application to public private partnerships', *Economic Journal*, **113**(485), C69–C76.

Hart, O., A. Shleifer and R.W. Vishny (1997), 'The proper scope of government: theory and an application to prisons', *Quarterly Journal of Economics*, **112**(4), 1127–61.

Haut Conseil du Service Public (1999), 'Quelle régulation pour l'eau et les services publics?', Technical Report, La Documentation Française, Paris, France.

Huet, F. and S. Saussier (2003), 'The provision of public interest services through private law contracts', *European Business Organization Law Review*, **4**(3), 403–28.

Kappeler, A. and M. Nemoz (2010), 'Public–private partnerships in Europe: before and during the recent financial crisis', Economic and Financial Report 4.

Levin, J. and S. Tadelis (2010), 'Contracting for government services: theory and evidence from US cities', *Journal of Industrial Economics*, **53**(3), 507–41.

Marin, P. (2009), 'Public–private partnerships for urban water utilities: a review of experience in developing countries', Washington, DC: Public–Private Infrastructure Advisory Facility, World Bank.

McAfee, R.P. and J. MacMillan (1987), 'Auctions and bidding', *Journal of Economic Literature*, **25**(2), 699–738.

Ménard, C. and S. Saussier (2002), 'Contractual choice and performance: the case of water supply in France', in E. Brousseau and J.-M. Glachant (eds), *The Economics of Contracts: Theory and Applications*, Cambridge: Cambridge University Press, pp. 440–62.

Ruester, S. and M. Zschille (2010), 'The impact of governance choice on performance: an application to the German water distribution sector', *Utilities Policy*, **18**(3), 154–62.
Williamson, O.E. (1976), 'Franchise bidding for natural monopolies: in general and with respect to CATV', *Bell Journal of Economics*, **7**(1), 73–104.

6. Water governance in Spain: a dynamic system in transition

Belen Olmos Giupponi

6.1 INTRODUCTION

Spain represents an interesting case in terms of water management in Europe. As a Mediterranean country with semi-arid regions and unequal distribution of water resources, Spain had to introduce various changes to the system in order to secure access to water for its 44 million inhabitants (as of 2008). Consequently, Spain's water policy and management have undergone a process of continuous change over the past 30 years.

Water governance in Spain has been categorized into three levels of institutional arrangements. The state (central government) plays a crucial role in the institutional coordination and allocation of water resources. At the same time, regional governments (Comunidades Autónomas) are vested with important powers in terms of water management, while local administrations are responsible for water supply and sanitation. Consequently, there are many cross-regional variations regarding the water governance architectures and the level of institutional development.

Various legislative and policy changes have occurred since the enactment of the 1985 Water Act. The core instrument is the National Hydrological Plan, a legally binding instrument which includes the main guidelines relating to water governance. Spanish legislation and policies on water have followed international trends and the recommendations and commitments arising from the European Union law. The implementation of the 2000 EU Water Framework Directive (WFD) introduced new challenges and further modifications in terms of water management legislation and policy. However, the implementation of the WFD has been an uneven process with differences across the various regions.

Currently, the water governance system in Spain is still in the process of developing a new paradigm to address water policy challenges. Because of its permanent evolution, one could define the Spanish water governance system as dynamic. At the same time, the system is in transition because it is in the process of switching to a new paradigm (Font and Subirats, 2010).

This chapter examines the main trends in water governance in Spain, with particular focus on the institutional aspects for public–private partnerships (PPPs). The chapter offers an overview of the Spanish water policy, highlighting the main turning points in its evolution and the features of the current water governance system. Section 6.2 provides an overview of the current status of water resources in Spain. Section 6.3 analyses the evolution of the legal framework and water policy in Spain. Section 6.4 addresses the question of water management highlighting the contractual and administrative frameworks, in terms of PPPs. Section 6.5 contains a general appraisal of urban water management in Spanish cities, while section 6.6 summarizes the author's conclusions.

6.2 WATER RESOURCES IN SPAIN: A BRIEF OVERVIEW

In terms of water resources, Spain is dependent on annual precipitation and surface water (FAO, 2009). Surface water provides around 76 percent of water services for extractive uses in Spain (FAO, 2011) and the level of use varies according to the river basin concerned. The main watersheds in Spain are the Duero, Guadiana, Guadalquivir, Ebro, Tajo, Segura and Júcar. The river basins in northern Spain provide more surface water compared to the Mediterranean and southern basins. Consequently, there is water imbalance between the basins which flow into the Atlantic–Bay of Biscay area, and the ones which flow into the Mediterranean, with the exception of the Ebro basin. These territorial imbalances concerning water availability are the basis of the water transfer policy among different hydrographical basins (MMA, 1998). Table 6.1 provides information about the water resources in Spain (as of 2009).

Different factors have aggravated this irregular distribution of resources. These include: (1) increase of water consumption in all its uses; (2) long periods of drought; (3) significant increase in the use of underground water; (4) progressive exhaustion of some aquifers, particularly on the Mediterranean coast; (5) water quality deterioration; (6) problems of salt intrusion; (7) restrictions on supply for irrigation; (8) soil erosion caused by inefficient land uses and deforestation; (9) problems related to the effects of climate change; and (10) pollution of water resources. These phenomenon have led to water scarcity in the country, especially in the Southern and South-Eastern areas, and resulted in an increasing pressure on the available water resources.

Table 6.1 Water resources in Spain, 2009

Water resources	Value	Unit
Average depth of annual precipitation	636	mm/year
Average volume of annual precipitation	321.4	km^3/year
Average internal annual renewable water resources	111.2	km^3/year
Average external annual renewable water resources	0.3	km^3/year
Total annual renewable water resources	111.5	km^3/year
Dependency ratio	0.2691	%
Total renewable water resources per capita	2443	m^3/year
Total dam capacity	52.65	km^3

Source: FAO (2009).

Over the past 50 years, water demand has increased due to growth of cities, extension of irrigation and increase in the demand for industrial uses. According to the World's Water Datasheet, the total freshwater withdrawal in Spain was 37.22 km^3 in 2002, of which agricultural use constituted 68 percent, industrial use 19 percent and domestic use 13 percent.

With regard to the use of water in agriculture, there are currently 3 700 000 hectares (ha) of land under irrigation. The public policy concerning the use of water for agricultural purposes has led to an increase in consumption to improve the conditions in rural areas and to encourage irrigation by private companies. In many cases, there are problems due to deficient conservation and maintenance of pipelines and channels and the use of old methods of irrigation.

In the current situation, water allocation in Spain is unsustainable, with problems arising from agricultural use. Many aquifers are subject to intensive use for irrigation, which has resulted in overexploitation. Spain is one of the few high-income countries in the Organisation for Economic Co-operation and Development (OECD) where water abstraction continued to increase through the 1990s (OECD, 2010). The future climate change scenario in Spain includes reductions of up to 70 percent in precipitation and more frequent occurrence of flash floods (Lopez-Gunn, 2009). In general, water availability will decrease, provoking a rise in water stress. At the same time, an increase in water consumption for agricultural use and domestic use in the cities is foreseeable. According to Fuentes (2011, p. 16), 'in view of the role of per capita income, urban water demand may be expected to continue rising if prices do not rise significantly in real terms'.

6.3 LEGAL FRAMEWORK AND WATER POLICY: A SYSTEM IN PERMANENT EVOLUTION

Water policy in Spain has undergone several changes over the past 20 years (Font and Subirats, 2009). Since the enactment of the Spanish Constitution in 1978 and the adoption of a new Water Act in 1985, various legislative and policy changes have taken place (on the features of contemporary domestic water resources legislation, see Burchi, 2004). These modifications of national legislation on water management have been also shaped by the European Union (EU) legislation since Spain became a part of the European Economic Community (now the European Union) in 1986 (Costeja et al., 2004). Different milestones can be traced during this evolution.

As a starting point, the Spanish Constitution of 1978 provided the legal basis for the public management of water resources (both surface water and groundwater), which were included within the 'hydraulic public domain'. This allowed government authorities to take measures in case of groundwater overexploitation. This system is based on the incorporation of water resources into the public domain of the state and the idea of water as a free natural resource (Garrido and Llamas, 2007). Accordingly, while tariff covers the cost of service, it is not a price paid for the resource itself.

As for water institutions, the 1978 Constitution established a new territorial organization with specific distribution of competences among the state and autonomous regions and the unique configuration of the 'Spanish Hydraulic Administration'.[1] The central government (Estado), through the Ministry of the Environment (Ministerio de Medio Ambiente) and the National Council of Water, is responsible for interregional basins and provides the main guidelines in terms of water policy.[2] The regional governments (17 Comunidades Autónomas) have increasing powers over the water resources situated in their territories, their procurement, water policies and sanitation and wastewater treatment. Finally, municipalities are responsible for water supply and sanitation in the cities. In accordance with the Law of Local Regime (Act 7/1985), water supply and the treatment of urban wastewater (extraction and purification) are public services falling within the competence of municipalities. The Constitutional Court (Tribunal Constitucional) is the judicial body in charge of settling conflicts of competences between the state and the different regions (Autonomous Communities) and between the regions themselves.

The next step in the evolution was the adoption of the Water Act in 1985, which represented a new paradigm in water management (Government of Spain, 2001).[3] The changes introduced by this new legislation contributed to the development of a new water policy.[4] Indeed, the 1985 Water Act reinforced the conception of all water resources as belonging to the public domain with a few exceptions regarding the use of groundwater. The 1985 Water Act underlined the watershed as the management unit (the river basin authority), attributing the management of watersheds covering the territory of several regions to the State Administration. The river basin authorities (RBAs) (*organismos de cuenca*) are autonomous organs which manage water resources (groundwater) and wastewater releases into natural water streams. The authorities develop drafts and implement river basin plans and manage water resources, including granting concessions. Moreover, under the Water Act users of aquifers can create users' associations.

Besides, the Water Act laid down the principles to be followed by the hydrological plans. From a financial point of view, this legislation provided economic support to water users, mainly those involved in irrigation uses. On the whole, the 1985 Water Act also reformed the institutional setting by granting financial and institutional autonomy to the basin authorities (*organismos de cuenca*), which were enabled to act within the basin boundaries.

The organizational model embodied by the Water Act in 1985 not only established the allocation of competences, but also created a complex organization involving the government (Junta de Gobierno) and planning bodies (Water Boards) and comprising members from regional governments, local governments and users, pursuant to the principle of organic and democratic cooperation.[5] In this respect, the 1985 Water Act represented a step forward in the configuration of a water management system in which direct users and the administrations should play a significant role in water management and planning at basin level (Mateos et al., 2003; Garrido and Llamas, 2007).

The 1985 Water Act was reformed in 1999, which led to the introduction of three main innovations (Embid Irujo, 2006). The first one was the modification of the water rights regime, allowing the holders of water rights to exchange them and the basin authorities to establish 'water banks' or 'trade centers' in the event of acute scarcity problems or droughts (Di Stefano and Schmidt, 2005). The second change consisted of granting a new status to public corporations, allowing them to build water infrastructure by signing financial arrangements. The third innovation was the new consideration of desalinated water and wastewater as belonging to the public domain on the same basis as other water resources. By introducing the aforementioned changes, the 1999 reform of the Water Act increased the role of water markets, and promoted water desalination and privately financed water projects.

The EU Water Framework Directive 2000/60/EC (WFD) represented a new shift in water management in Spain (Kaika, 2004). The implementation of the 2000 EU Framework Directive implied changes for the Spanish legal system, including the introduction of the concept of Integrated Water Resources Management. Besides introducing a focus on achieving better water quality, the Directive had economic implications (such as pricing of water based on cost-effectiveness analysis). The implementation of the Framework Directive also led to the inclusion of various channels of participation for civil society in water management (Harrison et al., 2001).[6]

In fact, the Water Framework Directive established that EU member states shall take into account the principle of cost recovery of water supply, comprising environmental costs. The enforcement of the WFD has been complex: the whole 1900s financial design of the water sector became outdated, leading Spain close to breaching the European legislation (del Moral, 2009). According to Garrido and Llamas (2009), 'the WFD set out a demanding agenda, enshrining ideas that were foreign to most traditional tenets of water management . . . Spain's long water-policy history was forced to change when the WFD was enacted . . . A more balanced mix of objectives replaced the traditional emphasis on supply policies, which had little concern about the quality of water and ecosystems.'[7] In sum, the WFD posed serious enforcement difficulties for Mediterranean countries such as Spain, which would otherwise never have passed such a national legislation (Garrido and Llamas, 2009).

Another important milestone was the adoption of 'National Hydrological Plans' at the national level (European Commission, 2009). The central government (Estado) decided to implement different water management plans in order to distribute and redistribute water resources amongst the regions (Autonomous Communities). Therefore, the National Hydrological Plans regulate the management and equitable distribution of water resources, providing for their inter-basin transfers. These plans became highly politicized and encountered strong opposition from certain regional governments. Between 1993 and 2004, various National Hydrological Plans were drafted and three different National Hydrological Plans were approved by laws passed between 1993 and 2005. These plans, however, failed to be effectively implemented and remained at the initial stages. The reason underlying this failure was mainly the refusal of certain regional governments (donor regions) to accept inter-basin transfers.

The initial 1993 draft version was based on the main idea of state regulation of water resources through the construction of large-scale infrastructure (dams and inter-basin transfers) to guarantee universal access to water supply. Additionally, the 1993 Plan foresaw significant water transfers from two rivers shared with Portugal, provoking tensions between the two countries (Garrido et al., 2009).[8] The successive draft issued by the government between 1993 and 1995 in order to adopt a water plan was highly criticized not only by 'donor regions' (such as Aragon) but also by environmental organizations and practitioners. The 2001 National Hydrological Plan was considered as 'the biggest water project' by the government and commentators but it also failed. The 2004 National Hydrological Plan focused on water quality and desalination and interrupted the Ebro transfer. Additionally, water issues became political in local, regional and national elections, and the debate about water resources took place with very little scientific basis. At the turn of the twenty-first century, water management has become an extremely controversial issue from a political point of view, giving rise to regional and environmental conflicts.[9]

Following these unsuccessful attempts, the National Hydrological Plan was modified, stopping the Ebro River transfer, and the central government adopted the Actions on Water Use and Management Program (AGUA program) in 2004 to introduce significant changes in water management policy (Downward and Taylor, 2007). The AGUA program was implemented during the period 2005–08. It aimed at solving the problems related to water transfers, establishing a 'better management of existing resources', including water desalination process and reuse of treated wastewater (AGUA, 2004).

These modifications in water management have been supported by changes in approaches to water policy. The AGUA program included more than 100 initiatives to change the water policy including construction of water desalination plants, proposals for alternative methods of water supply regulation and creation of public water banks in each hydro-graphical basin, which would allow historical water resources to be reallocated according to the criteria of equity, efficiency and sustainability (MMA, 2005). At the same time, the Ministry of the Environment launched a voluntary process to assess the environmental sustainability of urgent initiatives contained in the AGUA program in the Mediterranean basins. The AGUA program was well received because of the effort made in terms of transparency. However, the initial inclusion of water restriction was criticized. This new orientation of water management was inspired by the Water Framework Directive.[10]

After 2008, the Ministry of the Environment was integrated into the Ministry of Agriculture. This led to a change in the water governance paradigm towards a conventional supply approach to water management, and inter-basin water transfers have been included on the agenda. In recent years, regions have developed their own competences in water management. In the framework of the Statutes of Autonomy (special laws regulating the organization of the autonomous government, the electoral rules, the distribution of competences between different levels of governance and other regional-specific provisions), some regional governments (for example, autonomous communities such as Catalonia and the Basque Country) have included competences relating to water management.

Currently, water resources are not being used in a sustainable manner. The predictions about future climate change scenarios suggest that water resources will be seriously affected in Spain. Besides, an increase in water demand is already generating conflicts derived from water scarcity and it is foreseeable that these conflicts are going to be aggravated in the future. Between 2001 and 2007, urban water consumption grew at an annual rate of 4–5 percent. Among other factors which contributed to the increase in water consumption, rapid growth of tourist resorts, urban expansion and development of golf courses can be highlighted. There are also problems related to the irrigation sector where the resources are not enough to meet demand. Neither the liberalization undertaken by the 1999 reform of the Water Act nor national irrigation plans include measures to make the necessary water resources available to the agriculture sector.

From an environmental point of view, Spanish water ecosystems have suffered from severe deterioration with rising levels of water usage and harmful impacts on the environment (Schmidt and de Stefano, 2009). This has resulted in the irreversible loss of wetlands, biodiversity and habitats. One of the requirements for the implementation of the WFD is to develop river basin plans, initially scheduled to be applied in 2009 according to the WFD timetable. This will need to address initiatives including public participation processes (Menéndez-Prieto, 2009; Barreira, 2009). The environmental problems are likely to worsen and be aggravated due to the effects of climate change (Iglesias et al., 2009). After decades of unsuccessful attempts, the Spanish water management system is currently addressing the problems related to intensive use of groundwater resources. The main challenge is the scarcity of water resources, which is mainly due to the inefficient allocation of water resources and mismanagement in the agricultural sector, such as the intensive use of large amounts of 'virtual' water (water used in the production of a good or service).[11]

In the future, increasing water scarcity will make it difficult to meet current and future needs, including those related to environmental protection. Additionally, there are other interrelated processes linked to negative climate change impacts. Recurrent droughts and long-term climate projections appear to suggest that Spain's drought contingency planning will face the problems of reduced precipitation and stricter constraints on the development of new infrastructure (Iglesias et al., 2009). The models predict that run-off in all basins will diminish and extreme events will become more frequent (Bates et al., 2008). Spain will be among the EU countries which will suffer the most from climate change.

In this scenario, Spain's adaptation policy is in line with the adaptation policy adopted by the EU. The EU has already followed the Integrated Water Resources Management strategy for the impacts of climate change in the Water Framework Directive, the Floods Directive and the Communication on Water Scarcity and Drought. Many Spanish urban supply systems have already faced problems of water shortage during the early 1990s with water supply interruptions of various intensities (Cabrera et al., 2009). Commentators underline that the Spanish water supply services are unreliable because they lack preparedness to provide healthy and sustainable drinking water to one-third of the population (Garrido and Calatrava, 2009).

6.4 CONTRACTUAL AND ADMINISTRATIVE FRAMEWORKS FOR PPPS IN WATER MANAGEMENT IN SPAIN

As for the institutional and political structures concerning water management in Spain, there is a general institutional framework with variations depending on the territorial divisions indicated above: state, regions and municipalities. Additionally, it must be underlined that traditionally, the water sector in Spain has shown the structure of a natural monopoly (Mates Barco, 1991). Consequently, there was a predominance of public water management. This model has changed over the past 25 years with the emergence of various models of water management including public–private partnerships (PPPs).

In this context, PPPs refer to 'any contractual framework between a public party (government, regions or municipalities) and a private party for construction, financing, management and/or operation of a public infrastructure or equipment' (Deloitte, 2010; Marin, 2009; WEF, 2006). The essential feature of PPPs is private sector participation with a significant transfer of risks in the case of a long-term contractual relationship. Having said so, it is relevant to examine the legal and contractual framework related to PPPs in Spain. The 1985 Water Act is the main legal instrument which establishes the guidelines to be followed in the field. In addition, the different Autonomy Statutes provide regions with competences in the water sector, whereas local governments' attributions on water management are mainly established in the 1985 Act (Article 25.l). Apart from this core legislation, the legal framework is also composed of various administrative laws, such as the 2007 Public Procurement Act (Ley de Contratos del Sector Público).[12]

Among the array of administrative contracts in Spain, concessions are most frequently applied in the water sector. The national framework for concessions is established by the 1985 Water Act, which foresees that any concession must be granted under the provisions of the Hydrological Plans. Besides, according to Article 59 of the Water Act, concessions are temporary and cannot last for more than 75 years (Embid Irujo, 2008). Concessions are awarded through a discretionary decision but substantiated and adopted in the public interest (Article 59, Water Act).

A concession may be terminated due to one of the following reasons: revocation, review, expiration of deadline, expropriation of water rights or reallocations decided by law. The last two causes of termination (expropriation and reallocations by law) generate, as a logical consequence, the obligation to compensate. Compensation may also be payable in case of review of the concession where it is the outcome of a modification to a hydrological plan and causes harm (Article 65, Water Act). In principle, only the revocation of a concession excludes the idea of compensation (Article 66, Water Act).

In general, concessions can be transmitted as a right but the transmission is subject to public control, which involves the obligation to inform the citizens about the party which wants to transfer the concession. In certain cases, concessions that can be considered as a 'public service' (such as urban water supply) can only be transmitted after obtaining authorization from the competent authority. In other cases, transmission is possible with a notification to the competent authority, in order to incorporate the data of the new holder in the Water Register. This situation continues to reflect the current water legislation concerning the transmission of water rights (Article 63, Water Act).

Thus, the model followed in the implementation of PPPs in Spain is in line with the legal regime: whereas ownership of the assets continues to be public, certain functions are being delegated to a private company for a specified period.

The 1999 reform to the Water Act introduced a new contract regarding water rights alongside the concession. As stated above, this reform aimed at liberalization of water markets. In order to achieve this, the law provided for trading of water rights and authorized the creation of trading centers within a river basin. The trading of water rights implies that a holder can lease out water rights to the basin authority or another user for a period of time.[13] The law regulated the powers of trading centers to organize transactions of water rights of holders who are interested in subsequently transferring rights to other holders. The transaction takes place in the framework of a transparent process and the principles of public procurement are applicable. In practice, this reform has led to changes in the economics of water resources, including flexible allocation instruments and voluntary arrangements allowing new forms of PPPs. However, this process was complex and many voices were raised against 'water trading'.

In terms of implementation of contracts concerning water management, at the national level, the Administrative Contracts Advisory Board (Junta Consultiva de Contratación Administrativa) plays an important role in the implementation of different contracts (mainly concessions).[14] In case of conflicts concerning the granting process or the concessions, they can be solved through hierarchical review (appeal) or by the courts (Embid Irujo, 2007). In Spain, there are specific courts which deal with cases involving conflicts between private parties and governmental authorities (administrative jurisdiction, or *jurisdicción contencioso-administrativa*) which are competent to settle such disputes.[15]

On the whole, over the past three decades, there has been an increase in PPPs in the water sector in Spain. Among other drivers for the introduction of PPPs in the Spanish context, the reform of local governments, a new concept of efficiency, a broad framework of public sector and intervention in national monopolies to introduce private competition may be highlighted.

This new paradigm in terms of management of urban water utilities with the emergence of PPP projects can be seen as a trend. In the 2000s, various Spanish cities decided to incorporate private companies for domestic water supply (Bakker, 2002). According to this model, water was held in the public domain (ownership) but management and supply services were transferred to private companies. This specific type of water management is called indirect management (*gestión indirecta*) (Ordoñez and Bru, 2003). As a result, a significant proportion of the Spanish population is served by private companies.

As noted above, water and sanitation are included within municipal powers. In general terms, local governments can more or less autonomously regulate urban water supply by passing local laws. This generates differences among jurisdictions and variations in legal arrangements. Consequently, there are a variety of water management models at the municipal level. The traditional model is direct management (water supply was provided by the cities or through associations or consortia of cities). A further differentiation among direct water management models involves direct management without a specific body or with a specific body (autonomous municipal body or a public company).[16]

The most common forms of PPPs for urban water utilities in Spain are:

1. Lease contract: In this type of contract, assets are leased to the private operator, which receives a share of revenues. It implies higher commercial risk than under a management contract. Investment is completely or mostly financed and carried out by the public sector. The duration is typically 10–15 years.
2. Mixed-ownership company: In this case, a private investor owns a minority share in a water company with full management and responsibility vested in the private partner. This type of PPP is most common in Spain.
3. Concession: In this case, the private operator is responsible for water supply and sanitation including investment, which is mostly or fully financed and carried out by the private operator. The duration is typically 20–30 years.
4. Management contract: Under this type of PPP, the private operator is only responsible for running the system in exchange for a fee. Investment is financed and carried out by the public sector. The duration is typically three to five years.

The awarding of contracts for urban water management is implemented through public competition (open or restricted) or a negotiation. The law 57/2003 de Medidas para la Modernización del Gobierno local (LMMGL) reformed the legal framework for supply contracts (Romero Hernández, 2003; Martínez Pallarés, 2004). As for public control over private management, it includes measures such as the adoption of detailed regulation of management and control of water quality and price. The responsibilities for the private partner in water management remain and private companies owe property and damage liability. The interpretation of such contracts has been the subject matter of recent judicial decisions, at both national and regional level.

However, the assessment of the concession and the possibility of changing the private company if water supply does not meet efficiency standards are controversial issues. From a theoretical point of view, when the city council opens the tender to award grants to companies which presented offers in the competition, it is possible to introduce changes in the terms of the concession or to decide in favor of public management. However, in practice, there are few cases in which the private operator was changed through this mechanism. These obstacles together with the long-term nature of concessions (25 years) make it difficult to assess the level of competition in the sector. Therefore, it is difficult to establish any criteria to assess the performance of water management in Spain.

In recent years, indirect management of urban water supply has increased. This has been mainly implemented in the form of concessions to private companies or joint ventures (indirect water management).[17] This indirect water management model has gained ground in Spain.

This process runs parallel to the expansion of Spanish private companies at the international level. As a matter of fact, Spanish private companies (such as FCC) have been participating in water management tenders and building infrastructure in the water sector in different countries. Additionally, Spanish multinational corporations have actively participated in water privatization in Latin American countries.

On the contrary, public companies encounter many difficulties to be competitive in the water sector and to achieve internationalization. Indeed, public companies or entities originally created to provide a service locally are not prepared or designed to expand, except in the surrounding areas (such as in the case of the extension of Canal de Isabel II's activities in Extremadura or Consorcio de Aguas de Bilbao). Despite the attempts to achieve internationalization, it is clear that both Canal de Isabel II and Bilbao Consortium have serious difficulties in explaining the benefits and risks of their output abroad to their shareholders (the public).

This stage can be seen as a deceleration of the trend towards indirect management of water utilities in Spanish cities. On the one hand, the number of concessions and other forms of PPPs appears to be decreasing in recent years. On the other hand, some sections of civil society are increasingly more reluctant to allow new private projects in the water sector (Brennan et al., 2005).[18] Some commentators such as March Corbella (2010) even point out that there are signs of exhaustion in the expansion of indirect management. This is partly because, in some cities, private companies have failed to meet expectations; in other cases, they have not achieved expected profits.

6.5 URBAN WATER UTILITIES IN SPAIN: DIFFERENT SCENARIOS FOR PPPS

The urban water sector in Spain is relevant from an economic point of view. According to the Ministry of Environment, it represented 0.9 percent of gross domestic product (GDP) in 2002 (MMA, 2007).[19] In Spain, domestic water supply and sanitation are considered to be the responsibility of the municipalities (Antolin, 1991). They regulate, with some exceptions, the rights and obligations of users and supply companies by municipal ordinances. Such is the case of the domestic water supply rules in Andalusia, which are adopted at the regional level.[20] Even though in practice more than 8000 Spanish municipalities can exercise their competences in terms of water management, in many places some of the

Table 6.2 Water supply management systems, 2003

Management	%	Management	%
Public		Private	
Canal de Isabel II	15	Agbar	26
Consortium Aguas de Bilbao	3	Bouygues	5
Sevilla	4	Vivendi- FCC	3
Other municipalities	36	Other companies	4
Total	58		41

Source: Hispagua (2003).

water cycle phases (catchment, production and distribution) are assigned to supra-municipal associations (such as in the case of the Bilbao Water Consortium – Consorcio de Aguas de Bilbao – which brings together a large number of municipalities); a public company (Canal de Isabel II or Aguas de Sevilla SA) or concessionaires (Aquagest).

Traditionally, water management was public in the main Spanish cities (which represent approximately 40–50 percent of the total population).[21] From a historical perspective, during the period 1939–69, municipalities were in charge of water management and this resulted in the gradual withdrawal of private companies. Alternatively, during the period 1970–96, private companies returned to the water management sector (Matés Barco, 1991; P. Comín and Aceña, 1991). Among other reasons, this was due to efficiency reasons – local systems have a small size for economies of scale; hence, in cases where private operators had already begun water supply, they continued their operations through renewal of their concessions.

In this sense, as seen in the previous section, there are various public and private models for urban water management. Investment in water supply and sanitation involving transnational corporations has increased in the last 20 years and even in the cases where water supply remains public there is private funding for infrastructure. Table 6.2 highlights the percentage of public and private share in water supply management systems in Spain.

At present, water management is still public in small cities and villages (fewer than 10000 inhabitants) whereas there is an increasing participation of private companies in water management in medium-sized cities (more than 20000 inhabitants). Private companies are taking over urban water management by signing concession and lease contracts (*contratos de concesión o de arrendamiento/gerencia*). The largest private water companies in Spain are Aguas de Barcelona – Barcelona's Water (Agbar Group), and Aguas de Valencia – Valencia's Water (Group Agval), with concessions in more than 1000 municipalities involving water supply to over 20 million people. Agbar was initially created by French and Belgian investors, sold to Spanish investors in 1920, and gradually came back under French control in the early twenty-first century.

6.5.1 Urban Water Management in the Main Spanish Cities

In an appraisal of water management in Spanish cities, taking into account efficiency and costs (tariffs), a significant scattering can be observed. A brief overview of water management in five Spanish cities shows the complexity of this issue.

Madrid is the largest urban agglomeration in Spain with approximately 3 300 000 inhabitants. A state agency, Canal de Isabel II, was in charge of water services. The company was launched in the mid-twentieth century and transferred to the regional government (Comunidad Autónoma) of Madrid in the mid-1980s.[22] Canal de Isabel II is in charge of a large supply system that captures flowing water from the mountains and distributes it throughout the region. It is relevant to point out that civil society organizations are against privatization of the water supply in Madrid.[23]

The second-largest city in the country, Barcelona, shows a long tradition of private water management.[24] By the time the first Water Act was enacted in 1879, a number of companies were operating in the city (up to a total of seven cities on the then outskirts of Barcelona). Subsequently, they were all absorbed into Sociedad General de Aguas de Barcelona (currently, Agbar). Agbar, which was created in Paris on 20 January 1882, is the largest private water company, with a market share of about 50 percent of the private concessions. Agbar Water is the first private operator responsible for urban water management in Spain. It has operations in all the Spanish autonomous communities, and serves approximately 20 million people in over 1000 towns and cities, ranging from 1000 people to more than 3 million people.[25]

Valencia is the third-largest Spanish city in terms of population size. Aguas de Valencia, which was created in 1890 as Sociedad de Aguas Potables y Mejoras de Valencia, is in charge of water supply and sanitation. Water supply is regulated at the local level.[26] In general, there are significant differences in the tariffs applied in the same region (Generalitat Valenciana), which creates dispersion as well (Sevilla and Torregrosa, 2011).

Urban water supply in Zaragoza (675 000 inhabitants) represents an example of efficient public management. The complexity of the system is physically embodied in 14 treatment facilities, pumping and storage, more than 1000 miles of distribution network and more than 280 000 subscribers with large metering equipment. The production of water reaches more

Table 6.3 Urban water management in main cities in Spain

City	Population (2009)	Service/company	Management	Tariff (euro per 10 m³) (2009)
Madrid	3255944	Canal de Isabel II	Public	14.32
Barcelona	1621537	Aguas de Barcelona- (Agbar)	Private	12.66
Valencia	814208	EMVASA	Mixed	9.34
Zaragoza	674317	Aguas de Zaragoza	Public	9.95

Source: FACUA (2007).

than 200 million liters of drinking water per day that is distributed to users. The municipality is in charge of water supply and sanitation. The whole process is undertaken in Zaragoza by direct municipal management through various specific contracts aimed primarily at maintaining the metering equipment, the realization of investments and, partially, to do repair work. On the whole, the public management experience has been successful with an increase in efficiency in terms of water supply.[27]

The city of Cordoba (328000 inhabitants) is located in the region of Andalusia, in southern Spain. The public utility company EMACSA has been providing water service since 1969. The company provides good-quality services to users at low prices. Besides, EMACSA has successfully contributed to a substantial reduction in water consumption (Observatorio de los Servicios Públicos Córdoba, 2005). The company is responsible for water collection, planning and storage infrastructure and to help the city tackle seasonal differences in water consumption. As a result of these measures, Cordoba was the only city in Andalusia that did not have to restrict water delivery during a serious drought in 1995 (TI and CEO, 2009). Table 6.3 describes urban water management in four main cities in Spain.

The experience in water management in these cities shows that there have been significant but scattered changes. Indeed while assessing the management of water services at the national level, one can observe distinct tariffs for water supply which are applied in different cities. There are different arrangements for concessions (regional and local). This broad array of agreements governing concessions cause dispersal and make it difficult to establish a minimum quality of service and determine different tariffs among the cities (FACUA, 2007).

The large number of Spanish cities and towns and the lack of standardized measures on the criteria for the provision of services make it difficult to compare different systems. Even in the cases where municipalities are responsible for urban water management, it is not clear whether they follow criteria such as efficiency, quality and respect for the environment. This situation needs to be addressed: water is a scarce and sensitive resource, with consumers and users increasingly demanding efficient performance and less expensive service. There is no regulator of the water sector or equivalent body, except for the regional pricing commissions with powers to revise tariffs. These are 17 agencies which autonomously control the increase in prices charged by domestic water utilities in each region (Comisiones de Precios). As stated in the original rules of, and subsequent amendments to, the organization and powers of the commissions, their performance in relation to the hydrological cycle only relates to price increases in water tariffs. It is worth mentioning that these commissions have not taken into account other aspects such as quality of the service or water purification. Thus, the critical issues that are being raised outside the control of the regional pricing commissions remain within the jurisdiction of regional or municipal authorities. Finally, the functions of these price commissions depend on the degree of regulation at the regional level. There are regions with almost no regulation, and tariff increases are discussed between the municipality and the operator (Fernández Pérez et al., 2006). The most obvious outcome of this situation is the difference in tariffs applied in Spanish cities, with variations of more than three times (FACUA, 2007; Torregrosa, 2008). There is no doubt that the costs incorporated into drinking water supply vary from one area to another, due both to the cost of acquisition of water and local factors (topography, size of network, and so on). But often such oscillations cannot be correlated with better or worse water resource management. For instance, a very low tariff may be associated with an environmentally unsustainable use of water.

Furthermore, the information about each local service is usually provided by the concessionaires, except the data concerning the negotiation of the contract. In addition, each concession contract is usually different from others, with different commitments and variable contract durations. This creates a large asymmetry in the information available to the parties (Cabrera, 2008). An increase in transparency would generate information about the operative costs that are awarded to companies in the group (as in the case of Chile and the United Kingdom).

Access to information is crucial. Scholars recommend following the experience of the French system, which has similarities with the Spanish system, in terms of allocation of competences on water supply at the municipal level (Sevilla and Torregrossa, 2011). In 2007, the French government approved a decree which established the government's obligation to provide information about these services. Afterwards, the Arrêté

(Decree) No. 2007-675 of 2 May 2007 established a series of indicators that should be reported for the entire territory in which corporations provide water and sanitation.

Within this context, there have been various successful experiences of public participatory management. Cordoba also represents an interesting case study on the application of the participatory framework (De Miguel and Sanz Mulas, 2005). Since 1979, EMACSA has developed a widely accepted and well-functioning structure of participatory co-management. The board of directors is responsible for all main decisions in the company and has a diverse membership. Independent of municipal election results and majorities, each of the three political groups in the council nominates two members to the board. The two major trade unions each nominate an additional two members, and one is nominated by a council of civil society movements (neighborhood associations which organize around 13 percent of the citizens). The manager of EMACSA as well as the general secretary and the general financial controller of the city council take part in the board meetings without voting rights. EMACSA has thus included a participatory structure characterized by broad transparency that allows citizens to follow the decision-making process and intervene, for instance, by raising alternative opinions (TI and CEO, 2009).

Many participatory initiatives have been also undertaken in Catalonia. The Catalan Water Agency launched a broad process of participation and consultation and has designed a water management strategy with a focus on the improvement of water quality and the control of water demand (Esteban and Prat, 2006). Other regions like Navarra and Aragón have also launched participatory processes (Espluga and Subirats, 2007). In certain cases, such initiatives have remained in the deliberative phase, while in other cases, agreements have been reached and specific decisions have been made (Espluga and Subirats, 2007).

In sum, the exercise of local autonomy and the absence of clear guidelines that allow the establishment of efficient management objectives and environmental standards for urban water supply policy lead to a complex status quo in this field. There is a need for clear indicators to compare the efficiency gains of different forms of management. A regulator is also needed to implement best practices in the sector for the benefit of consumers, while promoting the sustainability of water resources.

6.6 CONCLUSION

This chapter describes the past and current situation of water management in Spain. After examining the evolution of Spanish water management, various conclusions on the weaknesses and strengths of the system can be drawn in order to contribute to the broader discussion on water management at a global level.

The evolution of water management in Spain shows the change of a deep-rooted water management paradigm (which is the result of different drivers). Water laws have adjusted to the new water policies. However, this transition has not been an even process. In particular, the implementation of the EU Water Framework Directive raised various complex issues relating to the limited capacity of the institutional set-up to face the new challenges.

As for best practices in water management, Spain illustrates a positive experience in increasing civil society participation in water management in certain cities (Pares et al., 2009). Spain also exhibits a significant experience in increasing the participation of stakeholders, users and regional and local government agencies in water institutions (basin authorities), taking into account their essential role in the process.

With respect to new challenges, water management in Spain faces problems such as environmental degradation, growing water demand, climate change and increasing demand for agricultural use. In this regard, the importance of regional initiatives in the area of water pricing and environmental policies must be underlined.

On the whole (and despite the difficulties), after the 'privatization wave', there have been significant changes in urban water supply, in particular concerning the emergence of various PPPs (Molina, 2001). Nevertheless, there is a need to provide information about the performance of private corporations in charge of water supply.

In terms of the weaknesses of the water management system in Spain, the failure to approve a national hydrological plan is widely accepted. The approval of the plans has become a highly political and controversial issue (Bukowski, 2007). Another main weakness of the current system relates to the financial implications and limited resources for the full enforcement of the WFD. The national water policy, which was traditionally considered a national and regional matter, must now meet the criteria laid down in the WFD.

The water management system also faces the problems of unsustainable groundwater management, despite the provisions of the 1985 Water Act and the initial attempts to establish a sustainable system. The water management system has also failed to diminish water pollution; on the contrary, the anthropogenic impacts on the environment have worsened.

In terms of urban water management, as seen in the previous section, many Spanish municipalities have delegated competence in these matters to private companies. The lack of information and opacity observed in some cases is an obstacle to conduct an assessment of PPPs.

Finally, it is difficult to access information on the provision of water supply by private companies, in order to assess their efficiency. It would be advisable to create a technical body in conjunction with regional governments and municipal organizations to determine the optimum standards of management performance in each case. This analysis should be as detailed as possible in order to assess the municipal management and to implement policies to address water management challenges.

NOTES

1. This system was created pursuant to Articles 148.1.10 and 149.1.22 of the Spanish Constitution. Water Administration units maintained the traditional name of Hydrographic Confederations.
2. The National Water Council consists of national and regional governments as well as representatives of user associations (agriculture, local governments), scientists and environmental non-governmental organizations. It provides recommendations on all policies affecting water resources at the national level.
3. The law was the first twentieth-century regulation on water, replacing the 1886 law.
4. After this law was passed, there have been many subsequent changes. The last reform of the Water Act was adopted in 2011.
5. The Spanish Constitutional Court, in a case concerning water management, held that the inclusion of entities and local users at basin level represents 'the most direct way' to participate in water management in accordance with the Spanish Constitution. See STC 166/1996 of 17 October.
6. The Directive was internalized through the Law 62/2003. Espluga and Subirats (2007) have developed an interesting analysis on different cases of public participation in integrated water resources management.
7. During the period 2000–2005, there were no experiences of participation. From 2006 onwards, there have been cases of bottom–up participation in the Basque Country and the Ebro basin.
8. Before the adoption of the WFD, the two countries were involved in negotiations which resulted in the adoption of the Albufeira Convention in 1998.
9. There were water conflicts during the political negotiations of new Autonomous Statutes for Aragón, Catalonia, Andalusia and Valencia (approved in 2006 and 2007), and Castille-la Mancha concerning the National Hydrological Plans of 2004 and 2005. Water issues were taken into account by voters as well as members of national and regional parliaments during the period 2004–08.
10. There are other initiatives concerning the management of specific rivers such as the Guadiana program or broader programs such as water banks according to new planning criteria and measures.
11. 'The Spanish water footprint should be analysed in its time and spatial dimension as well as considering the sectoral and geographical standpoints. Furthermore, we cannot forget about the multifunctionality of agriculture' (see Aldaya et al., 2009).
12. Ley 30/2007, de 30 de octubre, de Contratos del Sector Público. The 2007 Public Procurement Act contains specific regulations in Title II on Special norms for public works contracts, concessions, utilities management, supplies, services and collaboration between the public and the private sector (*Normas especiales para contratos de obras,*

concesión de obra pública, gestión de servicios públicos, suministros, servicios y de colaboración entre el sector público y el sector privado).

13. Different water centers were created, such as trading centers in the River Basin Júcar, Segura, Guadiana and Guadalquivir.
14. See Junta Consultiva de Contratación Administrativa, available at http://www.meh.es/, accessed 11 November 2011.
15. The tribunal of last resort could be either the Regional Supreme Court or the Supreme Court. See, for instance, Valencia Regional Supreme Court, Judgments 13 March 2003 (22703) and 19 May 2004 (79) in the case Aigües de L'Horta'.
16. The array of contracts includes: *Gestión directa: Sin órgano diferenciado: Con órgano diferenciado: organismo autónomo municipal; Sociedad mercantil municipal; Indirecta: Concierto; Concesión; Empresa mixta*. And for indirect water management: concession, *Concierto* and 'mixed corporations' (public–private company). See Delgado (2003, p. 1460).
17. Nevertheless, it should be pointed out that large cities are served by public companies (except Barcelona and Valencia), with Canal de Isabel II (in the metropolitan area of Madrid) being the largest public municipal company.
18. Cases of reversal have been noticed in some countries where municipalities have changed their initial decision to privatize water management and returned to public management. See 'Water Re-municipalisation Tracker', available at http://www.remunicipalisation.org/ (accessed 10 February 2012).
19. The MMA estimated that the services related to the water sector in Spain, taking into account the value of services provided by the different agents in 2002, was €6330 million a year. The most important urban services are distribution and wastewater treatment with a cost of about €4100 million while irrigation services have an approximate cost of €1285 million. An important aspect is the relatively low value of other services such as collection, extraction and transportation of surface water and groundwater as these only represent 6.64 percent (surface) and 8.25 percent (underground) of the total (MMA, 2007, p. 197).
20. See Reglamento de Suministro Domiciliario de Agua de Andalucía, los derechos y obligaciones de los usuarios y las empresas suministradoras, Decree 120/91, 11 June 1991, available at http://consorcioagua.malaga.es/documentos/normativa-7.pdf (accessed 10 February 2012).
21. According to the Asociación Española de Abastecimientos de Agua y Saneamiento (AEAS) – Spanish Association of Water Supply and Sanitation, water management was public in small-size cities with less than 10 000 population.
22. Ley 17/1984, de 20 de diciembre, reguladora del abastecimiento y saneamiento de agua en la Comunidad de Madrid. 'Aplicación a los concesionarios de contratos de gestión de servicios públicos de la Ley de Contratos de las Administraciones Públicas en los contratos de obra que deseen adjudicar'.
23. See, for instance, Plataforma en Defensa de los Servicios Públicos, ATTAC Madrid, Ecologistas en Acción-Madrid.
24. See March Corbella (2010).
25. See http://www.agbar.es/esp/b-2_ciclo_integral_agua.asp (accessed 10 February 2012).
26. Reglamento del Servicio de Abastecimiento de Agua Potable de la Ciudad de Valencia, approved on 30.07.2004. Publicación BOP: 14.09.2004.
27. Ibid.

REFERENCES

Actions on Water Use and Management Programme (AGUA) (2004), *Official Gazette*, BOE 2/2004, 19 June 2004.

Aldaya, M., A. Garrido, M. Ramón Llamas, C. Varela-Ortega, P. Novo and R. Rodríguez Casado (2009), 'Water footprint and virtual water trade in Spain', in A. Garrido and M.R. Llamas (eds), *Water Policy in Spain*, London: Taylor & Francis Group, pp. 49–59.
Antolín, F. (1991), 'Las empresas de servicios públicos municipales', in F. Comín and P. Martín Aceña (eds), *Historia de la Empresa Pública en España,* Madrid: Espasa-Calpe, pp. 283–330.
Bakker, K. (2002), 'From state to market? Water mercantilization in Spain', *Environment and Planning A*, **34**(5), 767–90.
Barreira, A. (2009), 'Public participation in developing and improving water governance', in A. Garrido and M.R. Llamas (eds), *Water Policy in Spain*, London: Taylor & Francis, pp. 185–94.
Bates, B.C., Z.W. Kundzewicz, S. Wu and J.P. Palutikof (eds) (2008), *Climate Change and Water*, Intergovernmental Panel on Climate Change (IPPC), available at http://www.ipcc.ch/pdf/technical-papers/ccw/frontmatter.pdf (accessed 10 February 2011).
Brennan, B., O. Hoedeman, S. Kishimoto and P. Terhorst (eds) (2005), *Reclaiming Public Water Managment*, Amsterdam: Transnational Institute and Corporate Europe Observatory, available also at http://www.tni.org/tnibook/reclaiming-public-water-book (accessed 11 March 2012).
Bukowski, J. (2007), 'Spanish water policy and the NHP: an advocacy coalition approach to policy change', *South European Society and Politics*, **12**(1), 39–57.
Burchi, S. (2004), 'Water laws for water security in the twenty-first century', in J. Trottier and P. Slack (eds), *Managing Water Resources: Past and Present*, Oxford: Oxford University Press, pp. 117–29.
Cabrera, E., R. Cobacho and E. Cabrera Rochera (2009), 'Water supply in urban areas', in A. Garrido and M.R. Llamas (eds), *Water Policy in Spain*, London: Taylor & Francis, pp. 77–84.
Cabrera, E. (2008), 'El suministro de agua urbano en España, Fundación Nueva Cultura del Agua', available at http://www.unizar.es/fnca/varios/panel/31.pdf (accessed 10 November 2011).
Costejà, M., N. Font and J. Subirats (2004), 'The evolution of water regime in Spain', in I. Kissling-Näf and S. Kuks (eds), *The Evolution of National Water Regimes in Europe: Transitions in Water Rights and Water Policies*, Dordrecht: Academic Publishers, pp. 235–63.
De Miguel, E. and A. Sanz Mulas (2005), 'Water management in Córdoba (Spain): a participative, efficient and effective public model', available at http://www.tni.org/sites/www.tni.org/archives/books/waterspain.pdf (accessed 11 March 2012).
del Moral, L. (2009), 'Changing water discourses in a modern society', in A. Garrido and M.R. Llamas (eds), *Water Policy in Spain*, London: Taylor & Francis, pp. 83–91.
Delgado, J.F.M. (2003), *Las Formas de Prestación de los Servicios Públicos Locales. En Particular, la Concesión*, Tratado de derecho municipal/coord. por Santiago Muñoz Machado, Vol. 2, pp. 1441–538.
Deloitte (2010), *La Contratación de Infraestructuras Mediante Paticipaciones Público Privadas*, Montevideo, Uruguay.
Di Stefano, L. and G. Schmidt (2005), 'Los bancos de agua: una oportunidad para los ríos y la gestión de sequías', *Ambienta* 72, available at http://www.mma.es/secciones/biblioteca_publicacion/publicaciones/revista_ambienta/n45/pdf/72_wwfadena45.pdf (accessed 10 March 2012).

Downward, S.R. and R. Taylor (2007), 'An assessment of Spain's Programa AGUA and its implications for sustainable water management in the province of Almería, southeast Spain', *Journal of Environmental Management*, **82**(2), 277–89.
Embid Irujo, A. (2006), 'El derecho al agua. Thomson-Aranzadi', Cizur Menor.
Embid Irujo, A. (2007), 'Conflictos jurídicos en la gestión y uso del agua', Madrid: Consejo General del Poder Judicial.
Embid Irujo, A. (2008), 'Asignación del agua y gestión de la escasez en España: los mercados de derechos de aguas', Zaragoza, España.
Espluga, J.L. and J. Subirats (2007), 'Participación ciudadana en las políticas de agua en España', Sevilla: Fundacion Nueva Cultura del Agua, available at http://www.unizar.es/fnca/varios/panel/53.pdf (accessed 10 March 2012).
Esteban, A. and N. Prat (eds) (2006), *Alternativas para la Gestión del Agua en Cataluña. Una Visión desde la Perspectiva de la Nueva Cultura del Agua*, Zaragoza: Bakeaz/FNCA.
European Commission – Environment (2009), 'Spanish National Hydrological Plan', available at http://ec.europa.eu/environment/water/spanish_hydrological_plan.html (accessed 10 March 2012)
Federación de Asociaciones de Consumidores (FACUA) (2007), 'Suministro de agua: Diferencias de hasta 422 euros anuales en la factura del suministro de agua de 28 ciudades', available at http://www.facua.org/es/estudio.php?Id=90 (accessed 11 February 2012).
Fernández Pérez, D., F. Morcillo Bernaldo de Quirós and A. Rodríguez Furones (2006), 'Análisis de los sistemas de retribución y evaluación de la gestión indirecta de los servicios del ciclo integral del agua', Jornadas Técnicas AEAS, XXVI, Proceedings w/p.
Font, N. and J. Subirats (2009), 'Spanish water management in transition: transition management watered down?', in D. Huitema and S. Meijerink (eds), *Water Policy Enterpreneurs: A Research Companion to Water Transitions around the Globe*, Cheltenham, UK and Northampton, MA, USA: Edward Elgar Publishing, pp. 272–86.
Font, N. and J. Subirats (2010), 'Water management in Spain: the role of policy entrepreneurs in shaping change', *Ecology and Society*, **15**(2), 25.
Food and Agriculture Organization (FAO) (2009), Aquastat Country Fact Sheet, available at http://www.fao.org/nr/water/aquastat/data/factsheets/aquastat_fact_sheet_esp_en.pdf (accessed 8 March 2012).
Food and Agriculture Organization (FAO) (2011), Spain's country profile, available at http://www.fao.org/countryprofiles/index.asp?lang=enandiso3=ESPandpaia=4 (accessed 10 March 2012).
Fuentes, A. (2011), 'Policies towards a sustainable use of water in Spain', OECD Economics Department Working Papers, No. 840, OECD Publishing, available at http://dx.doi.org/10.1787/5kgj3l0ggczt-en (accessed 17 March 2012).
Garrido, A., A. Barreira, S. Dinar and E. Luque (2009), 'The Spanish and Portuguese cooperation over their transboundary basins', in A. Garrido and M.R. Llamas (eds), *Water Policy in Spain*, London: Taylor & Francis, pp. 195–208.
Garrido, A. and J. Calatrava (2009), 'Trends in water pricing and markets', in A. Garrido and M.R. Llamas (eds), *Water Policy in Spain*, London: Taylor & Francis, pp. 131–44.
Garrido, A. and M.R. Llamas (2007), 'Water management in Spain: an example of changing paradigms', Paper presented at the Engelberg Academy Conference

on Water, Switzerland, 10–12 October 2007, available at http://www.rac.es/ficheros/doc/00311.pdf (accessed 17 October 2011).

Garrido, A. and M.R. Llamas (2009), 'Meeting future water challenges: Spain's strengths and weaknesses', in A. Garrido and M.R. Llamas (eds), *Water Policy in Spain*, London: Taylor & Francis, pp. 220–28.

Government of Spain, Water Act 1985, Ley 29/1985, de 2 de agosto, de Aguas. (Revised by the Decree 1/2001, 20 June 2001), available at http://noticias.juridicas.com/base_datos/Admin/rdleg1-2001.html. (accessed 11 March 2012).

Harrison, A., G. Schmidt, C. Avis and R. Hauser (2001), 'WWF's preliminary comments on public participation in the context of the Water Framework Directive and Integrated River Basin Management', Copenhagen: WWF European Freshwater Programme.

Hispagua (2003), 'Indicadores del agua', available at http://servicios2.marm.es/sia/indicadores/ (accessed 16 March 2012).

Iglesias, Alonso A., L. Garrote, M. Moneo and F. Flores (2009), 'Drought and climate risks', in A. Garrido and M.R. Llamas (eds), *Water Policy in Spain*, London: Taylor & Francis Group, pp. 62–75.

Kaika, M. (2004), 'Water for Europe: the creation of the European Water Framework Directive', in J. Trottier and P. Slack (eds), *Managing Water Resources: Past and Present*, Oxford: Oxford University Press, pp. 89–116.

Lopez-Gunn, E. (2009), 'Spain, water and climate change in COP 15 and beyond: aligning mitigation and adaptation through innovation', Working Paper 65/2009, Elcano Institute, Area: International Economy and Trade, 24/12/2009, available at http://www.realinstitutoelcano.org/wps/portal/rielcano_eng/Content?WCM_GLOBAL_CONTEXT=/elcano/elcano_in/zonas_in/dt65-2009 (accessed 10 March 2012).

March Corbella, H. (2010), 'Urban water management and market environmentalism: a historical perspective for Barcelona and Madrid', Phd Thesis, Universitat Autonoma de Barcelona, Spain.

Marin, P. (2009), *A Review of Experiences in Developing Countries, Public–Private Partnerships for Urban Water Utilities*, Washington, DC: Public–Private Infrastructure Advisory Facility, World Bank.

Martínez Pallarés, P.L. (2004), 'Los servicios públicos locales en la Ley 57/2003, de 16 de diciembre, de medidas para la modernización del gobierno local. Especial referencia a su incidencia en la legislación de Administración local de Aragón', *Revista Aragonesa de Administración Pública*, no. 24, pp. 289–320.

Mateos, L., I. Lorite, L. Lozano and E. Fereres (2003), 'Water governance and management in the water users' associations of Spain', *Options Méditerranéennes*, Series B, no. 48.

Matés Barco, J.M. (1991), '*El Servicio Público de Aguas Potables en España: Un Sector entre la Confluencia de los Intereses Públicos y Privados*', Revista de Obras Públicas 25, w/p, available at http://grupo.us.es/ciberico/archivos_acrobat/zaracomun1matesb.pdf (accessed 11 March 2012).

Matés Barco, J.M. (1998), *Cambio Institucional y Servicios Municipales. Una Historia del Servicio Público de Abastecimiento de Agua*, Granada: Editorial Comares.

Menéndez-Prieto, M. (2009), 'Facing the challenges of implementing the European Water Directive in Spain', in A. Garrido and M.R. Llamas (eds), *Water Policy in Spain*, London: Taylor & Francis, pp. 175–84.

Mestre Delgado, J.F. (2003), 'Las formas de prestación de los servicios públicos

locales. En particular, la concesión', in AA.VV, S. Muñoz Machado (ed.), *Tratado de Derecho Municipal*, 2nd edition, Madrid: Civitas.
Ministerio del Medio Ambiente (MMA) (1998), 'Libro Blanco del Agua en España' (Edición preliminar. MIMAM. Madrid), available at http://hercules.cedex.es/Informes/Planificacion/2000-Libro_Blanco_del_Agua_en_Espana/ (accessed 10 March 2012).
Ministerio del Medio Ambiente (MMA) (2005), 'Programa AGUA', available at http://www.mma.es/secciones/agua/pdf/programa.pdf (accessed 10 March 2012).
Ministerio del Medio Ambiente (MMA) (2007), 'Precios y costes de los servicios de agua en España'. Madrid, España.
Molina, A. (2001), *El Servicio Público de Abastecimiento de Agua en Poblaciones: El Contexto Liberalizador*, Valencia: Tirant Lo Blanc.
Observatorio de los Servicios Públicos Córdoba (2005), 'Water management in Córdoba: water time case study', available at http://www.watertime.net/docs/WP2/D17_Cordoba.doc (accessed 11 March 2012).
OECD (2010), 'OECD Environmental Data Compendium 2006–08 and OECD Economic Outlook: Statistics and Projections', available at http://www.oecd.org/document/57/0,3746,en_2649_37465_46478585_1_1_1_37465,00.html (accessed 17 March 2012).
Ordoñez, C. and L. Bru (2003), 'Análisis de la privatización y regulación del servicio de abastecimiento de agua en Málaga', in *Cuadernos del CC.EE. y EE.*, N. 44–45, pp. 81–98.
P. Comín, F. and M. Aceña dirs. (1991), 'Historia de la empresa pública en España', Madrid: Espasa-Calpe.
Parès, M., B. Gener, H. Cruz, and M. Martí (2009), *La Participació Ciutadana en la Gestió de l'Aigua*, Barcelona: Escola d'Administració Pública de Catalunya.
Romero Hernández, F. (2003), 'Los modos de gestión de los servicios públicos en el Proyecto de Ley de Medidas para la Modernización del Gobierno local (PLMGL)', *Cosital*, **69**, p. 15.
Saurí, D. and L. del Moral (2001), 'Recent developments in Spanish water policy: alternatives and conflicts at the end of the hydraulic age', *Geoforum*, **32**(3), 351–62.
Schmidt, G. and L. de Stefano (2009), 'Major processes degrading freshwater resources and ecosystems', in A. Garrido and M.R. Llamas (eds), *Water Policy in Spain*, London: Taylor & Francis, pp. 29–37.
Sevilla, M. and T. Torregrosa (2011), 'La cuestión de la regulación del sector del agua potable y saneamiento en España', Fundación Nueva Cultura del Agua, 2011, available at http://www.fnca.eu/congresoiberico/documentos/c0512.pdf (accessed 10 November 2011).
Torregrosa, T. (2008), 'El modelo socioeconómico de gestión de los recursos hídricos en la comarca de la Marina Baja (Alicante). Un enfoque de Gestión Integrada de Recursos Hídricos', Tesis Doctoral. Departamento de Análisis Económico Aplicado, Universidad de Alicante.
Transnational Institute and Corporate Europe Observatory (TI and CEO) (2009), 'Progressive Public Water Management in Europe', available at http://www.waterjustice.org/uploads/attachments/Progressive%20public%20water%20management%20in%20Europe.pdf (accessed 15 April 2012).
World Economic Forum/Financing for Development Initiative (WEF) (2006), 'Development-driven public–private partnerships in water', available at https://members.weforum.org/pdf/ppp_health_summary.pdf (accessed 10 March 2012).

7. Urban water challenges in the MENA region: integrating Islamic principles with demand management strategies

Arani Kajenthira and Sharmila Murthy

7.1 INTRODUCTION

In the Middle East and North Africa (MENA) region, the arid landscape has resulted in water playing a dominant role in human activities, socio-economic interactions and in the shaping of institutions. From the standpoint of water availability, the MENA region can be loosely categorized into three regions: those that are hyper-arid, those that rely on transboundary surface water, and those that have a high degree of water variability across their territories. Regardless of which category they fall into, almost all of these countries have focused on increasing the supply of water through large-scale infrastructure projects. In fact, as a proportion of the total freshwater resources available, the MENA region has more dams than any other part of the world. Energy-rich states have also pioneered desalination technologies as a way to augment supply. The perceived water scarcity, along with other historical, religious and cultural influences, has historically led to supply-centered water management regimes that are highly centralized and authoritarian in nature. Although the MENA region is often recognized as a water-scarce region, and one that is likely to become more arid in the face of climate change, a focus solely on resource availability obscures larger water governance issues.

Consistent with the international trend away from supply-side management and towards demand management, the MENA region has begun to enact reforms designed to lead to more efficient water management. These reforms include greater decentralization of the water governance structure, increased involvement of private actors, and the restructuring of tariffs to improve financial viability. At the same time, the cultural influence of Islam, which recognizes a 'right of thirst' (the right to drink when thirsty), helps to ensure that social equity and affordability in pricing structures are maintained. While many of these reforms are focused on improving the performance of water utilities operating in the urban water sector, effective water governance reform also requires reconsideration of how water in the agricultural sector is used. In the MENA region, the agricultural sector consumes, on average, nearly 70 percent of total water withdrawals (FAO, 2011), but contributes less than 20 percent (and in most cases less than 10 percent) to the total gross domestic product (GDP) (CIA, 2011). For example, Saudi Arabia has begun to reform its water institutions and engage in public–private partnerships, yet it relies primarily on energy-intensive desalination strategies to provide water to its urban consumers, while its aquifers are being rapidly depleted by unsustainable agricultural practices that consume approximately 85 percent of total water withdrawals, but only contribute 2.6 percent to GDP. Effective demand management reforms in the MENA region require consideration of both the urban and rural (that is agricultural) sectors.

This chapter explores the evolution of urban water governance in the MENA region and discusses recent water sector reform through a case study on Saudi Arabia. More specifically, section 7.2 considers the role played by Islam in the development of state-centered, centralized water ministries while section 7.3 examines the rise of supply-oriented policies by evaluating overall water availability, emerging supply-oriented technologies, and the potential impacts of climate change. Section 7.4 discusses recent trends in urban water utility reforms that rely on demand management principles and thereby promote greater decentralization, private sector participation and tariff increases to promote financial viability and foster water conservation. Section 7.5 focuses on Saudi Arabia, a hyper-arid country that has made significant progress in reforming its urban water governance framework but still faces major challenges in developing a more sustainable water management strategy. Section 7.6 concludes the discussion.

7.2 HISTORY OF WATER GOVERNANCE AND THE INFLUENCE OF ISLAMIC PRINCIPLES

Water governance in the MENA region has been influenced by both water scarcity and the highly centralized and authoritarian nature of many of the regimes. In most MENA countries, water is managed directly by national ministries, and the roots of the highly centralized nature of water institutions in the MENA region can be traced to Islamic traditions that promoted 'close community or state administrative control over water resources' (Civic, 1998).

Islamic law, or Sharia, literally means the 'way to water'. In Islam, water is of profound importance and is mentioned 63 times in the Koran (Civic, 1998; Faruqui et al., 2001). The Koran states at 21:30, 'Have not the unbelievers then beheld that the heavens and the earth were a mass all sewn up, and then We unstitched them and of water fashioned every living thing' (Faruqui et al., 2001). Water is the basis of life, and part of the purification process needed to enter into 'the state of divine grace' (Hughes, 2009).

Islam considers the sharing of water to be a 'holy duty' and preserves several fundamental water rights: a 'right of thirst' for humans, a 'right of thirst' for animals, and a 'right of irrigation' which means 'the right to provide water to land, trees, and plants' (Caponera and Nanni, 2007; Hughes, 2009).[1] Under traditional Islamic law, groundwater, natural wells, seas and great lakes, and rivers were held in the public trust and considered part of the public domain (UPI.com, 2008), leading to a communal approach to water resource management. Although the majority of countries in the MENA region no longer rely on Sharia law, Islamic principles have had a significant influence on water resources management (Mallat, 1995; World Bank, 2009), particularly the emphasis on affordability. The lack of water in the region and its importance to early agrarian and nomadic societies also prompted MENA countries to develop fairly complex systems of tribal law and Islamic doctrine to manage and share resources (Jagannathan et al., 2009).

In most MENA countries, water has been managed in the same manner as many other government functions: centrally, and most often by national ministries, with little transparency with regard to public expenditures and environmental impacts and almost no involvement by local actors (World Bank, 2009; Sowers et al., 2011). According to a 2007 World Bank report, water resource management in most MENA countries has been subsumed within ministries of irrigation, agriculture, energy/electricity, or environment/planning (World Bank, 2007) with very few nations (for example, Algeria and Iraq) having a dedicated ministry. In most cases, these ministries are responsible for both service delivery and quality control, but in some countries (for example Jordan, Morocco and Tunisia), operational and regulatory functions have been separated (World Bank, 2007).

Recent trends towards decentralization and increased private sector involvement have been motivated by internal and external pressures to improve efficiency and accountability, promote conservation, and raise capital. In addition, the problems posed by limited water resources and harsh climatic conditions have been exacerbated by high population growth rates, rapid urbanization and the continued use of nearly 70 percent of total water withdrawals by the agricultural sector (FAO, 2011), resulting in the number of people without access to water in urban settings increasing rather than decreasing over time (Pérard, 2008). To situate recent reforms within their proper context, the next section considers how water scarcity in the MENA region has historically promoted an emphasis on supply-oriented water management.

7.3 WATER SCARCITY AND SUPPLY-SIDE MANAGEMENT

Average per capita renewable water resources in the MENA region declined from 4000 m^3/person/year in 1950 to just 1100 m^3/person/year in 2007, and based on projected population growth, is predicted to decline further to just 550 m^3/person/year by 2050 (World Bank, 2007).[2] With this scarcity as the backdrop, this section examines the degree to which water availability varies across the MENA region, explores options for expanding water supply through technological innovation, and discusses the potential impact of climate change on future water resources.

7.3.1 Classifying the MENA Region into Three Categories

As water management strategies are contingent not only on absolute availability but also the variability and distribution of water resources, the World Bank has classified the MENA countries into three broad groups based on their primary water challenges: hyper-aridity, transboundary and high variability. Each of these is discussed below.

Hyper-arid countries

All MENA countries rely on groundwater to a certain degree, but hyper-arid countries are the most reliant on non-renewable groundwater and the desalination of saline or brackish water to augment their renewable freshwater supplies (World Bank, 2007). Managing groundwater extraction in these nations in a manner that sustains agricultural productivity without exhausting existing resources or causing saltwater intrusion is a careful balance that incurs different socio-economic challenges in those countries with relatively high per capita incomes – Bahrain, Kuwait, Oman, Qatar, Saudi Arabia, the United Arab Emirates (UAE), Libya and Israel – than those with lower incomes – Gaza, Yemen and Jordan (World Bank, 2007).

It is also worth noting that although many of the non-renewable aquifers

Table 7.1 Shared aquifers in the MENA region

Aquifer	Countries
Nubian Sandstone	Chad, Egypt, Libya, Sudan
North Western Sahara	Algeria, Libya, Tunisia
Mountain	Israel, Palestine (West Bank)
Disi	Jordan, Saudi Arabia
Rum-Saq	Jordan, Saudi Arabia
Great Oriental Erq	Algeria, Tunisia
Al-Kabeer Al-Janoubi	Lebanon, Syria

Source: World Bank (2007).

in the MENA region are transboundary in nature (Table 7.1), none are managed under a multi-country cooperative framework (World Bank, 2007). As a result, only countries with sufficient political and economic clout are able to exploit these finite resources effectively. Without robust regulatory systems, a 'tragedy of the commons' situation can result, with each country seeking to obtain as much water as possible (World Bank, 2007).

Transboundary surface water countries

Three rivers (the Nile, Tigris and Euphrates) account for the majority of the MENA region's surface water and are all shared between multiple riparian states (AFED, 2010) (for example Egypt, Iraq and Syria) (World Bank, 2007). Although not the focus of this chapter, much has been written about the challenges of transboundary water management and the resulting potential for conflict in the MENA region (Sadoff and Grey, 2005; Gryzbowski et al., 2010; Dolatyar and Gray, 2000).

Countries with high water variability

The final group of countries (Algeria, Djibouti, Iran, Lebanon, Morocco, Tunisia and the West Bank) suffer primarily from water variability and, despite possessing adequate quantities of renewable water at the national level, still face significant challenges with regard to internal distribution, both geographically and temporally (World Bank, 2007). Under these circumstances, the ability to store and transfer water is crucial and the World Bank notes that as a proportion of the total freshwater resources available, the MENA region has more dams than any other part of the world (World Bank, 2007). In addition to dams and surface water storage, several countries have also invested in bulk water transport systems: Libya's Great Man-Made River transfers non-renewable groundwater from the North Western Sahara Aquifer to population centers in the north, while

Morocco's inter-basin transfer systems are capable of redistributing 2.5 billion cubic meters of water per year. These cases also demonstrate two occasions where the state has assumed responsibility for strategic water allocations between basins (World Bank, 2007).

Although investments in water storage have brought significant economic growth to the MENA region, they have also caused significant hydrological and land quality challenges, including adverse impacts on soil fertility (World Bank, 2007). The variable nature of rainfall in the region has also affected dam operation: many recently constructed dams in Morocco, for example, have not been filling to design capacity, affecting water allocations for rural farmers (Jagannathan et al., 2009).

Finally, large-scale dams and infrastructure projects also have a tendency to engender controversy. One recent example is the construction of the Red Sea–Dead Sea Water Conveyance Concept,[3] which resulted in several Palestinian civil society groups filing a complaint with the World Bank's Inspection Panel over the Bank's feasibility study of the project, asserting that it did not seriously examine alternatives for restoring the flow of water to the Jordan River and favors Israel's current policies in the region (Bretton Woods Project, 2011).

7.3.2 Technological Options for Expanding Supply

With limited and variable freshwater sources, the MENA region has invested heavily in technological options for augmenting supply. Many countries are relying more and more on desalination, but as this section explains, the comparative contribution of desalination to total water availability is small, despite significant energy and environmental costs. Wastewater reuse also has great potential to enhance water availability, but a lack of social acceptance and political commitment remain obstacles to its expansion.

Desalination

From the perspective of capital costs and operating costs, the price of desalination is now competitive with water reuse and groundwater withdrawals. But even in Saudi Arabia, which possesses the largest desalination capacity in the world, the total water produced by desalination is dwarfed by the total demand on surface water and groundwater.

Considering economic costs alone, the World Bank estimates that desalination can be less expensive than exploiting more conventional water resources when major investments (that is, dams and inter-basin transfers) are required (World Bank, 2007, p. 38). However, there are several environmental externalities: most significantly, the large-scale

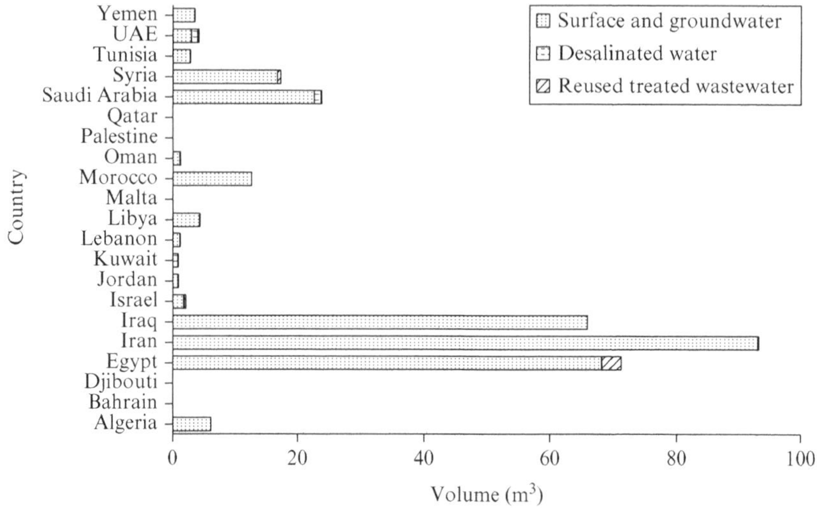

Figure 7.1 Total water withdrawals by source and country, MENA region

thermal desalination that is prevalent in the MENA region remains highly energy-intensive (Gude et al., 2010), an important consideration for those countries that do not have significant energy reserves. In addition, neither capital costs nor operation and maintenance costs fully account for the environmental impacts associated with desalination (that is, CO_2 emissions and the production of concentrated brine), or transportation costs that can actually exceed the cost of production (Kally, 1993; Zhou and Tol, 2005; Kajenthira et al., 2012).

Wastewater collection and reuse

There is significant potential for increasing wastewater treatment and reuse in the MENA region to augment surface and groundwater resources (Hamoda, 2004; Kajenthira et al., 2012). Recent estimates suggest that, on average, only 43 percent of the wastewater generated in the MENA countries is treated (FAO, 2011). However, there is also substantial disparity across the region in the percentages of total wastewater reclamation (Figure 7.1) (Qadir et al., 2010).

Exploring the use of treated wastewater as a non-conventional resource highlights the dilemma that MENA policymakers face in aligning Islamic law and customs with the realities of modern-day water resource management (Jagannathan et al., 2009). Given the importance of cleanliness in Islam, many Muslims had previously declared that wastewater reuse was undesirable, or even unlawful (*haraam*) under Islamic law (Faruqui et al.,

2001). However, in 1978, Saudi Arabia took the dramatic step of issuing a special *fatwa* stating that wastewater could be used for drinking and ablution if treated to a sufficient standard such that it presented no health risks. This *fatwa* led to a dramatic expansion of wastewater reuse in Saudi Arabia, with other Gulf Cooperation Council (GCC) countries (Bahrain, Kuwait, Oman, Qatar and the United Arab Emirates) soon following suit (Faruqui et al., 2001).

Unfortunately, the lack of political commitment and national policies or strategies to support wastewater reclamation (AFED, 2010), the incomplete economic analysis of wastewater treatment and reuse options, and lingering misperceptions about the cost of developing wastewater collection networks and treatment plants (Qadir et al., 2010) remain significant constraints to effective reuse in the region. Rapid increases in urban population and the resulting discharge of larger volumes of wastewater have also resulted in reclaimed water of variable quality as treatment plants are overloaded; many treatment plants were also not designed to treat the mixture of both domestic and industrial sewage, which is now the most prevalent form of wastewater in the region (Qadir et al., 2010). With good reason, many people in the MENA region remain suspicious of wastewater reuse as they are uncertain of the quality of treated water (AFED, 2010), particularly in reuse activities that involve a large degree of human contact (Hamoda, 2004). However, both municipal and industrial wastewater can be reused within the industrial sector with lower levels of treatment (that is, secondary treatment versus tertiary or advanced treatment), fewer public health implications than domestic reuse (Angelakis et al., 1999) and more consistent water demand than seasonal agricultural applications such as irrigation (Asano et al., 2007).

These remaining infrastructural and institutional challenges to wastewater reuse highlight the critical need for enforceable wastewater treatment standards and policy reform to spur further investment in treatment facilities. Without this, MENA countries will continue to invest in desalination (Desal Data, 2010) rather than reuse as a means of addressing water scarcity.

7.3.3 Climate Change

Finally, the combination of climate change and demographic growth are anticipated to profoundly affect the quality and availability of water resources in the MENA region. Acceleration in the hydrological cycle is expected to cause extended droughts and more intense and variable rainfall events, increasing probabilities of flooding and desertification (Sowers et al., 2011). More specifically, the majority of global climate models project a 25 percent reduction in precipitation, an increase in evaporation, and a subsequent reduction in both run-off and soil moisture for the northern part of the Arab region. Although the GCC countries are anticipated to experience reduced precipitation, it can be argued that this region is also the least vulnerable to climate change as internal renewable water resources are already extremely limited (AFED, 2009). By contrast, agricultural production in the Maghreb countries (Morocco, Algeria, Libya and Tunisia) is extremely vulnerable given their almost exclusive dependency on rainfall (AFED, 2009).

Despite the pursuit of supply augmentation strategies and the procurement of virtual water through agricultural imports and foreign direct investment (Allan, 2001), the fact remains that per capita municipal water withdrawals in excess of 100 m^3/year (as is the case in Bahrain, Kuwait, Qatar and the UAE) are unsustainable in this arid region, and there is a critical need for integrated water resource and demand management to promote conservation and increase efficiency (Sowers et al., 2011).

7.4 WATER UTILITY REFORM AND DEMAND MANAGEMENT

Despite a historical focus on centralized and supply-oriented water management, reforms promoted by international institutions (for example the World Bank) over the last decade have led to an increasing emphasis on demand management approaches to promote conservation and increase efficiency. This shift in focus was based in part on the realization that a system based on securing supply can create 'excess demand', which can increase costs and result in social tension. For example, farmers may plant more water-intensive crops in response to the widespread construction of irrigation canals; however, in water-scarce regions, that increased demand for water may not be met consistently (for example, in Morocco; Jagannathan et al., 2009), causing economic disruption. Similarly, and as mentioned previously, without appropriate regulation, a 'tragedy of the commons' situation can result (World Bank, 2007).

The concept of demand management is premised on increasing the efficiency of water by ensuring that it is going to its most valuable and economically productive uses. Demand management strategies have several key elements, including: (1) ministerial reform to promote greater financial discipline and decentralization; (2) the involvement of private sector actors to promote efficiency and financial sustainability; and (3) tariff increases to both promote conservation and improve the efficiency and financial sustainability of water utilities. However, the challenges of water management reform in the MENA region cannot be understood without considering the role of agriculture, which consumes on average nearly 70 percent of total water withdrawals (FAO, 2011), but contributes less than 20 percent of the regional GDP (CIA, 2011). This section considers each of these issues in turn.

7.4.1 Governance Reform and the Challenges of Greater Decentralization

Improving water governance is a key component of demand management strategies. The term 'governance' in this context refers to the 'range of political, social, economic and administrative systems that are in place to regulate the development and management of water resources and provision of water services at different levels of society, while recognizing the role played by environmental services' (AFED, 2010).[4] Although there is no agreed-upon definition of good water governance, it can be understood as requiring participation of all stakeholders, equity, transparency, anti-corruption, promotion of water legislation and regulations (AFED, 2010).

Most MENA countries suffer from a weak water governance structure (Pérard, 2008; AFED, 2010). Legacies of the centralized systems of planning, taxation and revenue distribution prevalent in the region have also continued to limit the extent to which subnational actors can play a role. Improving water governance in the MENA region therefore requires that governments have stronger organizational capacity. 'Analyses of water management problems in almost all countries of the region diagnose problems stemming from factors such as public agencies having overlapping and unclear functions, difficulties coordinating different uses of water, or the same organization working as a service provider, planner and regulator' (World Bank, 2007). With the growing risks associated with climate change, Sowers et al. argue that 'adaptive governance' strategies require that governments and water experts make more effort in involving local communities in decision-making, especially as they deal with water risks (Sowers et al., 2011). Moreover, if public–private partnerships become increasingly common, there will be a corresponding need for a stronger regulatory framework for these partnerships to be successful (Bakker, 2010). By emphasizing the involvement of local governments, civil society and other stakeholders, demand management reform requires that there be improved accountability, transparency and inclusiveness (World Bank, 2007).

Some progress has been made. For example, in countries such as Egypt, Jordan, Libya, Morocco, Oman, Tunisia and Yemen, community groups and user associations have begun playing a larger role in water management (AFED, 2010). The case study on Saudi Arabia included in this chapter also describes in detail the water utility reforms undertaken by the Kingdom.

7.4.2 Private Sector Involvement

Formal private sector

Private sector participation has been promoted as a way to improve water demand management and increase financial discipline. When compared to other regions of the world, formal private sector participation in the MENA has historically been low:

> Based on the World Bank Private Participation in Infrastructure (PPI) Database for the period 1990–2002, there were 106 such projects in Latin America and the Caribbean and 73 in East Asia and Pacific, but only seven projects in the Middle East and North Africa and 14 in sub-Saharan Africa. Latin America and the Caribbean and East Asia and Pacific together accounted for more than 95 percent of total investment. (Kirkpatrick et al., 2006)

More recently, however, countries and cities within the MENA region are turning to the private sector in order to gain access to large amounts of capital, achieve water utility reforms, and implement the demand management strategies promoted by international institutions (for example, the World Bank). In fact, according to the market research agency Global Water Intelligence, which maintains a global public–private partnership tracker for projects that are proposed, under bidding or signed, as of 2010, 16 out of 21 countries in the region have solicited private sector involvement in ongoing projects (GWI, 2010a). Private sector participation in water and sanitation varies according to asset ownership, responsibility for capital investments, assumption of risk, responsibility for operation and maintenance, and contract length, and the MENA countries also differ in their preferred arrangements (GWI, 2010a).

It is important to recognize that private sector involvement in the water sector is not necessarily a panacea. A private actor may be able to operate more efficiently than the state, but only if the proper incentives are in place. This depends on both the terms of the contract and the strength of the state's regulatory power over the private actor. Moreover, although the involvement of the private sector can provide the state with the political cover it needs to raise tariffs, this is not always the case, as discussed in more detail below.

Informal private sector

While the 'formal' private sector is playing an increasingly large role in the management of water services, the role of the 'informal', unregulated private sector cannot be overlooked. Despite strong central control over water planning, decentralized private actions have had a large impact on

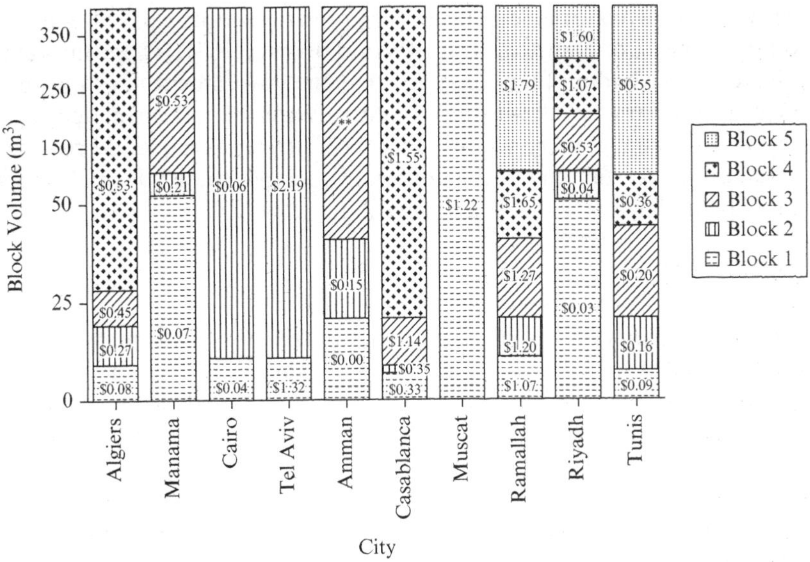

Notes: *All figures in 2010 US dollars; **volumetric charge then varies per m^3.

Source: GWI (2010b).

Figure 7.2 Block water tariffs for selected cities, MENA region

water resource management. For example, with the advent of low-cost drilling in the 1960s (that is, mobile diesel pumps), farmers across the MENA region have been able to extract water directly from aquifers, greatly increasing pressure on limited groundwater resources.

The 'informal' private sector has also played a key role in meeting the demands of those who do not have access to reliable sources of water. As in many parts of the world, the poor and marginalized communities in MENA countries are often not connected to municipal networks and, due to their reliance on private providers, often pay much higher rates than publicly supplied consumers (Sowers et al., 2011). In addition, during times of water scarcity, even networked consumers must rely on such private vendors. For example, during the summer of 1997, the city of Amman, Jordan experienced a severe water shortage that required the public to purchase water from private vendors. The black-market price for tanker truck water consequently reached up to US$14/$m^3$ (Faruqui et al., 2001), nearly two orders of magnitude more expensive than the block water tariff (Figure 7.2). Moreover, informal water markets in water-scarce regions (for example Jordan, Morocco and Yemen) create incentives for overextraction because the private vendors can sell as much water as they are able to extract (World Bank, 2007). These informal private water vendors fill a key need that the government does not, but often at a steep cost to consumers and to water management schemes. Improving water governance in MENA countries consequently requires a nuanced understanding of the important role that these informal actors can play in filling gaps in service provision.

7.4.3 Tariffs

Increasing water tariffs is a central tenet of many demand management strategies, intended to increase the financial viability of water service delivery and to promote water conservation. This section critically examines these underlying rationales, as well as the common assumptions that are made about the challenges of tariff reform.

Most water utilities in the MENA region do not have sufficient cash flow from water service provision and are therefore dependent on some form of government support, both to finance investments and to subsidize operations and maintenance (World Bank, 2007). However, MENA countries are not necessarily unique in this respect. In urban areas around the world, networked water and sanitation services have historically been subsidized to some degree. The rates paid by those connected to the water or sanitation networks may have covered the marginal operating costs but, in most cases, did not cover the maintenance costs or future marginal costs, which could potentially include the costs of obtaining new sources of water. Hidden underground, water and sanitation infrastructure in most parts of the world has also degraded over time, increasing leakage rates and the potential for contamination. While most governments now recognize the importance of rebuilding and maintaining these pipes, doing so is financially prohibitive in many situations. The trend towards full cost recovery, a large part of which includes tariff increases, has also only arisen in recent decades (Rogers et al., 2002).

Social equity and the importance of providing affordable access to household water are rooted in Islamic law; these factors, along with cost recovery and water conservation, play an important role in the setting of water tariffs in the MENA region (Magiera et al., 2006). The majority of MENA countries have escalating block water tariffs for municipal water withdrawals (Figure 7.2). Although they vary significantly across the region, these block tariffs generally help to subsidize the basic water needs of poor consumers while simultaneously sanctioning heavy water users (Magiera et al., 2006). However, as mentioned above, block tariffs can also inadvertently increase water rates for these marginalized individuals who do not have access to networked service and are forced to purchase water from their neighbors or from informal private vendors (Bakker, 2010; World Bank, 2009). Therefore, in addition to block tariffs, some countries have developed 'lifeline' policies based on Islamic law. For example, Iran provides 30 liters of water per capita per day for free to domestic customers in urban areas (Faruqui et al., 2001).

The recent emphasis on cost recovery and raising water tariffs has raised questions as to whether allocating water according to market forces is inconsistent with Islamic teaching (Faruqui et al., 2001). As noted by the World Bank, 'any legal reforms in the region must respect and attempt to incorporate the traditions of customary and Islamic law' (Jagannathan et al., 2009). Islamic scholars have, however, concluded that while water in its natural state cannot be bought and sold, individuals or groups have the right to sell and recover value-added costs of most categories of water. Therefore, if a public utility constructs a water supply, treatment and distribution system, then the utility has the right to recover its full costs (Faruqui et al., 2001). Such a qualified right of ownership of water must, however, generally be combined with the required distribution of surplus. In other words, Islamic values around water do not conflict with the idea of raising tariffs, but there is an emphasis on social equity. Nothing in Islamic doctrine specifically prohibits public–private partnerships: water can remain part of the public domain but still be managed in part by non-state, private actors. It is worth noting that even outside MENA, the full-scale privatization of water (including the divestiture of underlying assets) is rare, with England and Chile being the two most cited examples (Marin, 2009).

An interesting parallel can also be drawn between the Islamic influence on water and the human right to water and sanitation under international law, as well as existing misconceptions about what this human right means with respect to public–private partnerships. The United Nations (UN) recognition of a human right to water and sanitation means that everyone, without discrimination, should have access to adequate amounts of safe, accessible, affordable, acceptable water and sanitation (UN ESC, 2002; de Albuquerque, 2012). It does not require that these services be free, but requires states to implement policy measures to ensure that even the poor have affordable access, which in many cases requires some type of subsidization (UNHRC, 2010). The most famous example of such a policy is South Africa's free basic water policy, which although based on that country's constitutional human right to water (O'Reagan, 2009), is conceptually similar to Iran's lifeline policy, which is based on Islamic law. While the human right to water has often been linked to anti-privatization arguments (Prasad, 2006), the two issues are distinct. As the UN Special

Rapporteur on the Human Right to Safe Drinking Water and Sanitation has stated, 'human rights are neutral as to economic models in general, and models of service provision more specifically . . . [I]nvolving non-State actors requires, inter alia, clearly defining the scope of functions delegated to them, overseeing their activities through setting regulatory standards and monitoring compliance' (UNHRC, 2010). In other words, market-based approaches for allocating water are not necessarily incompatible with the human right to water and sanitation, provided that individuals still have equal, affordable and physical access to sufficient, safe and acceptable water and sanitation services.

Many commentators have observed that raising tariffs in MENA countries is difficult because of Islamic values (Prasad, 2006; World Bank, 2007). It is certainly undeniable that water 'evokes political emotions like no other issue' (Prasad, 2006). The challenge is, therefore, to reconcile these values with financial sustainability. One strategy may be to examine critically what affordability really means within a region, and to set tariffs accordingly across the population. In addition, in many MENA countries, urban populations have grown and become wealthier, and in many cases, are in a position to pay for the costs of services (World Bank, 2007). Subsidies to well-off consumers who are able to pay are burdensome on public budgets and prevent water utility companies from recouping costs (Magiera et al., 2006) and investing in much-needed infrastructure and comprehensive asset management. As discussed earlier, neither Islam nor the recognition of water as a human right require that the costs of delivering services to well-off consumers be subsidized; rather, the focus on social equity means that the poor may require subsidies to ensure affordability.

One thing that is important to recognize is that in the oil-rich countries of the MENA region, governments also provide widespread subsidies as a means of maintaining political stability (Jones, 2010), even if such subsidies make no sense from an economic or environmental point of view. The authoritarian nature of these regimes, combined with a lack of transparency and strong centralized rule, also mean that a realistic assessment of affordability, with a corresponding increase in tariffs, is politically unlikely. International institutions like the World Bank and the Arab Country Water Utility Network have consequently been promoting water utility reform with the goal of having a separate regulatory agency that can set water tariffs. Currently, water tariffs are commonly set by ministries, or at least reviewed by ministries when public or private operators exist (Magiera et al., 2006). By having an independent agency set prices instead, the goal is to reduce political interference with tariffs. Further, although demand management strategies emphasize increasing tariffs as a way to promote water conservation in addition to improving the financial sustainability of water utilities, it is not clear that this has actually materialized in countries where tariff increases have been implemented. For example, evidence from Palestine and Yemen suggests that in communities where water is used primarily to serve basic human needs, consumption does not decrease when the price increases (Magiera et al., 2006).

Awareness of the need for water conservation also varies widely across the region. Although most MENA countries appear to have some form of government-supported water conservation awareness campaign which in some cases target specific subsections of the population (for example, schoolchildren), they have differing degrees of effectiveness. For example, a 2006 pan-Arab public opinion survey found:

> that a relatively high percentage of respondents from countries poor in natural fresh water resources, and largely depending on sea water desalination, said that water was not a major environmental problem. Those who thought water was not a major problem or no problem at all included 64 percent in Bahrain, 52 percent in Tunisia, 41 percent in Qatar, 35 percent in Oman and Kuwait and 31 percent in UAE. (Sowers et al., 2011; Tolba and Saab, 2006)

In contrast, in countries with comparatively high amounts of renewable surface water, such as Iraq and Sudan, over 90 percent of the population indicated that access to freshwater was a major problem, which was attributed to the poor state of public infrastructure (Sowers et al., 2011; Tolba and Saab, 2006). These figures highlight some of the challenges of supply-oriented approaches to water management. Some commentators have suggested that to promote water conservation, MENA countries would do well to invoke Islamic principles, which favor water conservation and impose liability for polluting or degrading water. For example, the Koran instructs 'waste not by excess, for God loveth not the wasters' (Faruqui et al., 2001). Moreover, mosques and religious schools could play an increasingly important role in conveying messages for water conservation (Faruqui et al., 2001).

7.4.4 Agricultural Water

Although this chapter focuses predominantly on urban water service provision in the MENA region, it is impossible to overlook the significant role that agricultural water plays in developing a comprehensive demand management strategy. Until the early twentieth century, most MENA countries depended on agricultural production as their main source of wealth (Beaumont, 2002). However, in more recent years, the diverse fortunes of the region have been more heavily influenced by the discovery of oil and gas (Beaumont, 2002) as well as by the institution of economic

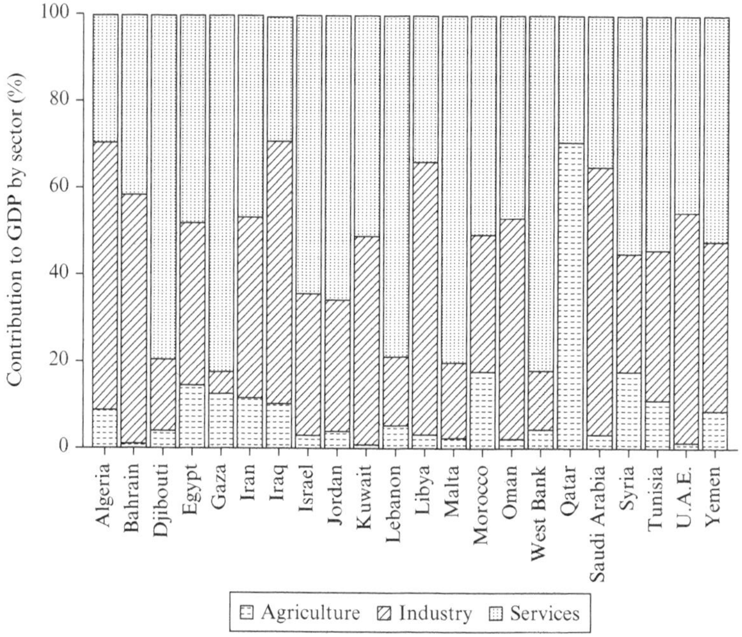

Sources: CIA (2011), FAO (2011).

Figure 7.3 Percentage contribution to GDP by sector, MENA region

structures and policies that emphasize the role of the state (World Bank, 2011). The influence of the oil industry, combined with rapid population growth and increased urbanization, has also resulted in the GDP of most MENA countries being dominated by the industrial and service sectors (Figure 7.3) (Beaumont, 2002). In contrast, the agricultural sector is by and large showing a decreasing contribution to GDP over time (*The Economist*, 2010) and makes a smaller contribution than the industrial and service sectors in all MENA countries. In addition, although increasing wastewater reuse in irrigation has the potential to reduce overall agricultural water withdrawals, the fact that in most MENA countries reclaimed water cannot be used for the irrigation of high-value vegetable crops also poses a significant barrier (AFED, 2010).

Yet despite the high volume and low economic productivity associated with agricultural water use, tariffs are incredibly low. In many MENA countries, irrigation water prices have only 'symbolic value', and in some places, irrigation water is free of charge (Magiera et al., 2006). Further, in the few countries that have used tariffs to recoup the costs of investing in wastewater treatment infrastructure, the historic availability of untreated wastewater free of charge has made it challenging to convince farmers to pay for treated wastewater of questionable quality (AFED, 2010).

While the need for reform is apparent, it may also be difficult to directly control prices for small-scale users. Instead, countries can reduce incentives for water extraction, for example, by removing subsidies on diesel fuel for pumping. At the same time, policymakers must be made aware that removing these subsidies could hurt small-scale farmers and exacerbate rural poverty (World Bank, 2007; Sowers et al., 2011). Charging agri-business holdings higher water tariffs may be a more viable option, though according to Sowers et al. this has not yet been widely explored in the MENA region (Sowers et al., 2011).

In summary, while urban water management in the MENA region has begun to undergo tremendous reforms that adhere to demand management principles, it remains critical to consider the role of the agricultural sector. For political reasons, it is often difficult for MENA countries to directly address the fact that agriculture consumes the vast majority of water, pays a fraction of the rates charged to urban consumers, and contributes comparatively little to GDP. However, as the case study of Saudi Arabia below illustrates, reforming the urban water sector in line with demand management principles requires an integrated approach that considers both urban and rural water use.

7.5 CASE STUDY: THE KINGDOM OF SAUDI ARABIA

In Saudi Arabia, harsh climatic conditions and limited renewable water sources have recently prompted a critical rethinking of the management of water and wastewater. This case study describes the recent evolution of the water sector in Saudi Arabia from its historical emphasis of supply-side management. Although the Kingdom has begun to adopt elements of a demand management strategy, such as restructuring the governance framework and involving the private sector, it has not addressed some key issues, such as improving water conservation, which are essential for a sustainable water management strategy.

7.5.1 Saudi Arabia's Supply-side Policies

Saudi Arabia is a paradigmatic hyper-arid country that depends heavily on its aquifers for water. Approximations of Saudi Arabia's groundwater resources vary wildly; however the majority agree that the economically abstractable volume of water is less than 1 trillion m^3 (FAO, 2008). Currently, the agricultural sector is responsible for 88 percent of total water withdrawals (FAO, 2010), which is primarily from non-renewable aquifers. At current withdrawal rates, the Kingdom's aquifers are expected to supply water for only about 25 more years (Kajenthira et al., 2012). In addition, when rates of groundwater withdrawal exceed recharge rates, saltwater intrusion can cause a serious deterioration in groundwater quality and land subsidence becomes an issue (Al-Ibrahim, 1991). For example, the Minjur aquifer, which provides approximately 33 percent of Riyadh's water supply, has been drilled to depths of about 1400 meters, and the water table decreased by 10 meters between 1960 and 1990 (Al-Ibrahim, 1991; Llamas and Custudio, 2003). Moving forward, the Kingdom needs to reconsider the continued high-volume extraction of water from non-renewable aquifers.

Like the other GCC countries, Saudi Arabia's vast oil revenues have enabled it to invest in large-scale infrastructure projects as a way to augment its supply of non-renewable groundwater. To maximize use of water from erratic rainfall events, over 200 dams were constructed between 1975 and 2009 with a total storage capacity of 1.35 billion m^3. Although the Saudi government aims to increase this storage capacity to 2.5 billion m^3 by 2014, this volume would still contribute less than 15 percent of the total annual water withdrawals of over 20 billion m^3 (FAO, 2011). To meet increasing water demands, Saudi Arabia has instead turned to producing freshwater through desalination.

Desalination plays a critical role in Saudi Arabia's urban water policy, serving primarily the municipal and industrial sectors. The Kingdom now boasts the world's largest desalination capacity at just over 1 billion m^3/year (SWCC, 2010). However, the Kingdom still needs to double its potable water supply over the next two decades to meet rising demand due to population growth and urbanization (US–SABC, 2010; CIA, 2011; World Bank, 2008). Desalination is expected to play an increasingly important role in meeting water demand but, as mentioned previously, the desalination process is highly energy-intensive. Saudi Arabia's Minister of Water and Electricity estimated that 25 percent of Saudi oil and gas production in 2009 was used domestically to generate electricity and produce water, while present demand rates suggest that this figure will reach 50 percent by 2030 (Al-Hussayen, 2009). In addition, the demand for natural gas in Saudi Arabia, used primarily in petrochemical and desalination plants, is expected to double from approximately 7.1 billion cubic feet (bcf) per day in 2007 to an estimated 14.5 bcf per day in 2030 (US EIA, 2011). While Saudi Arabia is in the fortunate position to possess large energy resources, the long-term environmental sustainability of the current approach is not clear. However, innovations in desalination technology could result in a more optimistic prognosis in the near future.

Wastewater treatment and reuse will also increasingly play a significant role in urban water management. Currently in Saudi Arabia, only 45 percent of the urban sewerage is treated and the existing wastewater systems are overloaded, resulting in inconsistent levels of treatment and potential health hazards (Qadir et al., 2010). Although treated wastewater is reused in the agricultural sector only in the irrigation of non-edible crops (FAO, 2008), it still has the potential to adversely affect soil and groundwater due to its high residual saline content (Hamoda, 2004). Treating wastewater to a much higher standard would reduce this problem and also provide Saudi Arabia with a much-needed source of water, but progress is impeded by both infrastructural and public perception challenges (Qadir et al., 2010).

7.5.2 Recent Trends towards Demand Management Strategies in the Kingdom

In response to pressure from international institutions, and to drive growth within the water sector, the Kingdom has begun to adopt some elements of a demand management strategy, including the restructuring of the water and sanitation sector to boost private sector participation. As this section will discuss, these innovations have improved operational efficiencies and provided opportunities to raise capital, but they have not addressed the underlying problem of unsustainable water use – which is at the heart of demand management theory.

Restructuring water governance framework

The restructuring of Saudi Arabia's water sector commenced with the establishment of a separate Ministry of Water Affairs in 2001, intended specifically to manage the Kingdom's water resources and to facilitate private investment into the sector. All of the water-related departments in the Ministry of Agriculture and Water and the Ministry of Urban and Rural Affairs, and all sewerage and sanitation departments, were integrated into the new ministry, which was also intended to increase water sales and review tariffs (US–SABC, 2009).

In 2003, the electricity sector merged with the water sector to form the Ministry of Water and Electricity (MoWE), which laid the foundation for ongoing reforms of both the water and power sectors by issuing guidelines for the operation of Saudi Arabia's Independent Water and Power Plants (IWPPs) (US–SABC, 2009). A limited liability company, the Water and Electricity Company, was created as the sole project manager of the IWPPs, which were to be developed by the private sector using build–own–operate (BOO) schemes (US–SABC, 2009). In addition, the MoWE aimed to reduce total water consumption by launching a four-phase conservation program in 2004 (US–SABC, 2009). The first phase of the program focused on reducing residential water consumption, and the distribution of more than 2.9 million water rationalization kits between 2004 and 2006 led to a 30 percent reduction in residential water consumption (US–SABC, 2009). Subsequent stages of the project have also succeeded in reducing water consumption in public buildings (hospitals, schools and mosques) and private enterprises (hotels, restaurants and commercial centers) by 38 percent and 25–35 percent, respectively (US–SABC, 2009).

Under the guidance of the MoWE, the Electricity and Co-Generation Regulatory Authority (ECRA) was created in 2004 as an independent body to help implement the government's water and power policies (US–SABC, 2009). ECRA issues licenses to any entity engaged in water desalination or power generation while also seeking to protect the interests of service providers and the rights of investors to see reasonable returns on their investments (US–SABC, 2009). Another independent entity guided by the MoWE is the Saline Water Conversion Corporation (SWCC), which is the primary source of water supply for the Kingdom (US–SABC, 2009). The SWCC is responsible for the 30 large-scale desalination plants that account for about 50 percent of the municipal water supply, and is also the Kingdom's second-largest supplier of electricity.

Between 2005 and 2009, the water sector in Saudi Arabia went through further significant institutional and organizational changes. In October 2006, the National Water Company (NWC) was created as a government-owned entity to oversee groundwater extraction, drinking water distribution, and sewage collection and treatment (US–SABC, 2009). The NWC is anticipated to gradually take over the control of all wells, sewage treatment plants and desalination plants from various government bodies and the SWCC, and to achieve a separation of water and wastewater operations on a city-by-city basis. The privatization and restructuring of the SWCC was also approved in July 2008 (US–SABC, 2009). Additional changes include the decentralization of water and sanitation administration to the 13 administrative regions, the adoption of a set of policies and measures aimed to raise the performance of the water and sanitation sector, and the approval of Executive Regulations for treatment and reuse of wastewater and the standardization of specifications for wastewater treatment plants (Ministry of Economy and Planning, 2010).

Private sector involvement

The restructuring of Saudi Arabia's water and sanitation sector has created opportunities for private sector involvement in three areas: (1) water and power production (desalination and IWPPs); (2) water distribution and management; and (3) wastewater collection and treatment.

Firstly, as mentioned previously, the Saudi Arabian government facilitated private sector involvement in water and power generation through independent water and power plant projects (IWPPs) developed on the basis of build–own–operate (BOO) and build–own–operate–transfer (BOOT) schemes. In 2004, the government launched an initiative to develop ten IWPPs by 2016 at a total investment cost of US$16 billion (US–SABC, 2009). In the wake of the financial crisis, these projects have been temporarily abandoned as a result of financing challenges (US–SABC, 2009). However, they will likely be undertaken in the future to meet the government's goal of doubling the total capacity of desalination plants to over 2 billion m^3 by 2014 (Ministry of Economy and Planning, 2010).

Secondly, in addition to encouraging private sector involvement in water production, beginning in 2008, the Kingdom has also sought to utilize the expertise of the private sector to upgrade and manage its water distribution infrastructure. Much of the water infrastructure in Saudi Arabia is over 25 years old. The desalination plants that supply much of the Kingdom's municipal water supply are linked to urban centers by over 4000 kilometers of pipelines with leakage rates ranging from 20–35 percent (US–SABC, 2009). Therefore, beginning with the Kingdom's five largest cities (Riyadh, Jeddah, Mecca, Medina and Dammam, which together account for 65 percent of the country's total water consumption), the MoWE has used management contracts to monitor levels of water loss and calculate occurrences of non-revenue water. In Riyadh, it was found that an astounding 60 percent of the total water supply was lost in the 10 000 kilometers of pipes that distribute water to the population of 5 million. Veolia Water is now aiming to reduce this leakage rate to 20 percent (US–SABC, 2010).

Thirdly, public–private partnerships have been employed in wastewater collection and treatment. Given Saudi Arabia's low urban sewerage rate of just 45 percent, the MoWE had previously estimated that approximately US$23 billion would need to be invested over the next 20 years to reach an overall sewerage rate of 100 percent (US–SABC, 2009). Jeddah, for example, despite having a much lower leakage rate in its water distribution infrastructure than Riyadh, has very limited sewerage capacity: less than 15 percent of Jeddah is currently connected to sewer pipes. As a result, Jeddah awarded Suez Environment a US$61 million contract in May 2008 that covers the production of groundwater, supply of potable water and collection of wastewater, with the goal of providing continuous drinking water and cutting emergency response time to repair leaks (US–SABC, 2010). In 2010, additional public–private partnerships were also established to improve the performance of water distribution and wastewater collection networks in Dammam, Medina, Mecca and Taif (GWI, 2010a); these five-year management contracts will subsequently convert into 25-year concession contracts, with a potential initial public offering (IPO) for each city's water network at a later stage (US–SABC, 2009).

Obstacles to effectively managing water demand in Saudi Arabia

Saudi Arabia has made significant progress in reforming its water sector, but it has not yet addressed some fundamental issues that are integral to a successful demand management strategy. Limited renewable water resources notwithstanding, Saudi Arabia is the second-highest per capita water consumer in MENA after Egypt, and the eighteenth-highest in the world out of 199 countries (FAO, 2011). The fact that a desert country with almost no surface water and primarily non-renewable aquifers is in the top 10 percent of world per capita water users underscores the need for greater demand management in Saudi Arabia.

Saudi Arabia's high per capita rate of water use is no doubt influenced by the agricultural sector's high water extraction rates. Agriculture consumes approximately 88 percent of total water withdrawals, but only contributes 2.6 percent to GDP (FAO, 2010). Recently, in response to the global commodity crisis of 2007 and growing awareness of the need to increase water productivity, the government has abandoned its support for water-intensive crops such as wheat and alfalfa, and instead devoted additional resources to organic farming and vegetables for human consumption (Lippman, 2010). The Ninth Development Plan aims to reduce agricultural water demand at a rate of 3.7 percent (Ministry of Economy and Planning, 2010). However, the government also seeks to improve annual growth rates of local production of vegetables, fruits, dairy, meat, fish and poultry, which indicates that the expansion of the agricultural sector still remains a priority. To create a sustainable strategy moving forward, Saudi Arabia needs to consider adopting better agricultural demand management strategies. However, in the wake of political instability in the MENA region following the Arab Spring, the Kingdom is less likely to implement any controversial reforms, including those relating to agricultural water tariffs.

In contrast to the agricultural sector, the urban water sector in Saudi Arabia consumes 15 percent of all water withdrawals and uses less than 100 m^3 of water per capita, which is lower than other countries in the region (FAO, 2011). While Saudi Arabia has engaged in some water conservation educational programs (Ministry of Economy and Planning, 2010) and also distributed water-saving devices (US–SABC, 2009), more can be done. For example, international institutions like the World Bank have been encouraging the Kingdom to raise block tariff rates as a way to curb demand.

Raising tariffs would also improve the financial viability of the newly restructured water utilities. For example, in 2001, it was estimated that a tariff of US$1.60/$m^3$ for water supply or a combined tariff of US$2.68–3.20/$m^3$ would be required to allow cost recovery and encourage investment, but urban water tariffs have remained at 2001 prices (GWI, 2010b). Although Saudi Arabia has transitioned to a block tariff system where water pricing varies based on total consumption (Figure 7.2), the water price does not approach the total economic cost of water production and transportation until the consumption volume is in excess of 300 m^3/month. Moreover, the tariffs do not account for the environmental costs associated with the production of desalinated water (Kajenthira et al., 2012). Although the Ninth Development Plan indicated the Kingdom's intent to raise municipal water tariffs (Ministry of Economy and Planning, 2010), the government may determine that such reforms are not politically feasible in light of recent unrest in the region.

While raising tariffs is a key component of demand management strategies, subsidies may still need to remain in place to help the poor. Given the important role that Islam plays in Saudi Arabia, this may be one way that the historic 'right of thirst' can continue to be respected. In addition, although significant improvements in water network management have been achieved over the past two decades, a proportion of the population is still forced to rely on water provided by tanker trucks (Ministry of Economy and Planning, 2010). Because such informal, private sources of water are more expensive than the subsidized networked water, the government also needs to consider how to expand network access and reduce this disproportionate impact on the poor.

Beyond urban household use, there are also substantial prospects for reducing industrial water withdrawals by increasing wastewater collection, treatment and reuse. In the petrochemical sector, for example, the institution of water conservation and reuse measures in oil and gas plants managed by the national oil company, Saudi Aramco, have the potential to reduce industrial water withdrawals by as much as 29 percent (Kajenthira et al., 2012). Financial support from the Saudi government (for example, loans, loan guarantees, tax credits) for renovations and process improvements to increase water efficiency could encourage water reclamation and reuse in this sector (Kajenthira et al., 2012).

In short, despite recent reforms in Saudi Arabia's water sector (such as restructuring the ministry framework to facilitate private sector involvement), more significant demand management measures are needed to create a sustainable and comprehensive national water management plan.

7.6 CONCLUSION

The chapter has explored the history of water governance in the MENA region, examining the historical basis for the centralized water management structure and the influence of Islam. After reviewing the different sources of water supply in MENA, the chapter analyzed more recent trends towards water demand management, focusing on tariffs, decentralization and private sector involvement. It then highlighted the importance of considering the role of agricultural water in developing an effective urban water reform strategy.

The chapter also presented a case study of Saudi Arabia. As a hyper-arid country with significant oil wealth, Saudi Arabia has been able to pursue a supply-oriented policy, with heavy reliance on desalination for municipal and industrial use. Although the Kingdom has implemented numerous reforms, such as restructuring its ministries to better engage in public–private partnerships, it has not addressed several issues that are critical to the long-term sustainability of its water management policy, such as curbing demand for water by the agricultural sector. In addition, it has been reluctant to engage in tariff reform that would improve the long-term financial sustainability of its water utilities. Moving forward, the Kingdom would benefit from consideration of a more holistic strategy for water management.

As a whole, water governance in MENA has made important strides in recent years to adopt reforms that involve non-state actors, promote efficiency and improve conservation. The challenge moving forward is to ensure that frameworks in place not only meet demand, but are also sustainable in the long term.

NOTES

1. The three main schools within Islam (Sunni, Shi'ite and Ibadite) all have slightly different approaches (see Jagannathan et al., 2009).
2. The Falkenmark Water Stress Indicator, which quantifies the fraction of total annual runoff available for human use, identifies a threshold of 1700 m^3/person/year of renewable water as being necessary to meet household, agricultural, industrial, energy and environmental needs. Countries with renewable water supplies in deficit of this figure are

said to be under 'water stress'. Countries with renewable water supplies below 1000 m^3/capita/year are 'water scarce', while those with less than 500 m^3/capita/year are said to be experiencing 'absolute water scarcity' (see Falkenmark, 1989).
3. It is a five-phase project originally designed to generate electricity and desalinate water near the Jordanian coast at Aqaba, distribute it throughout the country and also replenish the Dead Sea (see World Bank, 2007, p. 38).
4. This reference cites GWP (2000) and Rogers and Hall (2003).

REFERENCES

AFED (2009), *Arab Environment: Climate Change. Impact of Climate Change on Arab Countries*, M.K. Tolba and N.W. Saab (eds), Reports of the Arab Forum for Environment and Development, Beirut: Arab Forum for Environment and Development.

AFED (2010), *Arab Environment: Water. Sustainable Management of a Scarce Resource*, M. El-Ashry, N. Saab and B. Zeitoon (eds), Reports of the Arab Forum for Environment and Development, Beirut: Arab Forum for Environment and Development.

Al-Hussayen, A. (2009), 'Inaugural speech to Saudi Water and Power Forum', Jeddah: Saudi Water and Power Forum.

Al-Ibrahim, A. (1991), 'Excessive use of groundwater resources in Saudi Arabia: impacts and policy options', *Ambio*, **20**, 34–7.

Allan, J.A. (2001), *The Middle East Water Question: Hydropolitics and the Global Economy*, London: I.B. Tauris.

Angelakis, A.N., M.H.F. Marecos Do Monte, L. Bontoux and T. Asano (1999), 'The status of wastewater reuse practice in the Mediterranean basin: need for guidelines', *Water Research*, **33**, 2201–17.

Asano, T., F.L. Burton, H.L. Leverenz, R. Tsuchihashi and G. Tchobanoglous (2007), *Water Reuse: Issues, Technologies, and Applications*, New York, McGraw-Hill.

Bakker, K. (2010), *Privatizing Water: Governance Failure and the World's Urban Water Crisis*, Ithaca, NY: Cornell University Press.

Beaumont, P. (2002), 'Water policies for the Middle East in the twenty-first century: the new economic realities', *International Journal of Water Resources Development*, **18**, 315–34.

Bretton Woods Project (2011), 'Red-Dead sea plan in hot water', available at http://www.brettonwoodsproject.org/art-568924 (accesed 27 April 2012).

Caponera, D.A. and M. Nanni (2007), *Principles of Water Law and Administration: National and International*, London: Taylor & Francis.

Central Intelligence Agency (CIA) (2011), 'The World Factbook', 5 October 2009 edn.

Civic, M.A. (1998), 'A comparative analysis of the Israeli and Arab water law traditions and insights for modern water sharing agreements', *Denver Journal of International Law and Policy*, **26**, 437–62.

de Albuquerque, C. (2012), 'On the right track: good practices in realising the rights to water and sanitation', UN Special Rapporteur on the Human Right to Safe Drinking Water and Sanitation, available at http://www.ohchr.org/Documents/Issues/Water/BookonGoodPractices_en.pdf.

Desal Data (2010), 'DesalData.com', Global Water Intelligence and Water Desalination Report.
Dolatyar, M. and T.S. Gray (2000), *Water Politics in the Middle East: A Context for Conflict or Cooperation?*, New York: Palgrave Macmillan.
The Economist (2010), 'Economist Intelligence Unit', Economist Intelligence Unit Limited.
Falkenmark, M. (1989), 'The massive water scarcity threatening Africa – why isn't it being addressed', *Ambio*, **18**, 112–18.
FAO (2008), 'Irrigation in the Middle East region in figures: AQUASTAT Survey – 2008', FAO Water Reports. Rome: Food and Agriculture Organization of the United Nations.
FAO (2010), 'Saudi Arabia – AQUASTAT', Rome: Food and Agriculture Organization of the United Nations.
FAO (2011), 'FAO Aquastat', Rome: Food and Agriculture Organization of the United Nations.
Faruqui, N.I., A.K. Biswas, and M.J. Bino (eds) (2001), *Water Management in Islam*, New York: United Nations University Press.
Gryzbowski, A., S. Mccaffrey and R. Paisley (2010), 'Beyond international water law: successfully negotiating mutual gains agreements for international watercourses', *Pacific McGeorge Global Business and Development Law Journal*, **22**, 139–54.
Gude, V.G., N. Nirmalakhandan and S. Deng (2010), 'Renewable and sustainable approaches for desalination', *Renewable and Sustainable Energy Reviews*, **14**, 2641–54.
GWI (2010a), 'PPP Tracker', Global Water Intelligence.
GWI (2010b), 'Tariff Survey', Global Water Intelligence.
GWP (2000), 'Towards water security: a framework for action. Stockholm', Global Water Partnership.
Hamoda, M.F. (2004), 'Water strategies and potential of water reuse in the south Mediterranean countries', *Desalination*, **165**, 31–41.
Hughes, R.A. (2009), 'Pro-justice ethics, water scarcity, human rights', *Journal of Law and Religion*, **25**, 521.
Jagannathan, N.V., A.S. Mohamed, and A. Kremer (eds) (2009), 'Water in the Arab World: management perspectives and innovations', available at http://siteresources.worldbank.org/INTMENA/Resources/Water_Arab_World_full.pdf (accessed 19 November 2010).
Jones, T.C. (2010), *Desert Kingdom: How Oil and Water Forged Modern Saudi Arabia*, Cambridge, MA: Harvard University Press.
Kajenthira, A., L. Diaz Anadon and A. Siddiqui (2012), 'A new case for promoting wastewater reuse in Saudi Arabia: bringing energy into the water equation', *Journal of Environmental Management*, **102**, 1–9.
Kally, E. (1993), *Water and Peace – Water Resources and the Arab-Israeli Peace Process*, Westport, CT: Praeger.
Kirkpatrick, C., D. Parker and Y.-F. Zhang (2004), 'State versus private sector provision of water services in Africa: a statistical, DEA, and stochastic cost frontier analysis', Working Paper No. 70, University of Manchester.
Kirkpatrick, C., D. Parker, and Y.-F. Zhang (2006), 'An empirical analysis of state and private-sector provision of water services in Africa', *World Bank Economic Review*, **20**(1), 143–63.

Lippman, T. (2010), 'Saudi Arabia's quest for food security', *Middle East Policy*, **17**, 90–98.
Llamas, R. and E. Custudio (eds) (2003), *Intensive Use of Groundwater: Challenges and Opportunities*, Lisse, The Netherlands: A.A. Balkema Publishers.
Magiera, P., S. Taha and L. Nolte (2006), 'Water demand management in the Middle East and North Africa', *Management of Environmental Quality: An International Journal*, **17**, 289–98.
Mallat, C. (1995), 'The quest for water use principles: reflections on Shari'a and custom in the Middle East', in J.A. Allan and Chibli Mallat (eds), *Water in the Middle East: Legal, Political and Commercial Implications*, New York: I.B. Taurius, pp. 127–37.
Marin, P. (2009), *Public–Private Partnerships for Urban Water Utilities: A Review of Experiences in Developing Countries*, Washington, DC: World Bank.
Ministry of Economy and Planning (2010), 'The Ninth Development Plan (2010–2014)', Riyadh, Kingdom of Saudi Arabia.
O'Reagan, J. (2009), 'Case CCT 39/09', Johannesburg.
Pérard, E. (2008), 'Private sector participation and regulatory reform in water supply: the southern Mediterranean experience', Development Centre Working Papers, OECD Development Centre, Paris.
Prasad, N. (2006), 'Privatization results: private sector participation in water services after 15 years', *Development Policy Review*, **24**, 669–92.
Qadir, M., A. Bahri, T. Sato and E. Al-Karadsheh (2010), 'Wastewater production, treatment, and irrigation in Middle East and North Africa', *Irrigation and Drainage Systems*, **24**, 37–51.
Rogers, P., R. De Silva and R. Bhatia (2002), 'Water is an economic good: how to use prices to promote equity, efficiency, and sustainability', *Water Policy*, **4**, 1–17.
Rogers, P. and A. Hall (2003), 'Effective water governance', Background Paper No. 7, Stockholm: Global Water Partnership Technical Committee.
Sadoff, C.W. and D. Grey (2005), 'Cooperation on international rivers: a continuum for securing and sharing benefits', *Water International*, **39**, 420–27.
Saline Water Conversion Corporation (SWCC) (2010), 'Working plants', available at http://www.swcc.gov.sa/default.asp?pid=96 (accessed 4 November 2010).
Sowers, J., A. Vengosh and E. Weinthal (2011), 'Climate change, water resources, and the politics of adaptation in the Middle East and North Africa', *Climatic Change*, **104**, 599–627.
Tolba, M. and N. Saab (2006), 'Arab public opinion and the environment conference report of 18-country survey', available at http://www.afedonline.org/en/inner.aspx?contentID=88 (accessed 4 November 2010).
UN ESC (2002), Committee on Economic, Social and Cultural Rights, 'Substantive issues arising in the implementation of the International Covenant on Economic, Social and Cultural Rights; General Comment No. 15, The right to water (arts. 11 and 12 of the International Covenant on Economic, Social and Cultural Rights, U.N. Doc. E/C.12/2002/22)', United Nations Economic and Social Council.
UNHRC (2010), 'Report of the independent expert on the issue of human rights obligations related to access to safe drinking water and sanitation, Catarina de Albuquerque', United Nations Human Rights Council A/HRC/15/31, available at http://www2.ohchr.org/english/issues/water/iexpert/docs/A-HRC-15-31-AEV.pdf.

UPI.com (2008), 'Sadr City residents get potable water', 4 December 2008, available at http://www.upi.com/Top_News/Special/2008/12/04/Sadr-City-residents-get-potable-water/UPI-23281228420503/ (accessed 4 November 2010).
US EIA (2011), 'Saudi Arabia energy data, statistics and analysis – oil, gas, electricity, coal', in *Country Analysis Briefs*, Washington, DC: US Energy Information Administration.
US–SABC (2009), *The Water Sector in the Kingdom of Saudi Arabia*, Vienna: US–Saudi Arabian Business Council.
US–SABC (2010), 'Public–private agreements upgrade Saudi water infrastructure', US–Saudi Business Brief, Vienna: US–Saudi Arabia Business Council.
World Bank (2007), *Making the Most of Scarcity: Accountability for Better Water Management Results in the Middle East and North Africa*, Washington, DC: World Bank.
World Bank (2008), 'Urban Challenges in the MENA Region', available at http://go.worldbank.org/AGTL4WQE40 (accessed 19 November 2010).
World Bank (2009), 'Water in the Arab world: management perspectives and innovations', available at http://siteresources.worldbank.org/INTMENA/Resources/Water_Arab_World_full.pdf (accessed 19 November 2010).
World Bank (2011), 'Middle East and North Africa: Countries', available at http://go.worldbank.org/7UEP77ZCB0 (accessed 28 September 2011).
Zhou, Y. and R. Tol (2005), 'Evaluating the costs of desalination and water transport', *Water Resources Research*, **41**, W03003.

8. Implementing PPP programs in the urban water and sanitation sector: some insights from the Indian experience in selected states

Ashwin Mahalingam

8.1 INTRODUCTION

The Indian Planning Commission's draft approach paper to the 12th Five Year Plan (2012–17) notes that India's urban population has grown from 285 million in 2001 to 380 million in 2011 (Planning Commission, 2011). Urban infrastructure will need to be augmented considerably in order to accommodate this growing population. In particular, considerable investment must be undertaken to rehabilitate, create and operate water supply and sanitation systems due to the critical role that these systems play in promoting healthy and productive lifestyles, and spurring economic growth. The 11th Five Year Plan (2007–12) pointed out that 27 percent of the population in Class I cities (cities with population greater than 100 000) did not have direct access to drinking water, while for cities with population lower than 20 000 this figure increased to 42 percent (Planning Commission, 2008). Only 63 percent of urban residents have access to sewerage and sanitation facilities. Based on these figures, the 11th Five Year Plan estimated an outlay of nearly US$28 billion for water and sanitation projects over the plan period (Planning Commission, 2008). These numbers are expected to rise for the 12th Five Year Plan period (2012–17). Executing and financing such large volumes of projects poses a major challenge and it is unclear whether government departments possess the technical, human and fiscal resources to meet this challenge on their own. In response to this quandary, the draft approach paper to the 12th Five Year Plan further goes on to note that 'A massive push is needed to attract private investment . . . for drinking water supply, waste water recycling, treatment of MSW (Municipal Solid Waste) and treatment of urban sewerage' (Planning Commission, 2011). Public–private partnerships (PPPs) are therefore being conceptualized as a key tool for the delivery of urban water and sanitation services in India.

The Department of Economic Affairs (DEA) housed within the Ministry of Finance, Government of India, has been tasked with crafting the PPP agenda in India across a variety of infrastructure sectors including water and sanitation. The DEA has set in place a series of measures aimed at assisting various government departments in conceptualizing and executing PPP projects. A draft national PPP policy has been circulated that defines the process of awarding PPPs and the enabling frameworks and institutional mechanisms that are present within the Indian administrative system (Department of Economic Affairs, 2011). The DEA and Planning Commission provide a series of toolkits, manuals, guidelines and model concession agreement templates on their websites that departments can use as a basis to create their project agreements. The DEA, in partnership with the Asian Development Bank (ADB), has also created PPP cells in several Indian states. These PPP cells are staffed by PPP experts from the ADB, who work alongside state government representatives to identify and award PPP projects, provide policy and documentation support, and conduct capacity building and training programs. The DEA has recognized that various government departments might not have the capacity to draft PPP concession agreements and monitor the project award process. While PPP cells can certainly help in this regard, the DEA has empanelled a set of Transaction Advisers who can be directly and quickly selected (without going through a competitive bidding process), who in turn can assist government departments in structuring PPP projects (Department of Economic Affairs, 2008a). In order to finance the project development phase, the DEA has instituted the Indian Infrastructure Project Development Fund (IIPDF) to provide funds to meet project development expenses such as the cost of hiring Transaction Advisers (Department of Economic Affairs, 2008b). In addition, a Viability Gap Fund (VGF) has been constituted that can provide grants to projects (up to a maximum of 40 percent of project costs) in order to encourage private sector investment and participation in project delivery (Department of Economic Affairs, 2008c). These and other measures are a part of the institutional fabric that supports the delivery of infrastructure projects through PPPs in India.

With specific reference to the water and sanitation sector, the Ministry of Urban Development (MoUD), Government of India, has constituted the Jawaharlal Nehru National Urban Renewal Mission (JNNURM) to grant partial funding for urban infrastructure creation (including the construction of water and sanitation systems) in selected cities (Ministry of Urban Development, 2006). JNNURM explicitly encourages cities to seek recourse to private investment and participation in order to bridge the gap between project costs and funds devolved through JNNURM. In addition, the MoUD is also in the process of creating toolkits and templates for PPPs in the water and sanitation sector and has established 'Centers of Excellence' to assist urban local bodies (ULBs) in identifying, structuring and awarding PPPs.

All of these frameworks and institutions will enable 'PPP-able' projects in the Indian water and sanitation sector to proceed from the project conceptualization to the project award stage. However, amidst all of these efforts, very little discussion has been undertaken on two issues:

1. What factors would constitute an ideal enabling ambient environment for PPPs? Are the current measures taken by DEA and other departments sufficient to encourage widespread adoption and success of PPP programs? What potential interventions have thus far been ignored?
2. Given the existing institutional environment for water and sanitation PPPs in India, when should PPPs be undertaken? What kind of PPP models should be adopted, and under what circumstances?

This chapter aims to address these two questions. The first question assesses the actual strength of the institutional environment for PPPs in the water and sanitation sector in India. The second question seeks to unravel the variables that influence the decision on which PPP model to adopt. Currently, the rationale for the selection of, say, a build–operate–transfer (BOT) model or an operation and maintenance model over a public delivery model for a given project is unclear. Furthermore, this question explores the interplay between mode choice, project characteristics and the institutional environment. The rest of the chapter is organized as follows. The next section summarizes a research study that identifies the various factors that constitute an enabling environment for PPP projects. Following this, two states in India – Tamil Nadu and Karnataka – are selected and the strength of their institutional environments for water and sanitation PPPs is assessed based on the framework described previously. Then the chapter moves on to the second research question and the literature on public versus private delivery choice is summarized to arrive at a set of guidelines that influence when PPPs should be considered over public sector delivery, as well as the choice of PPP modes. Following this, case studies from the state of Tamil Nadu are presented to illustrate how wrongly chosen or wrongly structured PPPs run into problems, whereas PPPs that 'fit' within the existing institutional environment are more likely to succeed. The final section concludes by linking rational decision-making on PPPs with the strength of the institutional environment for PPPs, and provides some insights that may be useful to policymakers in the developing country context.

8.2 CHARACTERISTICS OF A ROBUST AMBIENT ENVIRONMENT FOR PPPS

A recent study undertaken by IIT-Madras and the Center for Development Finance (CDF), Madras, led by the author, identified a series of 12 capabilities that constitute a robust institutional environment for PPPs (IIT-Madras, 2011). The study combined a survey of relevant theoretical and empirical literature on institutions and processes that enable PPPs, interviews with practitioners from the public and private sectors, and observation of PPP projects and programs undertaken (some of which were and some of which were not delivering their intended benefits) under varied institutional conditions. Data collected was carefully coded, and compared in order to understand the underlying institutional factors that contribute to successful PPP programs. Twelve 'capabilities' were identified as a result, along with strategies to build them. Neither the capabilities nor the accompanying strategies are prescriptive or normative in nature. Rather, each capability is a construct represented by a series of options. While government departments wishing to undertake PPPs should possess each of these capabilities, their approach to implementing these capabilities can vary, and they can pick options best suited to their environments. The 12 identified capabilities are: (1) the presence of a mature rationale for PPPs; (2) the presence of legitimacy for PPPs; (3) political willingness for PPPs; (4) transparent, credible and equitable processes for project award; (5) ensuring public sector commitment to decisions; (6) ensuring private sector commitment to decisions; (7) the capability to identify, structure and award projects; (8) the capability to govern projects; (9) private sector capacity to finance and execute projects; (10) financial incentives; (11) risk management frameworks; and (12) frameworks to create stakeholder awareness and acceptance towards PPPs. Each of these capabilities is briefly explained below.

First and foremost, the need and rationale for PPPs should be clearly articulated. Traditionally, the lack of financial resources within the public sector has been cited as the driving factor behind the decision to turn towards the private sector for the provision of infrastructure services. However, this rationale is flawed since the private sector's cost of borrowing is almost always higher than that of the public sector. Therefore, unless the private sector can provide greater value for money (VfM) invested in the project, a PPP project is likely to be a more expensive proposition as compared to public sector delivery. Efficiencies in construction and operation brought in by the private sector and not fiscal stress should be the guiding rationale behind selecting PPPs. Next, developing projects through PPPs should be seen as a legitimate mode of project delivery. This will persuade public officials to enter into PPP agreements without fear of enquiry from anti-corruption and independent audit agencies. Similarly, the private sector will also derive confidence from the legitimacy of the arrangement, since they are assured of legal recourse to protect their interests. This can ensure better participation from the private sector. Governments can adopt a variety of measures to create legitimacy around PPPs. PPP-enabling legislation can of course be enacted, enshrining the functioning of PPPs within law. However, governments can also issue PPP policy statements describing the role and functioning of PPPs, or enact executive orders for specific PPP initiatives. Third, political willingness to implement projects through PPPs is crucial to the success of PPP programs since elected representatives are key decision-makers in the project approval process. Very often myths and uncertainties surrounding PPPs (for example, questions such as whether the public will actually be willing to pay for a service) lead to a reluctance to enter into PPPs even when they are the most economically viable alternative. Political willingness to consider PPPs must therefore be carefully nurtured and cultivated through training and mentoring programs, publicizing of success stories and so on.

Although legitimacy can be viewed as a necessary precondition for effective private sector participation, it is not sufficient. The predictability of the tendering process, the fairness of the decision-making criteria, the transparency of the bid criteria and award, and the fairness of the contract with regard to risk allocation are other considerations that determine effective private sector participation and competition. This represents the fourth capability that should be present in an ambient environment for PPPs. Enabling the speedy issuance of clearances, land and building permits, as well as the creation of model concession agreements that have fair and standardized risk allocation clauses, are some strategies that can strengthen this capability. The existence of legislation that ensures transparency in procurement can also help provide a fillip to private sector participation in PPPs. The fifth and sixth capabilities relate to strategies that can ensure the prevention of two chief failures that are inherent to PPPs – government expropriation and private sector hold-up. Measures must be present in the government's administrative environment that can ensure that the public sector will honor and fulfill its obligations with regard to contractual preconditions and payments. A speedy, independent and effective judiciary can perform this function. Sector-specific or cross-sectoral regulators and project-level dispute resolution boards are other strategies through which the capability to ensure public sector commitment to decisions can be strengthened. Simultaneously, PPPs are long-term transactions and replacing the private operator is often costly. Given the monopolistic nature of most PPPs, there exists the risk that the private sector, once entrenched, may focus on furthering its profit motive at the expense of the societal obligations of the project. The capability to rein in private sector opportunism must therefore be present. Once again the judicial or regulatory regimes can perform this function. In addition, strategies such as public sector shareholding in project vehicles can also ensure that private sector behavior is held accountable.

Capacity building is an oft-uttered term when PPPs are discussed in the context of developing countries that have traditionally engaged in public delivery for the procurement of infrastructure services. This forms the foundation for the next three capabilities in the framework. The public sector should have the capacity to identify project opportunities that can be viably executed through PPPs. Following this, public agencies should also have the capability to perform initial financial and economic analysis, to structure the terms of the transaction, prepare the concession agreement, and manage the bid process that will result in the selection of a concessionaire. These skills are fundamentally different from the skill set required to procure construction services and as a result may not be present across government departments. Therefore the capability to identify and award projects must be built either through training government officers, or through outsourcing these competencies to trained consultants who can advise governments throughout the procurement process. In addition to the capability to structure projects, government departments should also have the capability to monitor and govern projects over their lifecycle. Contract management is a key skill in this regard. However, it must be noted that PPPs are long-term contracts executed under conditions of uncertainty. These contracts are therefore necessarily incomplete (Hart, 2003). The management of PPP contracts is thus not restricted to the mere oversight of whether obligations are being executed. Rather, unfolding eventualities may lead to the need to renegotiate contractual terms and to manage the contractor–client relationship (Klijn, 2010). The public sponsor of a project must therefore be able to deal with such *ex post* uncertainties and govern PPP projects through turbulences. Training as well as flexibility in negotiating or renegotiating agreements as part of the PPP law or policy can help strengthen this capability. In addition, while the above discussion focuses on public sector capacity, augmenting the private sector's capabilities is no less important. Private sector firms need to integrate the capability to design, build and operate infrastructure

– skill sets that are typically separated due to the fragmented nature of the building and construction industry. Furthermore, the private sector also needs to have the bankability to finance infrastructure development. Governments therefore need to proactively provide incentives and measures to help private firms develop the capacity to execute complex long-term infrastructure transactions. Otherwise, the outcomes of a PPP may be far inferior to traditional public procurement.

The tenth capability is the creation of financial incentives that will encourage private participation. Schemes such as the VGF and IIPDF programs in India mentioned earlier are steps in this direction. Grants can allow socially desirable but financially unviable PPPs to be executed, while also lowering project risks and thereby the private sector's premium. A related capability is the presence of risk mitigation tools in an environment. Credit guarantee funds at the state or national level that can alleviate the effects of payment defaults by a department are likely to instill confidence in the private sector, lower their risk premiums and ensure a more competitive bidding environment, leading to better price discovery. The twelfth and final capability that should be present in an enabling ambient environment is awareness amongst project stakeholders and the general public about the need for and the mechanisms of PPPs. Very often public consultations are bypassed and social protests result from stakeholders who are unable to see value in the project, object to raised tariffs and charge the government with having 'sold the family silver' or with having abdicated its responsibilities to provide services to the taxpayers. While some of these arguments can serve to check indiscriminate privatization and the introduction of crony-capitalism, many of them are often ideological and misguided due to the lack of awareness of the need for a PPP, and the resulting efficiencies. Campaigns sensitizing stakeholders to the notion and advantages of PPPs can help overcome such criticisms and lead to a stable environment for PPPs.

Taken together, these 12 capabilities constitute a framework that describes the various characteristics that must be present within an ambient environment to foster and sustain PPPs. This framework can be used as a guide by governments in evaluating the strength of their environment for PPPs in the water and sanitation sector, and as a tool to decide on policy interventions that will augment the ambient environment for PPPs. The stronger these capabilities are, the more likely that a large number of PPPs can be successfully awarded. To illustrate the use of this framework for water and sanitation PPPs, the next section demonstrates its application to two Indian states – Tamil Nadu and Karnataka. Both states feature among the top quartile of Indian states in terms of economic growth and demographic indicators, making them more likely to adopt

PPPs to develop water and sanitation infrastructure. Public and private officials from these states were interviewed, and several projects were visited in order to obtain the data upon which this analysis was performed.

8.2.1 Institutional Environment for Water and Sanitation PPPs in Tamil Nadu

Assessing Tamil Nadu's institutional environment for water and sanitation projects using the framework described above indicates that the ambient environment for PPPs is quite weak. The rationale for PPPs in the water and sanitation sector in Tamil Nadu appears to be a combination of the technical expertise of the private sector, consequent efficiency gains and cost reductions as well as access to external capital. Hence the rationale for PPPs appears to be quite mature in the state as it emphasizes efficiency considerations as the driving force behind PPPs, as opposed to a need for finances.

There are no policies or laws that govern the use of PPPs in the water and sanitation sector in Tamil Nadu. Also, in general, the political and bureaucratic willingness to undertake PPP projects in the water and sanitation sector appears to be relatively low, especially in undertaking large-scale BOT model projects. Individual project champions are often the drivers of such projects. The process of issuing project bids and contract awards is fairly transparent in this sector since it is governed by the Tamil Nadu Transparency in Tenders Act 1998, which mandates competitive bidding for public procurement of any goods or services. However, predictability and equity remain significant concerns.

The institutional framework in this sector comprises a number of public sector departments, and processes followed at the pre-award and post-award phases are entirely at the discretion of each of these agencies. Further, model concession agreements do not exist, nor are there risk management frameworks, nor effective dispute resolution mechanisms. There are no specific institutional forms or measures to prevent or counter issues arising from public sector expropriation or private sector hold-up. Regulation is done solely through the contract. Performance considerations are also not imposed on the private contractors in many operation and maintenance projects. In terms of capacity to identify and structure projects, institutions such as the Chennai Metropolitan Water Supply and Sanitation Board (CMWSSB) and the Tamil Nadu Water and Drainage Board (TWAD) have their own internal training cells where some training is imparted on select project delivery modes. However, the sector does not have specific guidelines for identifying and selecting PPPs. Due to these shortcomings, state institutions in this sector often do not undertake PPP projects. The sector also lacks the capacity for carrying out the governance of certain kinds of PPP projects such as build–operate–transfer (BOT). More often, line departments do not have the capacity to clearly define project risks and performance parameters and hence are wary of undertaking complex PPP projects. Line departments do not have any well-defined, standardized set of performance indicators to measure contractor performance and monitoring of contracts is largely absent. Private sector capacity is also lacking. Larger companies do not take up projects below INR50 crores (US$10 million) while the capacities of smaller players are not noteworthy. There are no specific risk mitigation mechanisms outside those stated in the concession agreement that exist in this sector in the state. No specific funding incentives to the water and sanitation sector exist. Finally, beyond a handful of workshops, very little has been done in the state to create awareness amongst stakeholders on the concept of PPPs (IIT Madras, 2011).

Not surprisingly, relatively few PPPs have been undertaken in the water and sanitation sector in Tamil Nadu. A water supply project in the city of Tirupur and a sewerage PPP in the town of Alandur were some of the earliest PPPs to be attempted in India. Both of these projects were implemented through the BOT model. As will be discussed later, these projects faced considerable challenges post-award. However, beyond these initial experiments, few PPP projects have been undertaken outside of service contracts for the operation of pumping stations and the like. Recently, the BOT model was adopted for building a desalination plant in Minjur for bulk water supply to the CMWSSB. The CMWSSB guarantees an off-take from the desalination plant at an agreed-upon cost. For a water treatment plant in Chembarambakkam, the CMWSSB decided to explore the design–build–operate (DBO) model to avail of this model's inherent benefits of cost reductions and newer technologies. This DBO model will be discussed in detail later and is essentially a combination of an engineer–procure–construct (EPC) and an operation and maintenance (O&M) contract, whereby a private operator is first asked to design, build and then operate a facility for a period of time. However, the operator does not bear demand risks. Further, construction costs are recouped upon completion of the construction of the asset. Following this, the private operator then operates the asset and is paid periodically on the basis of performance. The CMWSSB and TWAD have implemented a few other projects based on this model, which is in essence a hybrid variety of PPP that combines aspects of the BOT model with those of EPC contracts.

8.2.2 Institutional Environment for Water and Sanitation PPPs in Karnataka

In comparison with Tamil Nadu, the ambient environment for PPPs in Karnataka appears more robust, although it is not without deficiencies. The rationale for PPPs in the water and sanitation sector in Karnataka partially rests on the need to augment financial capabilities to execute projects. However, operational efficiencies and innovations brought in by the private sector seem to be the primary drivers behind PPPs thus far, leading to a relatively mature rationale for PPPs.

The water and sanitation sector falls under the purview of the Infrastructure Policy brought out by the state of Karnataka. The policy's advocacy of PPPs therefore also influences and encourages PPPs to be undertaken in this sector across the state. In addition, Karnataka has brought out a Karnataka Urban Drinking Water and Sanitation Policy in 2002, which specifically states that PPPs and private sector participation should be encouraged in the pursuit of achieving sectoral goals. The political and bureaucratic willingness to undertake PPP projects in the water and sanitation sector appears to be relatively low except in certain pockets in the state. PPPs in the water and sanitation sector in Karnataka have been based on the Karnataka Transparency in Public Procurements Act 1999. The process of calling for bids and award of the contract have by and large been transparent. Both the Infrastructure Policy and the Karnataka Urban Drinking Water and Sanitation Policy have not been enacted as legislation and as a result, there is no separate framework for enforcement of public sector commitment. Traditional dispute resolution measures are followed, and no specialized structures have been set up for PPPs. As a result of this, private operators have often felt the need to add a risk premium on to their bids to price in the lack of this capability. Also, no specific measures exist to combat the private sector hold-up problem. Despite this, PPPs in Karnataka have not reported significant reneging by either party. The Karnataka Urban Water Supply and Drainage Board (KUWSDB) and the Bangalore Water Supply and Sewerage Board (BWSSB) are some of the key government departments involved in the water and sanitation sector. The Government of Karnataka possesses a well-staffed and vibrant PPP cell that augments the capacity of these government departments to structure water and sanitation PPPs. Several training programs on PPPs have also been conducted for the staff from these departments.

However, there appears to be a lack of project governance capacity within the state and some projects have run into trouble post-award. Private sector capacity to finance and execute projects has not appeared to be a bottleneck in this sector. Most of the projects that have been undertaken in Karnataka have attracted a reasonable number of bidders. Several international firms such as Suez, Veolia and Degremont have entered the Indian market and have a strategic focus on Karnataka. There are no specific risk mitigation mechanisms that exist at the sectoral level for water and sanitation in Karnataka. Most risks are mitigated or allocated through the concession agreements. Karnataka also does not have separate PPP-focused financing or funding mechanisms in the water and sanitation sector. However, several schemes exist that promote or accept the use of PPPs, and which provide funds for project implementation and payment to private operators. The Karnataka Urban Water and Sanitation Improvement Program (KUWASIP) funded by the World Bank is one such example. Finally, Karnataka has undertaken some efforts to raise awareness about PPPs. The CII Water Summit in 2009 was attended by government and private sector representatives and attempted to promote awareness on PPPs in the water sector. The Karnataka Urban Infrastructure Development and Finance Corporation (KUIDFC) engaged in project-specific public consultations as part of the KUWASIP project in 2003 and held a series of city-level consultation workshops. Some public consultations with non-governmental organizations (NGOs) and government representatives have also been undertaken (IIT-Madras, 2011).

Not surprisingly, Karnataka's experience with PPPs in the water and sanitation sector has been more numerous and vibrant as compared to Tamil Nadu. Three key sets of projects underlie Karnataka's experiences with PPPs in this sector thus far. The first are a set of PPPs to deliver 24×7 water supply in three towns in Northern Karnataka: Hubli Dharwad, Belgaum and Gulbarga. The second is a similar attempt to deliver 24×7 water supply in Mysore. Third, a few PPPs have been attempted in the city of Bangalore, including PPP projects in the wastewater treatment sector. In 1996, a 50 MLD recycling plant was developed in Bangalore through a PPP. Other effluent treatment plants in Yelehanka and Vrishabhavathi Valley have also been developed on similar models. Another PPP project that has been implemented was part of a water supply project to lift water from the River Cauvery for supply to the city of Bangalore. The BMWSSB has had more success in terms of smaller projects, particularly in the water treatment and sanitation sectors. A management contract was awarded for water supply in some peripheral areas of Bangalore, wherein the BMWSSB supplied bulk water that was distributed and billed for by a private operator. Some sewage treatment plants have been set up across the city on a design–build–operate–transfer mode. For instance, a 60 MLD plant has been set up by Degremont in TK Halli. Besides these, a few other PPP projects in bulk water supply, operation and maintenance of distribution networks and so on are being envisaged and developed. Project reports are being prepared for the provision of 24×7 water supply systems to coastal towns such as Mangalore, Udipi, Puttur and Kundapura. Bulk water supply through PPPs is being contemplated in Bijapur and Chitradurga and so on. Most PPP projects that have been undertaken in Karnataka thus far in the water and sanitation sector feature operation and maintenance types of agreements, as opposed to concessions or affermage types of contracts.

This discussion indicates that the more robust the institutional environment for PPPs in a region, the likelier it is that larger numbers of PPP projects can be awarded and will flourish. The next section addresses the second research objective of this chapter: under what circumstances should PPPs be adopted in the water and sanitation sector in India?

8.3 PUBLIC VS PRIVATE DELIVERY CHOICE

Vives et al. (2006) provide a framework with which the optimal modality of project delivery for water and sanitation projects can be assessed. The framework consists of eight primary factors: (1) the presence of a legal framework for PPPs; (2) the presence of political willingness (or a lack of political risk) to undertake PPPs; (3) the presence or absence of fiscal strength within a government organization to undertake projects themselves; (4) macroeconomic factors and economic stability that will influence project costs and revenues in the future; (5) institutional capacity to execute PPPs; (6) willingness to pay for services; (7) sustainability or affordability of tariffs charged; and (8) the size and profitability of the project. A simple result that emerges from this framework is that when weak legal frameworks are present, political willingness is lacking, governments are fiscally able to develop their own infrastructure, macroeconomic factors are unstable, institutional capacity and willingness to pay are poor, tariffs are not sustainable and project sizes are small, projects are best undertaken directly by the public sector. However, given different configurations of these factors and the presence of other risk mitigation tools, various forms of PPPs from management contracts to leases to concessions can be considered as viable project delivery modes. Vives et al.'s (2006) framework provides a matrix of modalities versus project conditions to help decision-makers choose between project delivery options. Sinha (2009) and Manohar (2010) augmented this framework by adding factors such as government stability, political willingness to charge tariffs, trust between the public and private sectors, and so on, and assigned weights to these factors. Using this system, one is able to arrive at a 'score' for each project delivery mode for a given project, and objectively select the mode with the greatest score.

Although these frameworks are helpful in providing practitioners with easy-to-use tools for delivery mode selection, they are limited by the fact that rigorous empirical validation of these frameworks has not been carried out, theoretical grounding for these frameworks is absent and the data required to use these frameworks may either be difficult to gather, or may require significant preliminary analysis before a decision on the mode of project delivery is possible. The theoretical literature on infrastructure projects provides some insight into when PPPs should be attempted. There are two main approaches in this area: the first originates from the literature on project management, while the other draws upon economic theory. The literature on project delivery and management relates the complexity of a project to its delivery mode. As Garvin and Mahalingam (2011) note, when the technology and organizational structure required to execute a project is simple and there is sufficient competition in the market, government departments can consider competitively tendering out the design and construction phases of a project separately. PPPs may not be necessary. On the other hand, as technological and organizational complexity increases, construction input into the design phase can lead to lower costs, time savings due to greater constructability of the design, and the ability to overlap the design and construction phases. In such cases, client organizations may consider bundling design and construction services into engineer–procure–construct or design–build contracts. As operational complexity increases, 'operability' of the asset will need to be built into the design and construction phases, and there is a case to be made for the integration of operations and management with design and build, leading to a design, build and operate model. Public sector clients with the capability (both financial and human-resource related) to execute such projects can consider bundled execution within the public sector domain. However, when this is not possible, PPPs wherein the private sector provides integrated design, construction and operations services emerge as an obvious alternative. Project complexity can therefore be related to the choice of project delivery mode. Iossa and Martimort (2008) and Martimort and Pouyet (2008) provide a more nuanced but similar viewpoint by noting that construction and operations can be bundled when positive externalities result (that is, when better construction leads to a lowering of O&M costs), while these contracts should be awarded separately when such externalities are not present. As argued here, positive externalities are likely to accrue in highly complex projects, thereby positioning them as candidates for PPPs.

Interestingly, conventional economic theory contradicts the above conclusion. Transaction cost economics (for example, Williamson, 1979) suggests that in cases where asset specificity is high (often the case with complex infrastructure in water and sanitation) and transaction frequency is low, neoclassical contracting or trilateral governance is a viable option. While this does not preclude the use of a PPP-type arrangement, it does indicate that robust third-party dispute resolution mechanisms must be present in order to undertake such projects via PPPs. De Bettignies and Ross (2010) review the extensive literature on public versus private delivery choice and offer a related but extended decision model for PPPs. In their model, assets that are relationship-specific and do not involve an uncertain or complex exchange environment are best undertaken through PPPs, whereas public delivery might be a more appropriate choice if the complexity of the transaction is high. If assets are not relationship-specific, then spot market solutions can prove to be efficient through the virtues of competition. The work of Hart et al. (1997) aligns with this opinion, as they show that the private sector has a propensity to reduce costs and thereby 'cut corners'. Thus, unless contractual outcomes can be clearly specified and monitored (that is, the exchange is not complex), the overall quality of the service can be lowered in the case of a PPP. There also exists empirical support for this line of argument. For instance, in his review of the Private Finance Initiative (PFI) experience in the UK, Corry (2004) notes that several PPP projects failed under conditions where specification of project outcomes was difficult. Further factors such as the strength and credibility of the private sector, the ability to structure equitable contracts, and so on, all contribute to the success of PPP projects. Therefore, in cases where project complexity is high, and where this leads to difficulties in the ability to clearly specify project outcomes that the private sector must deliver, PPPs are likely to lead to stakeholder dissatisfaction and project failure.

Other studies provide a different viewpoint. Schmidt (1996) shows that public provision provides greater allocative efficiency, whereas private provision provides greater productive efficiency. This implies that in cases where allocation and equity concerns are paramount, then public provision might be a more appropriate choice for infrastructure service delivery. A similar view is endorsed by other commentators. Besley and Ghatak (2001) submit that the more 'caring' party should own the asset. If social equity concerns are paramount, then it is likely that the public sector will be the more 'caring' party and should therefore retain control over service provision. Flinders (2010) describes how stakeholders resort to 'voice'-based responses (protests against projects) when confronted with public delivery failure and 'exit'-based responses (switching to other service providers) when confronted with private delivery failure. Since water and sanitation projects are likely to be monopolistic in nature, stakeholders may neither be able to 'exit' nor exercise 'voice' against a private provider when confronted with a PPP failure. Societal concerns are thus difficult to manage under PPPs.

Two key propositions emerge from this debate, as summarized below:

Proposition 1: When social equity considerations are paramount, projects are best kept within the public sphere.

Proposition 2: When project complexity increases, and when the public sector does not have the capacity to undertake such a project, then the strength of the institutional environment will determine the complexity of the PPP that can be adopted. Complete integration (for example, BOT contracts) may be possible in the case of complex projects enacted within strong institutional environments, while simpler forms of PPPs must be contemplated in weaker institutional environments.

The next section investigates support for the second proposition through the empirical analysis of PPP projects in the water and sanitation sector. It may be recalled that Tamil Nadu was found to be a state with a weak institutional environment for water and sanitation PPPs. The next section describes the experiences of three PPP projects: the Alandur sewerage project, the Tirupur water supply project and the Coimbatore sewerage treatment plant. The first two projects were undertaken through the BOT model, while the third was undertaken through the comparatively simpler design–build–operate (DBO) model.

8.4 CASE STUDIES IN TAMIL NADU

8.4.1 Alandur Sewerage Project

Alandur is a town in the south of India with a population of nearly 150 000. A relatively low cost of living combined with a proximity to several industrial and information technology units led to an increase in Alandur's population in the early 1990s. However, the state of sewerage infrastructure was problematic. No integrated sewerage system existed, and residents built individual septic tanks that often overflowed and resulted in health hazards. The municipality turned its attention towards building world-class sewerage infrastructure in Alandur.

An initial feasibility study estimated the cost of construction at US$10 million. The Alandur municipality's annual revenues were of the order of

US$750 000. It was therefore clear that the municipality could not finance the project. The Alandur municipality hired the Tamil Nadu Urban Development Fund (TNUDF), a public sector organization focused on helping municipalities finance and develop urban infrastructure. TNUDF suggested splitting the project into two packages. The first package would involve the construction of an integrated sewerage network, financed through user deposits as well as debt, and procured through a traditional EPC contract. In the second package a private operator would finance, build and operate the sewage treatment plant (STP) and receive regular payments directly from the municipality based on the amount of sewage that was processed, subject to certain performance standards. The municipality agreed to guarantee a minimum level of sewage flow into the STP. Furthermore, the TNUDF recommended that both packages be awarded to the same firm in order to ensure that the construction of the sewerage network would align with the development of the sewage treatment plant.

The mayor personally spent a considerable amount of time in canvassing the project to the residents. He organized meetings at street corners, met with residents' welfare associations, extolled the project's benefits through a loudspeaker as he toured the neighborhoods, and made the financial details of the project transparent so that the citizens could understand the rationale behind project costs and financing charges. This effort yielded fruit and the value of the user deposits collected exceeded expectations. As a result, the debt component and the resultant debt servicing expenses for the project decreased.

A concessionaire was selected through a non-controversial bid process. A normalized combination of the construction cost for the sewerage network and the proposed concession period for the STP was used to select the winning bidder. There were several hiccups and disputes post-award. As per the contract, the sewerage network would run close to the boundary of a house. The municipality would provide the connection from the boundary to the network. The municipality attempted to renegotiate this clause and asked the private operator to provide these connections, at a cost of US$0.50 per meter – a price that was not appealing to the private contractor. Another private firm was then selected for this portion of the project. Next, even as parts of the sewerage network and the pumping stations were built, municipal officials were slow to take them over, citing a lack of capacity to manage these assets. A change in government then led to the replacement of project officials from the government's side. The new team had to go through a learning curve, which caused further delays on the project.

The TNUDF then withdrew from the project, even as the municipality questioned the readiness of the STP. Without the TNUDF's presence, the municipality was unsure as to how to deal with the operator. When the STP was finally commissioned, the sewage inflow was much lower than expected. The municipality refused to pay for the minimum guaranteed amount of sewage and only relented when the private operator threatened to stop operations. Arbitration proceedings are currently under way as the operator seeks compensation for the delayed inflow of sewage into the system.

8.4.2 Tirupur Water Supply Project

Tirupur's textile industry is water-intensive. The demand for textiles and consequently for water grew rapidly in the early 1990s. In addition, a lack of rainfall prevented the Tirupur municipality from providing a sufficient quantity of water to its residents and the nearby textile units. Urged by the Tirupur Exporter's Association (TEA), the Government of Tamil Nadu (GoTN) partnered with Infrastructure Leasing and Financial Services (ILandFS) – a leading infrastructure consulting firm – to provide a water supply solution to Tirupur. Initial studies indicated that the cost of supplying water to Tirupur would be of the order of US$100 million, an amount that was beyond the budgetary capacity of the Tirupur municipality. ILandFS therefore proposed a BOT approach. The Government of Tamil Nadu accepted this proposal on the condition that water would be delivered at subsidized rates to the residents of Tirupur and nearby villages.

The GoTN, ILandFS and TEA then formed a special-purpose vehicle called the New Tirupur Area Development Company Limited (NTADCL) to supply water to Tirupur's industries, the town itself and selected neighboring villages at subsidized rates, and to provide a municipal sewerage system free of cost in the town of Tirupur. In order to make the project financially viable, NTADCL decided to implement a differential tariff system, wherein industrial consumers would pay higher rates and thereby subsidize residential consumers who would pay lower rates for water. The TEA was not in favor of this strategy since it was part of the consortium and was faced with having to pay high rates for water. However, the project would not have been feasible without these differential tariffs and the textile industry was in dire need of water. Further, the Chief Minister of Tamil Nadu personally had a discussion with TEA representatives, after which TEA relented. A concession agreement on these terms was then signed between the GoTN and NTADCL. NTADCL could earn a return on its investment of up to 20 percent, and the overall duration of the concession was 33 years from the award of the contract.

Raising finances for the project took longer than expected since this was a first-of-its-kind project in India and potential lenders were unsure of the risks that the project entailed. Once this was done, construction contracts were awarded on a typical EPC basis, and the construction was completed in 2005, ahead of schedule and within budget.

As operations commenced, it appeared that the project was proceeding smoothly. However, by 2008 it was clear that water consumption by the industrial users was far less than expected. This was due to economic recession that had lowered demand for textiles, technological advances that had reduced the need for water in the textile industry, and stringent policing by the Tamil Nadu Pollution Control Board that shut down several plants (who were potential customers of water) for not complying with environmental standards. In addition, good rainfall in recent years led to a rise in the groundwater table. Several small and medium textile units (which made up a significant fraction of the total textile industry) were exceedingly cost-conscious, not strict on their quality standards and preferred to draw groundwater as opposed to paying a large amount of money for the higher-quality treated water supplied by NTADCL. Although a law against extraction of groundwater had been enacted, it was never enforced. These factors combined to result in a revenue shortfall that was nearly 60 percent lower than estimated revenues.

NTADCL pursued many strategies to address this situation. It first lowered its prices in order to attract more customers, in the hope that these customers would continue to be loyal to NTADCL even as prices were increased later, due to the superior quality of water supplied by NTADCL. It also conducted and disseminated several scientific studies that showed the relative purity of NTADCL water over groundwater, and the consequent financial impact on the textile industry. However, these efforts did not bear fruit. In 2009, with NTADCL's reserves nearly exhausted and on the verge of bankruptcy, the GoTN infused nearly INR80 million into the project in order to sustain it.

It follows from these two case studies that although the BOT model seemed an appropriate choice when confronted with the project's parameters alone, this model has faced challenges in Tamil Nadu due to the weaknesses in the state's institutional environment for PPPs. For instance, the lack of project governance capacity led to disputes in the post-award phase of the Alandur sewerage project. A lack of project development capacity led to the bundling of additional services such as the sewage treatment plant into the Tirupur Water supply concession, leading to a rise in project costs. Also inadequate legal enforcement mechanisms and poor stakeholder awareness led to the local textile units searching for alternative

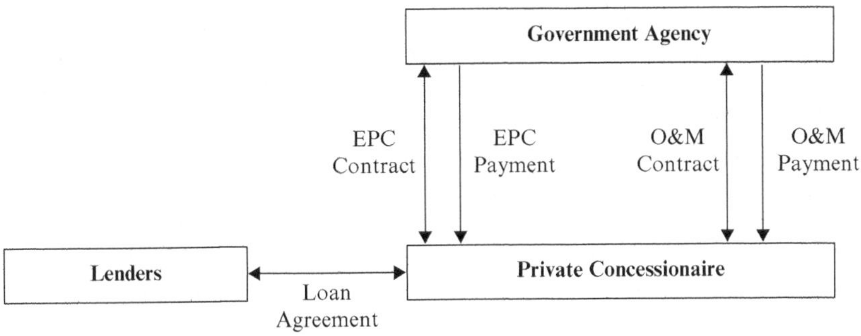

Figure 8.1 The design–build–operate (DBO) model

water sources. The next section describes a third case study where a less sophisticated PPP model was used for the delivery of water and sanitation services in Tamil Nadu.

8.4.3 Design–Build–Operate Model in Delivering Water and Sanitation Services

As mentioned earlier, a sewage treatment plant in Coimbatore and several other projects have been undertaken through a mode of PPP that is referred to as the design–build–operate (DBO) model. Figure 8.1 graphically depicts this model. The concessionaire is tasked with designing, financing, building and operating the asset. However, unlike conventional PPP concessions where the private sector's revenues (whether through user charges or availability payments) are received over the life cycle of the project, here the private provider's capital investments (asset creation costs) are reimbursed upon completion of the construction phase of the project. The private provider is then tasked with operating and maintaining the asset as well for a fixed period of time, in return for a performance-linked fee that is in turn linked to the direct costs incurred by the operator. The project sponsor (government agency) in effect signs two interrelated contracts with a private provider: the first contract is an engineer–procure–construct contract, followed by an operation and maintenance contract with the same provider.

This model has several advantages. First, it ensures that the same provider will design, build and operate the asset, thus providing the private provider with wide latitude for technological innovation, while also ensuring the constructability of the design and the ease of maintenance of the asset. Second, the juxtaposition of an EPC and an O&M contract often leads to this mode of contracting being considered within the bounds of conventional procurement, as opposed to a PPP. As a result, there is virtually no need to create additional legitimacy around this form, and political willingness to implement DBO is likely to be high. The ability of government departments to structure and monitor these contracts is also high since they correspond to traditional contracting practices. Further, the revenue or payment risk for such projects is quite low. The short time horizons within which the private provider can expect to recoup its investment (essentially the duration of the construction period) implies that the risk premium placed on such projects will be quite low, leading to lower overall costs. Further, this is likely to ensure wider private sector participation, in particular the inclusion of smaller firms in the tendering process, leading to heightened competition and price discovery. Finally, O&M contracts can be of short duration – the Coimbatore project is five years long – since the private provider's cost recovery is not based on the operation phase. This implies that the likelihood of private sector 'hold-up' is small. Perhaps the chief disadvantage of this model is that once construction costs are recovered, the private provider may have no incentive to operate and maintain the asset to high quality standards, since a large portion of its profits might be tied to the construction phase. However, a minor modification of this structure, where some of the construction costs will be 'retained' and distributed across the O&M period, can ensure that the private provider remains incentivized in service delivery as well as asset creation. Such modifications have been enacted in other sectors in India, such as in rural electrification and road construction.

This chapter has earlier noted several weaknesses in the institutional environment for implementing PPPs in Tamil Nadu, the chief among these being a lack of legitimacy for PPPs, a predominant lack of political will, as well as a lack of public and private sector capacity. The DBO model, being a milder version of the traditional BOT version of PPP, addresses most of these limitations. Explicit policy directions or legislation are not needed to enact such projects, and they fit within the ambit of the existing limited public and private sector capacity for PPPs in the state. In addition, they allow for efficiencies brought in by the private sector to be brought to bear on infrastructure service delivery. The DBO model thus fits well within an ambient environment with weak PPP capabilities. The experience in Tamil Nadu bears testimony to this. Projects implemented through DBO are operational and appear to be meeting societal objectives. Construction is often done according to schedule, and very few disputes between the sponsor and the concessionaire have been reported in the construction or operation phase.

8.5 CONCLUSION

Very often, the choice between public and private delivery and the selection of a particular private delivery mode are made ideologically and not rationally. The objective of this chapter was to infuse some rationality into this debate on the selection of project delivery mode. The first contribution of this chapter is to illustrate the close linkage between project delivery mode choice and the robustness of the institutional environment to support PPPs. The choice of project delivery mode is not based on the features of the project alone. Rather, the ability of the institutional environment to support the delivery mode is a key consideration. The examples from Tirupur and Alandur in Tamil Nadu have shown that although the projects themselves were not unlikely candidates for BOTs, the inability of the institutional environment to support the complexity of the project structure led to their downfall. In Tirupur, an inability to enforce public sector commitment, poor project structuring capacity (as evinced by the inclusion of a sewage treatment plant with no revenue streams attached under the concessionaire's scope) and a lack of stakeholder awareness were some of the capability gaps in the ambient environment that led to project failure. In Alandur, the inability of the project sponsor to monitor and govern a concession led to turbulences during the operation phase of the project.

The second, related contribution of this chapter is to indicate that governments should select PPP models that can survive within their ambient environments, and 'upgrade' them while simultaneously undertaking measures to strengthen the institutional environment. For instance, the DBO model requires a much less mature institutional environment as compared to a BOT project, and thus fits well within Tamil Nadu currently. However, should the state strengthen its ambient environment by announcing a PPP policy or enacting enabling legislation, and strengthen its dispute resolution apparatus and the capacity of the public sector with regard to PPPs, it is conceivable that BOTs and other forms of PPP can thrive within the state. These lessons are particularly important in the context of developing countries with growing economies. The rate of infrastructure growth that is required to sustain economic growth in these countries is so high that governments are likely to have no choice but to turn towards the private sector to deliver infrastructure services. However, if governments choose forms of PPP that cannot be supported within their administrative environment, they risk jeopardizing their overall service delivery goals. It is here that the framework to assess institutional capabilities can be used as a tool for governments to assess their environment, and to then iterate between enacting PPPs that fit, while simultaneously strengthening their institutions.

REFERENCES

Besley, T. and M. Ghatak (2001), 'Government versus private ownership of public goods', *Quarterly Journal of Economics*, **116**(4), 1343–72.
Corry, D. (2004), 'New Labour and PPPs', in A. Ghobadian, D. Gallear, N. O'Regan and H. Viney (eds), *Public Private Partnerships: Policy and Experience*, Basingstoke: Palgrave Macmillan, pp. 24–36.
De Bettignies, J.-E. and T. Ross, (2010), 'The economics of public–private partnerships: some theoretical contributions', in G.A. Hodge, C. Greve and A.E. Boardman (eds), *International Handbook on Public–Private Partnerships*, Cheltenham, UK and Northampton, MA, USA: Edward Elgar Publishing, pp. 132–58.
Department of Economic Affairs (2008a), 'Panel of Transactions Advisors for PPP Projects: a guide for use of the panel', New Delhi: Ministry of Finance, Government of India.
Department of Economic Affairs (2008b), 'Scheme and guidelines for India Infrastructure Project Development Fund', New Delhi: Ministry of Finance, Government of India.
Department of Economic Affairs (2008c), 'Scheme and guidelines for financial support to public private partnerships in infrastructure', New Delhi: Ministry of Finance, Government of India.
Department of Economic Affairs (2011), 'National Public Private Partnership Policy: draft for consultation', New Delhi: Ministry of Finance, Government of India.
Flinders, M. (2010), 'Splintered logic and political debate', in G.A. Hodge, C. Greve and A.E. Boardman (eds), *International Handbook on Public–Private Partnerships*, Cheltenham, UK and Northampton, MA, USA: Edward Elgar Publishing, pp. 115–31.
Garvin, M.J. and A. Mahalingam (2011), 'Project delivery and financing: conventional and alternative methods', in P. Chinowsky and A. Songer (eds), *Organization Management in Construction*, London: Spon Press.
Hart, O. (2003), 'Incomplete contracts and public ownership: remarks, and an application to public–private partnerships', *Economic Journal*, **113**(486), 69–76.
Hart, O., A. Shleifer and R.W. Vishny (1997), 'The proper scope of government: theory and an application to prisons', *Quarterly Journal of Economics*, **112**(4), 1127–61.
IIT-Madras (2011), 'Creating an enabling ambient environment for PPPs', IIT-Madras, Chennai, India.
Iossa, E. and D. Martimort (2008), 'The simple micro-economics of public–private partnerships', Centre for Market and Public Organization, Department of Economics, University of Bristol, UK.
Klijn, E.-H. (2010), 'Public–private partnerships: deciphering meaning, message and phenomenon', in G.A. Hodge, C. Greve and A.E. Boardman (eds), *International Handbook on Public–Private Partnerships*, Cheltenham, UK and Northampton, MA, USA: Edward Elgar Publishing, pp. 68–80.
Manohar, S. (2010), 'Developing a framework to determine the appropriate delivery mode for urban infrastructure projects', graduate thesis, Indian Institute of Technology, Madras.
Martimort, D. and J. Pouyet (2008), 'To build or not to build: normative

and positive theories of public–private partnerships', *International Journal of Industrial Organization*, **26**(2), 393–411.

Ministry of Urban Development (2006), 'Jawaharlal Nehru National Urban Renewal Mission: modified guidelines', MoUD, New Delhi, available at http://jnnurm.nic.in/wp-content/uploads/2011/01/modified_guidelines3.pdf (accessed 4 October 2011).

Planning Commission (2008), *Eleventh Five Year Plan (2007–2012)*, Vol. 3, New Delhi: Oxford University Press, available at http://www.planningcommission.nic.in/plans/planrel/fiveyr/11th/11_v3/11th_vol3.pdf (accessed 4 October 2011).

Planning Commission (2011), 'Draft approach paper for the 12th Five Year Plan (2012–17)', Planning Commission, New Delhi, available at http://www.planningcommission.nic.in/plans/planrel/12appdrft/appraoch_12plan.pdf (accessed 4 October 2011).

Schmidt, K.M. (1996), 'The costs and benefits of privatization: an incomplete contracts approach', *Journal of Law, Economics and Organization*, **12**(1), 1–24.

Sinha, R.K. (2009), 'Developing a framework to determine the appropriate delivery mode for urban infrastructure projects', graduate thesis, Indian Institute of Technology, Madras.

Vives, A., A.M. Paris, J. Benavides, P.D. Raymond, D. Quiroga and J. Marcus (2006), *Financial Structuring of Infrastructure Projects in Public Private Partnerships: An Application to Water Projects*, Washington, DC: Inter-American Development Bank.

Williamson, O.E. (1979), 'Transaction-cost economics: the governance of contractual relations', *Journal of Law and Economics*, **22**(2), 233–61.

9. Urban water governance in China: legal and policy framework

Lovleen Bhullar

9.1 INTRODUCTION

Home to one-fifth of the world's population (1.34 billion people in 2010), the People's Republic of China (PRC or China) has witnessed a rapid increase in its urban population since 1978, when less than 20 percent of the population was living in cities. By the end of 2008, there were 607 million urban residents, accounting for 45.7 percent of China's total population (ADB, 2010). Based on current trends, China's urban population is projected to cross the 1 billion mark in 2030 and there will be eight megacities by 2025, each with a population of over 10 million (Woetzel et al., 2009).

This population growth and rapid urbanization has led to an increase in the demand for water and water-related infrastructure. But China has only 7 percent of the world's water and 26 percent of the world's per capita water resources, unevenly distributed in spatial terms. The total annual renewable water resources of China are 2830 km^3/year and the per capita renewable water resources stand at a mere one-fifth of the world average (that is, 2100m^3). Around 100 million urban residents do not have access to safe water. Over 400 of the 667 cities suffer from water shortage. Half of the ground water in urban areas is polluted in various degrees (Wang, 2008).

The authorities have been unable to meet the increasing demand for water. This can be attributed to several factors including low levels of water tariffs, inefficient management and operation of water service facilities due to lack of expertise, outdated technologies, mediocre level of management skills and facilities, lack of investment, poor service coverage and water quality, absence of economic incentives and demand side management and the highly centralized government structure (Lee, 2010; Browder et al., 2007; Fu et al., 2008; Zhong et al., 2008).

The situation in China is not unique. Other developing and transition countries, as well as several developed countries, face a water crisis.

Effective water governance has been advocated to address this problem. This has led to the adoption of different mechanisms of water governance, involving the public sector, the private sector and the end-users (individuals and communities). While there are several examples of private sector participation in the urban water supply sector in Africa and Latin America, limited private activity has been observed in Asia. China represents an anomaly as it is one of the few Asian countries where private sector involvement in the urban water supply sector has expanded.

China opened its urban water sector to private investors on a pilot basis between 1992 and 2001. Private sector participation in the water sector was formally allowed on a nationwide basis in 2002. By 2007, one-third of 208 urban water utilities across China had private shareholders and private investors held majority ownership in two-thirds of these utilities (Jiang and Zheng, 2010). In 2008, private water companies served 38 percent of the urban population (Lee, 2010). In 2009, China represented 80 percent of new infrastructure projects and 25 percent of investment in the water sector (PPIAF, 2010). The players include foreign and domestic private investors and developers. In recent years, however, the type of private players involved in the urban water sector has undergone a transformation with an increase in the share of domestic and regional private water companies. While China encourages private sector involvement in water, most of the investors are semi-public entities.

This chapter explores the role of the private sector in the provision of water to Chinese cities, and provides a detailed account of the applicable legal and regulatory framework and water governance mechanisms. The next section introduces the urban water sector reform process in China. Then, some important aspects of the legal and regulatory framework governing private sector participation in the urban water sector are discussed and analyzed with a focus on water supply. The penultimate section evaluates the prevailing forms of private sector participation in China and discusses the participants (foreign and domestic). This is followed by some concluding remarks.

9.2 URBAN WATER SUPPLY REFORM IN CHINA

The People's Republic of China was created in 1949 and a command economy system (which is planned and controlled by the central government) was established. During the pre-reform period, the communist regime consolidated traditional state monopoly in urban water management. The responsiblility for investment, construction and operation of water services vested in the housing and construction departments and bureaus of the local municipalities. As water was viewed as an abundant natural resource (and a public good), water supply services were subsidized and provided as a form of social welfare at low or no cost. The central government provided exclusive financial support (through fiscal transfers) to the municipalities, which handed over the profits to the central government for reallocation. During this period, private sector involvement was discouraged.

Economic reforms followed the implementation of the open-door policy and the introduction of the core policy of 'Separation of Governments and Enterprises' in 1978. The urban water sector was restructured from local government departments into state-owned corporations and enterprises. Local governments had to invest in urban water infrastructure as a result of rapid urbanization and an increase in water demand from domestic and non-domestic consumers. For this purpose, they relied on intergovernmental loans or loans from international financial institutions, which were guaranteed by the central government. However, local governments continued to be plagued with infrastructure development bottlenecks and water services deteriorated further.

The central government and local governments resorted to private sector participation in the provision of urban water services from the late 1990s. One of the main drivers was the central government's 1994 tax sharing reform, which decentralized the responsibility for financing and developing urban infrastructure, including urban water supply, to local governments (Cheng and Wang, 2009). However, most local governments lacked the fiscal capacity to deliver these services since domestic capital markets were underdeveloped and the national policy had curtailed access to credit. In these circumstances, in order to overcome the infrastructure development bottlenecks without further adding to their consolidated municipal debt, local governments relied on the private sector as a source of new investment capital (Leman, 1996). In order to attract private investors, local governments offered guarantees and incentives (such as tax exemptions) (Cheng and Wang, 2009; Lee, 2010).

The other major reason for the Chinese government's support for private sector participation was its inability to pursue other sources of funding (Chen, 2009). China used to be one of the biggest recipients of loans from international development agencies and private banking consortiums for water sector development (Lee, 2003). But as the repayment of these loans became a budgetary problem for local governments, the government began to consider infrastructure financing through alternative channels, such as private sector participation. China's accession to the World Trade Organization in 2001 has also contributed to foreign private sector involvement in the urban water sector (Lee, 2007a).

The private sector's interest in the urban water supply sector in China is strategic and it is driven mainly by long-term factors. They envisage an increase in water demand and tariffs and significant future returns in view of China's rapid urbanization and economic development, as well as increasingly constrained water resources (Jiang and Zheng, 2010).

9.3 LEGAL AND REGULATORY FRAMEWORK FOR URBAN WATER SUPPLY

9.3.1 Institutional Framework

The central and local governments in China initiated the process of creating the necessary institutional and regulatory frameworks in order to accelerate the reform process and to encourage private sector participation. The responsibilities are delegated through a four-tiered system (Carmody, 2010), which involves:

1. Central government
2. Provincial governments (33 divisions):
   a. Five autonomous regions with own local government but more legislative rights (Guangxi, Inner Mongolia, Tibet, Xinjiang, and Niingxia).
   b. Four municipalities directly under the central government (Beijing, Shanghai, Tianjin and Chongqing)
   c. Two special administrative regions (Hong Kong and Macao).
3. Municipal governments (333 prefectures).
4. County governments (2872 county districts and 41 636 town divisions).

Within the central government, the National People's Congress exercises the power of legislation, makes decisions on important national issues, and elects and chooses the leading personnel of the highest state organs. The State Council is the chief executive organ of the Chinese government, which drafts and implements policies and/or laws and regulations adopted by the National People's Congress. Under the State Council, several commissions and ministries are involved with water resources (Lee, 2010). These include the following.

Commissions:

1. The Environmental and Resources Protection Committee of the National People's Congress is responsible for generating legislation

on the environment and resources and supervising the implementation of these laws.
2. The State Environmental Protection Commission is the chief interagency coordinating body for setting broad guidelines on environmental policy in view of other national priorities.
3. The main functions of the National Development and Reform Commission (NDRC) (previously the State Development and Planning Commission or SDPC) include 'to approve, authorize, and review key construction projects, foreign-funded key projects, key investment projects for overseas resources development, and investment projects utilizing large amount of foreign exchange as mandated by the State Council; to guide and supervise the utilization of foreign loans in construction projects; to guide private investment; to study and put forward strategies and plans for foreign capital utilization and overseas investment'.

Ministries:

1. The Ministry of Water Resources leads most of the water resource management issues.
2. The Ministry of Housing and Urban–Rural Development (the MOHURD, previously the Ministry of Construction) oversees public–private partnership projects in the urban water sector. It examines and reviews project proposals in relation to foreign investment prior to submission to the NDRC for approval. The Ministry is also involved in water pricing issues.
3. The Ministry of Health is responsible for supervision and management of drinking water quality and its suppliers.
4. The Ministry of Environmental Protection and local Environmental Protection Bureaus are responsible for water quality control and water conservation activities.

The framework is complex, hierarchical and decentralized, and it has resulted in inconsistent policy-making and implementation at the central and local levels (Lee, 2006). The political power of the central government over local governments is undermined because the latter is responsible for water supply. For example, although the central government supports increasing water tariffs to cost recovery levels (which would encourage private sector participation), local governments are responsible for tariff-setting and the central government cannot intervene directly. Similarly, although the central government forbids government guarantees to attract private investment, some local governments have used them (Cheng and

Wang, 2009). The fragmentation and frequent overlap of responsibilities and authority, between central and local level, as well as at the local level, has resulted in disorganized management, inefficiency and unclear administrative accountability (Lee, 2010).

9.3.2 Legal and Regulatory Framework

The hierarchy of regulations in China, as defined in the *Legislation Law of the People's Republic of China* (promulgated on 15 March 2000),[1] is as follows:

1. The Constitution.
2. National laws promulgated by the National People's Congress or its Standing Committee.
3. Administrative regulations issued by the State Council.
4. Local regulations promulgated by the local People's Congress or its Standing Committee.
5. Rules formulated by ministries and commissions of the State Council, the People's Bank of China, the State Audit Administration as well as the other organs endowed with administrative functions directly under the State Council, and rules formulated by local governments.

Private sector participation in the provision of basic infrastructure services has been allowed through various legal instruments, to varying extent. The 1982 Constitution permits foreign individuals, enterprises and economic organizations to invest in China and to enter into several forms of economic cooperation with Chinese enterprises (Hanley, 1999). For a long time, joint venture laws and related regulations provided the exclusive legal framework for foreign investment in infrastructure projects in China (see section 9.4.1 below).

The PRC Ordinance on Urban Water Supply (No.158 Policy Paper of the State Council, 1994) was the first piece of national legislation for water supply management in China. The Ordinance emphasized the importance of economic instruments for water management and the water tariff was raised substantially. Following its promulgation, the Chinese government accelerated the creation of a legal framework for the urban water sector. In December 2001, the former State Planning Committee published its guidance on non-government investment to encourage foreign and domestic non-government investors to participate in constructing and operating infrastructural facilities and public utilities in various business models (Jiang and Zheng, 2010). The marketization reform in the urban water sector, among others, commenced in 2002 and enabled state-owned enterprises to act like market-oriented firms. It also led to the opening of public utilities to foreign and domestic investors. The third Plenary Session of the 16th Central Committee in 2003 proposed the opening of China's public utilities, including the urban water sector, to private investors (Fu et al., 2008).

However, there are no specific laws and regulations with regard to foreign investors in the Chinese water industry (Lee, 2007a, 2010). Instead, private sector participation is based on a complex framework of fragmented administrative regulations, including opinions, circulars, rules, measures, provisions or orders, which carry different legal force, depending on the issuing authority, and deal with different aspects. This also reflects the ad hoc, piecemeal, problem-oriented approach of the central government. The following section will discuss and analyse some of these regulations.

The Build–Operate–Transfer (BOT) Circular of 1995

China's urban water sector was opened to private sector participation, on a pilot basis, during the early 1990s. However, in the absence of guidance on, and rules or regulations governing, private sector participation, this was primarily done through build–operate–transfer (BOT) contracts and capital share transfer arrangements. The project sponsors were heavily dependent on government debt guarantees and they were not permitted to own or operate these facilities without forming a joint-venture company with a Chinese partner.

The enactment of two legislative circulars in 1995 provided the legal basis for public–private partnerships and foreign capital investment in urban infrastructure. These circulars set forth the legal procedures for experimental and step-by-step introduction of the BOT approach in China. The former Ministry of Foreign Trade and Economic Cooperation (MOFTEC) (now Ministry of Commerce) issued the Circular on Several Issues on Attracting Foreign Investment through BOT Approach (No. 89 policy paper of 1994, issued on 16 January 1995).[2] The other related circular, the Circular on Major Issues of Approval and Administration of the Franchise Pilot Projects with Foreign Investment (No. 208 policy paper of foreign investment, issued on 21 August 1995) (the 'BOT Circular'), was jointly issued by the former SDPC, Ministry of Electric Power (which no longer exists) and the Ministry of Communication.[3]

The provision for the creation of wholly foreign-owned enterprises permitted foreign companies, for the first time, to finance, construct, operate and maintain major infrastructure facilities without a Chinese joint-venture partner, at least until the project is transferred to the government at the end of the concession period. The rate of return on the sponsors' investment was not fixed. In addition, concession contracts were awarded after an international competitive bidding process, central-level approval was required for all projects, conversion of foreign exchange was guaranteed, and the project company could increase tariffs or tolls or extend the concession period if its profits fell because of a change in government policies (see Hanley, 1999). However, the circular is written in general terms and it does not include any specific guidelines for foreign investors (Li, 2007; Tjoa and Zhou, 2005).

Pursuant to the BOT Circular, five pilot BOT projects (in the power, transport and water supply sectors) were implemented by wholly foreign-owned enterprises. The Chengdu No. 6 Water Treatment Plant B project in Tianjin, Sichuan Province was the first water supply BOT project awarded under competitive public bidding in 1998. A consortium of Veolia Group (then, Générale des Eaux) (France) and Marubeni Corporation (Japan) (49 percent) and the Chengdu Municipal Waterworks General Company (51 percent) owned the project company. The agreement provided for a 'take-or-pay off-take agreement', which is a 20-year undertaking by the municipal government to buy a set volume of water from the company, whether it was needed or not. The company was thus guaranteed a fixed return on its investment. However, the forecasted growth in water demand could not be reached and the project triggered economic losses to the local government (Wang and Zhao, 2004; Lee, 2006; Fu et al., 2008; Chen, 2009). In an attempt to minimize its losses, the local government unsuccessfully approached Veolia to renegotiate the contract in the mid-2000s. The term of the project is 18 years and it is operational.

The second trial of this BOT model, the Beijing No.10 water project, was awarded to a consortium comprising Anglian Water (United Kingdom) and Mitsubishi Corporation (Japan) in 2002. However, the project failed to achieve financial closure because the foreign investors failed to secure debt-financing through local and international financial institutions. The project was subsequently revived in August 2007, and awarded to a local investor, Beijing Anling Water Technology Company Limited (Beijing Enterprises Water Group Limited and Golden State Environmental Group (50:50)). The operating period of this project is 23 years.

The SDPC was responsible for organizing the drafting of and issuing the Foreign Investment Concession Projects Tentative Provisions (the 'Tentative Provisions') to end the experimental projects and to provide the legal basis for the general implementation of concessions in all areas (Vernay and Gasne, 2004). However, it was not approved by the State Council and the experimental foreign investment franchise project approval procedure under the circular has since been canceled by the State Council (Li, 2007).

Foreign investment in urban water supply projects

The Catalogue for Guiding Foreign Investment in Industry (the 'Catalogue') (first promulgated on 20 June 1995, and updated in 1998) encouraged, permitted, restricted or forbade different types of foreign-invested infrastructure projects.[4] The 'construction or management of urban networks of water supply' was expressly prohibited and foreign companies were not permitted to own and manage the right of water distribution. Projects not specifically listed (such as water treatment projects) were deemed to be 'permitted'.

The Catalogue was revised in March 2002 (and since then, in 2004 and 2007) to remove the prohibition on foreign investment in urban water distribution networks; now, it is 'restricted'. In order to provide political comfort, joint ventures (not wholly foreign-owned enterprises) can be approved (Mahony and Fong, 2002) and the Chinese party must hold the controlling interest (51 percent or more). It does mean, however, that foreign developers can now construct and operate water supply and drainage networks in large and medium-sized municipalities. This reduces some of their revenue risk as they can secure revenues from management of water supply and drainage networks (Lee, 2010).

Pursuant to the 2002 revision of the Catalogue, Veolia was the first foreign company to be given the responsibility for drinking water production, network distribution, selling running water for domestic use, and customer services (Lee, 2007b). Veolia entered into a cooperative joint venture with the Shanghai Pudong Water Supply Corporation to create the Shanghai Pudong Water Company in May 2002. The concession period is 50 years and the project is operational.

Guaranteed fixed rate of return

The uncertain and underdeveloped policy environment in China meant that the private investors were extremely concerned about investment risks. In order to assuage their concerns, the 21st century Urban Water Management Pilot Scheme 1997 scheduled the liberalization of water tariffs for projects funded with foreign capital and allowed foreign investors to receive favorable rates of return for water projects in China (Rozner, 1998). In order to attract capital during the early stage of the reform process, the PRC Urban Water Price Administration Regulations (1985, revised in 1998) guaranteed a fixed investment rate of return of 8 to 12 percent for foreign investors (Lee, 2007b).

However, this practice was banned by the Measures on the Guarantee of Fixed Profit Margins for Foreign Investment Projects (issued on 10 September 2002 by the General Office of the State Council).[5] The ban was based on the argument that the practice was not in accordance with the principle of sharing benefits and jointly bearing risks by Chinese and foreign investors. Further, it violated the relevant laws and regulations on joint ventures and Sino-foreign cooperatives (Li, 2007). This resulted in modification of the relevant terms of the pre-existing contracts with guaranteed fixed rates of return from foreign investment, buy-back of all the shares of foreign investors, transfer of foreign investment into foreign loans, or dismantling of contracts with often severe losses (Zhong et al., 2008). Future contracts were required to include fair terms and conditions for foreigners as well as Chinese partners (Lee, 2010).

Guarantees

Due to political uncertainty in China, foreign investors have traditionally sought some form of (informal) guarantees from politically influential central government ministries and bureaus related to water projects in order to minimize their risks in complex negotiations and contracts (Lee, 2007a). This type of guarantee, which is not legally binding, was called a 'Government Support Letter/Comfort Letter' and it was based on a Chinese tradition of 'gentlemen's honour and agreement' (Turner III and Seem, 1999). However, with increasing decentralization, the central government is reluctant to provide such 'guarantees' to foreign investors.

Insofar as sovereign guarantees are concerned, except under special situations, Article 8 of the Guaranty Law and Article 4 of the Measures on Administration of Domestic Organizations' External Foreign Exchange Guaranty issued by the China People's Bank renders government guarantees on loaning from foreign governments or international economic organizations existing in BOT water projects in practice to be void (Choi et al., 2010).

National concession framework

The process of marketization reform of the Chinese urban water sector was initiated by the Opinion on Accelerating the Marketization of Urban Utilities (No. 272 Policy Paper of the former Ministry of Construction, 27 December 2002) (the '2002 Opinion').[6] The Opinion officially acknowledges the eligibility of foreign and domestic capital to participate, on the basis of fair competition, in urban water supply projects. It encourages local governments to open urban water service contracts to both foreign and domestic private investors through a competitive public bidding process, but the use of the term 'privatization' is avoided.

Pursuant to the 2002 Opinion, some local governments have enacted specific local laws and regulations for urban utility concessions (see, Vernay and Gasne, 2004; Yuan and Qin, 2004). The Urban Infrastructure Concession Operation Method of Beijing 2003 and the Interim Provision on Administrating Concession Right of Chengdu 2001 (No. 131 Policy Paper of Chengdu Municipality) prescribe a concession agreement. The Measures on Public Utilities Concession Operation Method of Shenzhen 2003 (No. 124 Policy Paper of Shenzhen Municipality) prescribe a specific government permit or ordinance. Tianjin has adopted a less formal government circular for granting concessions whereas a specific approval sets out the detailed terms of the concession in Chongqing. The grant procedure may be through bidding or direct negotiations, depending on the local government.

The discrepancies at the local level highlight the importance of standardization of procedures. In order to address this concern, the Measures on Public Utilities Concession Management (No. 126 Policy Paper, issued by the former Ministry of Construction on 19 March 2004 and effective from 1 May 2004) (the '2004 Measures') created the general legal framework for any type of concession project in local urban public utilities, including water supply.[7] The 2004 Measures are different from the 1995 Circulars, which were restricted to specific experimental and large-sized projects, and they seek to fill the regulatory gap resulting from the non-issuance of the Tentative Provisions by the State Council. The 2004 Measures specify the procedure for private sector participation as well as the main aspects of concession projects and contractual documentation. The concession projects will be awarded (to foreign or domestic companies) through a public bidding process (Vernay and Gasne, 2004). However, the 2004 Measures are 'general' in nature and they refer to local regulations for the actual implementation of concession projects (as urban utilities are a matter of local jurisdiction).

Following the 2004 Measures, the former Ministry of Construction issued the Draft Standard Concession Agreement for Urban Water Projects. Its purpose is to facilitate the effective application of concession agreements for urban water utilities and to ensure more consistency across local governments (Cheng and Wang, 2009; Vernay and Gasne, 2004). However, this document is not legally binding and it favors the interests of local governments over investors (Jensen and Reid, 2007).

This standardization of procedures has removed some market risks and accelerated foreign investment in urban water services in China (Lee, 2010). However, in the 2004 Measures, 'concession management' refers to all forms of private sector participation and it relies heavily on the BOT model (Zhong et al., 2008). This means that the terminology used to describe different forms of private sector participation is used interchangeably, without regard for their differences. Further, in relation to tariffs, the 2002 Opinion states that 'the reasonable investment returns to be obtained by an enterprise through its legal operation shall be guaranteed'.

This contradicts other regulations that prohibit guaranteed fixed returns on foreign investment. Neither the 2002 Opinion nor the 2004 Measures address the 'take or pay' mechanism where the government promises the lowest water quantity and agreed prices, a key condition for financing of BOT projects in the water sector as long as the scope of the concession does not extend to the operation of public networks (Vernay and Gasne, 2004). It is argued that this provision may fall under the prohibition of fixed return guarantees (Li, 2007).

The PRC Administrative Licensing Law (promulgated on 27 August 2003 and effective from 1 July 2004) requires that the grant of permits in special industries affecting the public interest, such as concessions in urban water utilities, shall be determined through fair competition such as bidding or auction. This would largely restrict the local government from directly negotiating a concession agreement. However, this is subject to the inclusion of other methods for grant of permits in laws and the State Council's administrative regulations.

Regulation

The government cannot withdraw itself from public water utilities once the water supply has been privatized without establishing a proper and independent monitoring system. However, during the early stages of reform in China, the local authorities focused on opening the urban water market and attracting private investments. The issues of regulation, such as monitoring of private water companies' services and operating procedures, and public participation received scant attention. This oversight was partly rectified by the Opinions on Strengthening Regulation of Public Utilities (No. 154 policy paper, issued by the former Ministry of Construction, 2005), which emphasized the importance of governmental regulation for the water sector. In practice, however, 'a systematic and comprehensive regulatory framework for the Chinese urban water sectors' remains absent (Zhong et al., 2008). Similarly, although there is a public hearing when water firms want to raise water tariffs, participation is restricted by the companies and related government departments (Globalization Monitor, 2009).

9.3.3 Institutional and Regulatory Risks

Private sector participation in the urban water sector in China can be affected by a number of institutional and regulatory risks, which may partly explain the retreat from, or the limited operations of, transnational water corporations in China.

Foreign investors are circumspect about the efficacy of the Chinese legal system. The absence of a 'uniform supervisory legal system', which can provide coherence, has allowed the local governments to interpret laws and regulations differently depending upon local socio-political circumstances (Blackman, 2001). The tendency to consider the laws as guidelines rather than the ones that should be implemented strictly has also led to inconsistent legal practices at the local level (Lee, 2007a, 2010; Li, 2007).

In this context, the 'change in law risk' is a major concern. It includes the adoption, promulgation, modification or reinterpretation of any Chinese laws by a governmental authority; and the imposition of any material condition by a governmental authority in connection with the issuance, renewal or modification of any approval that establishes requirements that render the performance by the project developer according to its terms illegal, after the signing of the concession agreement (Wang et al., 2000). For instance, the revocation of the guaranteed fixed rate of return for foreign investment by the State Council in 2002 discouraged foreign investors from expanding their operations in the urban water supply sector.

The marketization reform of, and private sector participation in, the Chinese urban water sector is based on various governmental policy papers rather than a mature legal system. In the absence of a well-established legal framework, the current legal codification of public–private partnerships in water services is largely a reactive process, where various policy papers address specific problems in the reform process (Zhong et al., 2008). Given the long duration of some concession (including BOT) contracts, foreign investors require consistent policies and regulatory programs but these are absent and the legal system is in a constant state of flux. New regulations are introduced without withdrawing older regulations (Jensen and Reid, 2007). The patchwork of existing regulation means that in some cases the rules overlap and may even seem incoherent.

Further, state dominance, a legacy of the planned economy in China, means that the government fully controls the right to decide on a project (Cheng and Wang, 2009). In the urban water supply sector, the delay in approval risk represents a major obstacle. It means that the central or local government authority does not approve the project-related issues in time or even cancels approved projects (Wang et al., 2000). The approval process in China is lengthy and time-consuming and it frequently involves protracted negotiations between several fragmented ministries, governmental agencies and departments at the central, provincial and local level, and the project partners (Wang et al., 2000; Mahony and Fong, 2002). The bureaucratic power of policy decisions overwhelms legislative and judicial powers and fails to inspire investor confidence (Cheng and Wang, 2009). The lack of objectivity and transparency in the approval process also hinders private sector participation in the urban water supply sector.

It is clear from the above discussion that private sector participation in the urban water supply sector in China faces several hurdles. Nevertheless, the Chinese government continues to promote private sector participation in the urban water supply sector. The more recent Opinions on Implementing Policies and Measures to Accelerate the Development of the Service Sector (issued by the General Office of the State Council on 13 March 2008) emphasize that the government should continue to promote the privatization of public utilities projects and grant concessions to private entities to operate certain types of projects, including urban water supply (Li, 2009). As a result, the legal and regulatory framework may be strengthened to promote private sector participation in China's urban water supply sector. However, the government is expected to continue to exert considerable influence on the sector, as illustrated by an examination of the type of participants and their respective share, and the agreements governing them.

9.4 ALTERNATIVE GOVERNANCE ARCHITECTURES

Chapter 1 has provided an overview of some of the alternative water governance architectures. It is generally possible to categorize and distinguish the different forms of public–private partnerships based on certain key parameters. However, in the absence of a mature institutional and regulatory framework, they have evolved in an ad hoc and experimental manner in China. In some cases, private parties (domestic or foreign) are the 'private' player. In other cases, state-owned enterprises are the 'private' partner in a public–private partnership; in other words, the government is the regulator and the regulated. Such state-owned enterprises could increasingly crowd out local private sector firms as well as foreign participation. Further, while the Chinese government is encouraging public–private partnerships in the urban water supply sector, the scope is heavily restricted to certain types of projects. This section highlights the salient features of the public–private partnerships in the urban water supply sector in China.

9.4.1 Joint Venture

A joint venture is an arrangement between a private company and the public sector, in which both the parties share responsibilities and (investment) obligations (Zhong et al., 2008). The local party injects equity (that is, the right to use the land, labor force, raw materials, services and cash) and the foreign party injects cash, according to the proportion of its shares. As a result, joint ventures are suitable for larger projects that require much more direct capital investment from the private sector (Zhong et al., 2008). Public sector entities assume most of the design, construction and operation responsibilities and risks. The private company may be given full management and operational control, or this may be shared with the local government, according to the terms of the contract and the equity contributions of the parties. At the end of the project period, the share of the foreign party is re-transferred to the local party (Chen and Doloi, 2008; Zhang and Kumaraswamy, 2001).

The joint venture approach has the largest share in China's water supply sector and, historically, the joint-venture laws and regulations have provided the exclusive legal framework for foreign investment in infrastructure projects. The majority of joint ventures are still exclusively for bulk supply, but a growing number of integrated supply and distribution projects have been signed. These include joint-venture projects between private companies and state-owned enterprises, such as government-supported investment companies or privatized local water companies, and joint-venture projects between state-owned enterprises without private participation (Choi et al., 2010).

In China, joint ventures take the form of cooperative and equity joint ventures.

Cooperative joint venture

The parties to the cooperative joint venture share profit, control and risks according to negotiated contract terms (and not necessarily in proportion to their capital contribution). The investment contributions are not limited to financial capital. They may also include non-financial assets, such as intellectual property rights, buildings, materials or machinery. A cooperative joint venture represents a flexible approach as it can either be registered as a limited liability company or operate as an unincorporated entity under which partners bear unlimited liability. The foreign investor returns the majority or full ownership of the enterprise to the domestic partner (without charge) after obtaining a fair or premium return on investment. Thus, cooperative joint ventures are ideal when the foreign investor is only looking for a short-term project because the profit, control and risks of the parties are shared according to negotiated contract terms (and not necessarily in proportion to their capital contribution). Further, transfer or withdrawal of investments is not subject to government approval and foreign investors can deploy their registered capital as they see fit. Subject to government approval, the venture can be renewed at any time and for any extended period.

The structure of cooperative joint ventures is similar to BOT contracts. But unlike BOT contracts, they often incorporate ownership and financial obligations (cash and assets) for local partners, in addition to the foreign partners.

During the early 1990s, cooperative joint ventures between foreign investors and local partners were the preferred arrangement in China's urban water sector in the absence of appropriate guidelines. The local partners understood the local market, facilitated negotiations with local governments and generally allowed the foreign investors to hedge risks (Lee, 2010). Cooperative joint ventures are governed by the Law of the PRC on Sino-foreign Cooperative Enterprises or the Sino-Foreign Cooperative Joint Venture Law (adopted on 13 April 1988 and amended in 2000), which is administered by the former MOFTEC (now Ministry of Commerce), and the PRC Sino-foreign Cooperative Joint Venture Law Implementing Rules (promulgated on 4 September 1995). An example of a cooperative joint venture is the Sino French Water Development between Suez Environnement (France) and NWS Holdings Ltd (Hong Kong).

Equity joint venture

An equity joint venture is a partnership between a foreign and domestic individual, company or enterprise, or financial organization, which has been approved and registered by the government as a limited liability company. The participation of a domestic partner is mandatory and the minimum level of foreign participation is 25 percent of the registered capital (which may encompass financial as well as non-financial assets, subject to government approval).

The parties are jointly responsible for investment and management and the profits, risks and losses are divided in proportion to the equity shares invested by each party (capital contributions or ratio of investment or registered capital). Unlike a cooperative joint venture, government approval is required to transfer or withdraw equity.

Foreign investors in China adopted this model in order to avoid risks, especially during the pilot BOT phase. The governing law is the Law of the People's Republic of China on Sino-Foreign Joint Equity Enterprises (adopted in July 1979 and amended in 1990), which was also the first legislation to permit direct foreign equity investment in China. Examples include the Dachang Water Plant in Shanghai, the Chengdu No. 6 Water Treatment Plant B project in Tianjin and the Beijing No. 10 water project.

9.4.2 Wholly Foreign-owned Enterprises

Foreign companies are the only investors in, and owners of, wholly foreign-owned enterprises. They are also solely responsible for profits and losses; there is no domestic equity or management involvement. The non-involvement of the public sector avoids a potential conflict of interest between the government's role as the regulator and as the (indirect) owner of project companies, as is the case in joint ventures. A wholly foreign-owned enterprise is required to register as a legal person and it is restricted to certain businesses. There is a minimum capital contribution (registered capital) requirement. Any extension is subject to government approval.

The applicable law is the Law of the PRC on Sole Foreign Investment Enterprises (adopted on 12 April 1986 and amended in 2000) and the Detailed Implementation Rules for the Wholly Foreign-Owned Enterprise Law (issued by the former MOFTEC on 12 December 1990).

9.4.3 Joint Stock Company

The municipal government may convert a state-owned enterprise or water utility into a joint stock company by share transfer, that is by selling or transferring a part of its stock or equity to private investors (Fu et al., 2008). The domestic water company already holds some form of concession, which is modified to accommodate the private investor upon equity transfer (Yuan and Qin, 2004). The government retains control over strategic assets while the private sector entities can bring in additional capital and management expertise. The primary difference between a share transfer and joint venture is that the funding brought by the private investors will go to the local government in a share transfer and remain in the joint company in a joint venture (Jiang and Zheng, 2010).

Initially, the Chinese government did not permit the incorporation of joint stock companies with foreign investment, thus preventing any kind of financing through the issuance of stocks on the domestic market (Vernay and Gasne, 2004). However, equity or share transfer in an existing water company has emerged as one method to attract foreign investment. The joint stock company model was first launched in Shanghai, in 2002, when Veolia won the contract to acquire a 50 percent equity share of the Shanghai Pudong Water Supply Corporation (Jiang and Zheng, 2010). In 2004, the Beijing Capital Veolia Water Investment Company, a joint venture of Veolia and the Beijing Capital Corporation, was awarded the 50-year contract for the Shenzhen Water Group Co. Ltd. The municipal government owns the majority stake (55 percent). Veolia completed a similar deal in Lanzhou in 2007.

9.4.4 Concession

There are a number of concession-based water supply projects in China involving both foreign and domestic investors. Examples include the Tanzhou (Guangdong) concession, which was signed in 1992, through a joint venture with Sino French Water Development, and marked the entry of foreign water companies into the Chinese market. While a large number of concessions are operational (given their long duration), the Macau concession of 1985 serves as an example of a concession that has been concluded and recently renewed in 2010.

In terms of market access, the concessions were initially limited to plants; but recently distribution networks have also been opened to outside investment (Yuan and Qin, 2004). However, the number of integrated water supply and distribution concession contracts in China is low because tariffs are low and the expansion of capital investment is prioritized over service efficiency (Jensen and Reid, 2007). Further, the 2004 Measures use the term 'concession' loosely, and in practice, concessions have been designed around the features of BOT contracts.

9.4.5 Build–Operate–Transfer (BOT) contracts

A BOT contract incorporates most of the features of concessions but it is oriented towards construction of new water supply infrastructure. This approach is viable where the investment environment is less favorable (Chen and Messner, 2005). The 1995 BOT Circular led to the implementation of pilot BOT projects in the urban water sector in China. This was followed by more BOT projects during the early 2000s with the re-emergence of a bottleneck effect of infrastructure on economic development. In most cases, a franchise agreement is signed between a local government department and the foreign investment enterprise, including Sino-foreign joint ventures and cooperatives and wholly foreign-owned enterprises. The BOT model is also known as 'franchise investment and financing mode' (Li, 2007).

9.4.6 Other Forms of Private Sector Participation

Management contracts and leases are the two forms of private sector participation that are recommended for systems with weak enabling institutional environments, such as China. However, one of the primary reasons for the introduction of private sector participation for urban water supply in China is the injection of capital. Further, the performance of local officers is often judged and evaluated by the investment share of the private partner (Choi et al., 2010). As management contracts and leases do not impose any investment obligations or commercial responsibility on the private operator, they are not the popular forms of public–private partnership in China. One exception is the management contract for the Shenyang Water Supply Project.

There are very few cases of full-scale privatization or divestiture in the Chinese urban water sector. Full-scale privatization has occurred mainly in small projects in specific provinces because the government will not sell profit-making water firms or plants (Globalization Monitor, 2009).

9.4.7 Corporatization

The failure of the state to provide water services to its citizens has been cited as one of the main reasons for recourse to private sector participation in several developing and transition countries. However, it is the discourse on efficiency, capital investments and service coverage that led to private sector participation in China's urban water supply sector (Zhong et al., 2008). As a result, in many cases, private sector participation in China is restricted to projects that cannot be otherwise operated and managed by government water utilities. In other cases, rather than opting for privatization, former state-owned enterprises have been transformed into profitable corporatized entities.

The corporatization of government departments that are responsible for urban water supply into state-owned corporations and enterprises involves detaching their asset management side from the government and introducing corporate management techniques to their administration. The state-owned enterprises can trade freely and assume sole responsibility for their profits and losses (Globalization Monitor, 2009). The advantage of such 'privatized' water companies lies in their ability to mimic the structure and efficiency of private corporations while assuring that social objectives are still emphasized through public ownership (Harding and Preker, 2000). This, it is argued, makes corporatization a suitable model for increasing efficiency of governance in public water utilities and a viable alternative to total privatization.

One example of corporatization is the Shenzhen Water Group Co. Ltd, which was formed in 2001 after the merger of the former Shenzhen Water Company and the drainage sector in Shenzhen city. It is also an example of share issue privatization as it became a listed company in 2004. The Shenzhen government holds 55 percent of the company's shares, thus retaining control. In other cases, the government may retain control by introducing different ownership classes of shares and constraining agents who hold the shares (Globalization Monitor, 2009).

9.4.8 Regional Distribution of Public–Private Partnerships

The form of urban water governance is usually determined by the level of development of water infrastructure as well as the local economic, social and political conditions. The regional distribution of China's urban water supply industry has a close relationship with urban development. Private investors choose utilities and cities with the greatest potential for profit. As a result, private sector participation is mainly concentrated in the comparatively urbanized (Shandong, Guangdong and Zhejiang) and populous (Shandong, Sichuan, Beijing and Shanghai) areas. For instance, the joint-venture approach is more often adopted in the southern coastal provinces (for example, Guangdong and Fujian) and the eastern coastal provinces (for example, Jiangsu) provinces with richer markets, more open economic policies and higher payment capacity of local residents (Zhong et al., 2008; Globalization Monitor, 2009). The corporatization model is more common among larger projects because of larger capital demands and the huge labor redundancies within such projects. It has been adopted more widely in inland provinces (for example, Gansu, Heilongjiang, Jilin, Sichuan, Xinjiang, Yunnan) than in coastal provinces (Zhong et al., 2008).

9.5 PRIVATE SECTOR PARTICIPATION IN THE URBAN WATER SUPPLY SECTOR

This section highlights the foreign and domestic water companies involved in the urban water sector in China.

9.5.1 Foreign Water Companies in the Urban Water Sector

Private sector participation in China's urban water market has been spearheaded by foreign water companies, with international experience and investment, advanced know-how, technologies and innovative management skills (Lee, 2007a). Their entry can be traced back to the year 1992, when a water production project was undertaken by the Sino-French Water Development Company Limited (SFW), a joint venture between Suez Environnement (France) and NWS Holdings Ltd (Hong Kong), in the city of Zhongshan. As of 2010, SFW operated 25 contracts in 18 municipalities in China, and served approximately 14 million domestic and industrial consumers (SFW, 2010).

Suez's main competitor, Veolia Water (France), established its strong position in the Chinese urban water market by winning the contract for the Chengdu No. 6 BOT water supply project in Tianjin, Sichuan Province, in

1996. Veolia has deployed the strategy of offering high premiums in order to win bids (with the expectation that water tariffs will increase in the future). As of 2011, Veolia was involved in 22 municipal projects in China, and provided urban water services to over 38 million people, including 27 million customers via full service concessions (Veolia, 2011).

Local governments have played an important part in facilitating the entry of foreign water companies into China's urban water market. They have pursued aggressive foreign investment and local economic growth, and cooperated in expediting approval processes (Wang et al., 2000). Following the 2002 revision of the Catalogue, foreign water companies have entered into joint ventures with local municipal water authorities in order to hedge against risks. For instance, the SFW has entered into six joint ventures with local municipal water authorities to deliver full water services in Sanya (Hainan province), Tanzhou (Zhongshan province), Chongqing (Sichuan province), Tanggu (Tianjin province), Changshu (Jiangsu province) and Qinhuangdao (Jiangsu province) (SFW, 2010). Similarly, Veolia Water has partnered with Beijing Capital Co. and some local water companies. In 1997, Veolia (then known as Vivendi) and the local government of Tianjin set up Tianjin General Water Company, a joint venture with 55 percent of shares held by Vivendi. This was the first time that the government allowed a foreign company's franchise to operate existing waterworks in China (Li, 2007).

The other less influential foreign companies in China's urban water sector are Bouygues (SAUR), Thames Water (now part of the German RWE Group) and Anglian Water. In 1995, the consortium of Thames Water (UK) and Bovis Construction entered into a cooperative joint venture with the Shanghai Waterworks Shibei Company for the construction and management of the Dachang water treatment plant in Shanghai. The 20-year concession agreement was the result of private negotiation between the parties. It was the first urban water supply project to be wholly financed by foreign capital and, therefore, financial risks at the municipal level were limited. The central government did not issue any symbolic letters or guarantees for the project. At that time, the success of the project was attributed to investor confidence in the political and economic stability and strength of the Shanghai government as well as favorable economic conditions for China's infrastructure projects (Donoghue et al., 1999). However, the project also illustrates the uncertain and risky nature of China's urban water market. In 2002, the State Council revoked the guaranteed rates of return for foreign investment and the contract provision that guaranteed a fixed return of 16 percent per annum became illegal. The parties failed to renegotiate the terms of the contract and the contract was terminated. In June 2004, the plant was sold back to the

Shanghai Shibei (Northern City) Water Treatment Corporation (Lee, 2006, 2007a; Fu et al., 2008).

Specialized water operators from other Asian countries (mostly Chinese expatriates from Hong Kong, Malaysia and Singapore) have also entered the urban water market in China. These include subsidiaries of Darco (Singapore) in Deqing (Zhejiang province); Salcon Engineering Berhad (Malaysia) in Yizheng (Jiangsu province), Linyi and Weifang (Shandong province) and Kunming (Yunan province); Puncak Niaga Holdings Bhd (Malaysia) in Yuanshi county (Hebei province); and Ranhill Utilities (Malaysia).

External agencies, such as the World Bank's Multilateral Investment Guarantee Agency (MIGA), have guaranteed foreign investment in China's urban water sector against risks. Examples of such projects and foreign investors include Shanghai Pudong – Veolia Water Corporation and Shenzhen Water (Group) Company Ltd (Veolia); Chongqing Water Project (Suez); and Zhenjiang Golden State Water Co. Ltd (Golden State Water Group Corporation) (MIGA, 2011).

9.5.2 Domestic Private Companies in the Urban Water Sector

The involvement of domestic companies in China's urban water sector has undergone a dramatic shift. Earlier, domestic private companies having a background in construction and environmental engineering focused on the wastewater treatment sector alone. Over the years, their close association with the water-related bureaus of local governments has facilitated their smooth transition to the urban water supply sector. Examples include the Beijing Sound Group and Tsinghua Tongfang Water Engineering Corporation.

However, the dominance of these domestic private companies gradually diminished after the lifting of the ban on private investment in the water distribution network in 2002. This period witnessed the emergence of Chinese investment developers and financial institutions in collaboration with Chinese water service companies (Lee, 2010). Choi et al. (2010) categorize them as quasi-private developers because these organizations are generally state-owned enterprises, either at the central or local level. The Xiamen Water Group, which was formed in 2004, as a result of a cooperative joint venture between Zhong Environmental Protection Water Investment Company, a local investment company, and Xiamen Municipality, is one example.

Additionally, in large cities, such as Beijing, Tianjin and Shanghai, rapid economic development has resulted in a large influx of taxes and public money. As a result, local governments have established their own subsidiary water companies and financial vehicles for the provision of urban water services. These include the Beijing Capital Group, Tianjin Capital Environmental Protection, and Shanghai Industrial Holdings. Some of them have begun to be partially privatized and in some cases, such as the Beijing Capital Group, these entities are competing with foreign water companies for market share.

The Chinese government has supported the development and expansion of these different domestic water companies. Many companies have some form of relationship with governmental bureaus and agencies: they are either former state-owned enterprises or the state is their majority shareholder. Usually, the managerial personnel are former government officers, and this facilitates access to market information and good relationships with local governments. It may be argued, therefore, that the 'private' nature of these domestic water companies is different from that of foreign or transnational water companies.

9.6 CONCLUSION

China has adopted a problem-oriented and 'develop first, control later' approach towards the urban water supply sector. As a result, it lacks a robust regulatory regime that is suited to its specific institutional characteristics. The development of the legal and policy framework illustrates these limitations. Therefore, interested private investors will have to adjust their expectations concerning China's urban water supply sector. Further, as the sector is driven by local provincial policies and approvals rather than the decisions of the central government, it is important to abandon the focus on the water sector as a whole in favor of local and regional approaches.

The government is pursuing urban water sector reforms. However, the scope of private sector participation is restricted and the government retains majority control. The nature of private sector participation is also different.

The government possesses limited financial capacity and know-how to proceed with the required investment to modernize and develop its water supply plants and networks, and it is technologically backward in water treatment and production. It relies on international companies to provide the best and most up-to-date equipment and the latest innovations for a competitive price. This is expected to lead to a shift in the nature of private sector participation – from water supply and wastewater to development of advanced water technologies.

NOTES

1. The full text is available at http://www.npc.gov.cn/englishnpc/Law/2007-12/11/content_1383554.htm.
2. It is also referred to as the Circular on Issues Relating to Attracting Foreign Investment to BOT Projects, Notice on Relevant Issues Concerning Attraction of Foreign Investment with BOT Mode, Circular on Several Questions concerning the Use of BOT as a Form to Attract Foreign Investment or the Circular Concerning the Issues of Absorbing Foreign Investment through BOT.
3. It is also referred to as the Circular on Several Issues Concerning the Examination, Approval and Administration of Experimental Foreign-Invested Concession Projects or Notice on Relevant Issues Concerning Approval Administration of Foreign Investment Franchise Projects.
4. The Catalogue was prepared by the State Planning Commission, the State Economic and Trade Commission and MOFTEC. It is also known as the Foreign Investment Industrial Guidance Catalogue.
5. It is also referred to as the Notice on Appropriate Handling of Current Issues Concerning Guarantee of Foreign Investment Project with Fixed Return or Notice on Appropriate Handling of Existing Problems in Guarantee of Foreign Investment with Fixed Return Projects.
6. It is also referred to as Circular on Accelerating Marketization/Privatization of Public/Municipal Utilities, Quickening the Process of General Adoption of Market Principles for the Municipal Public Utilities Sector, Opinions on Acceleration of Privatization Process of Public Facilities or Opinions Concerning the Acceleration of the Marketization of Urban Utilities Industries.
7. It is also referred to as Measures for the Administration of Concessionary Operation of Urban Utilities Industries, Administrative Rules for Concession of Municipal Facilities, Management Measures on Concessions of Municipal Public Utilities, Method of Managing Urban Public Utility Concessions, Administrative Measures on the Concession of Municipal Public Utilities or Administrative Method of Urban Utilities Concessions.

REFERENCES

Asian Development Bank (ADB) (2010), 'Urban water supply: case study of public–private partnerships (PPPs) in Maanshan', available at http://www.adb.org/documents/urban-development/urbandev-prc-nov2010-watersupply.pdf (accessed 10 October 2011).

Blackman, C. (2001), 'Local government and foreign business', *China Business Review*, **28**(3), 26–31.

Browder, G., S. Xie, Y. Kim, L. Gu, M. Fan and D. Ehrhardt (2007), *Stepping Up – Improving the Performance of China's Urban Water Utilities*, Washington, DC: World Bank.

Carmody, L. (ed.) (2010), *Issues for Responsible Investors – Water in China*, Singapore: Responsible Research.

Chen, C. (2009), 'Can the pilot BOT Project provide a template for future projects? A case study of the Chengdu No. 6 Water Plant B Project', *International Journal of Project Management*, **27**(6), 573–83.

Chen, C. and H. Doloi (2008), 'BOT application in China: driving and impeding factors', *International Journal of Project Management*, **26**(4), 388–98.

Chen, C. and John I. Messner (2005), 'An investigation of Chinese BOT projects in water supply: a comparative perspective', *Construction Management and Economics*, **23**(9), 913–25.

Cheng, C. and Zhengxu Wang (2009), 'Public private partnerships in China: making progress in a weak governance environment', Issue 56, Briefing Series, China Policy Institute, University of Nottingham.

Choi, Jae-ho, J. Chung and Doo-Jin Lee (2010), 'Risk perception analysis: participation in China's water PPP market', *International Journal of Project Management*, **28**(6), 580–92.

Donoghue, N., S. Nelson and D. Smith (1999), 'Water projects: growth for foreign investment, financing problematic', in B. Sorabs and D. Smith (eds), *Project Finance Models for Greater China*, Hong Kong: Asia Law and Practice.

Fu, T., M. Chang and L. Zhong (2008), *Reform of China's Urban Water Sector*, London: IWA Publishing.

Globalization Monitor (2009), 'The reform of the urban water supply in Southern China – water privatization in China', available at http://www.globalmon.org.hk/en/wp-content/uploads/2009/04/water-privatization-in-south-china_mar09.pdf (accessed 10 October 2011).

Hanley, Thomas P. (1999), 'The BOT Circular: an evaluation of the new regulatory framework governing privately-financed infrastructure projects in the People's Republic of China', *Standford Journal of Law, Business and Finance*, **9**, 59–98.

Harding, A. and A.S. Preker (2000), 'Understanding organisational reforms: the corporatization of public hospitals', Health, Nutrition and Population (HNP) Discussion Paper, World Bank Human Development Network.

Jensen, O. and S. Reid (2007), 'Privatization and regulation of China's water sector', in *Water Markets Perspective*, London: ICF International, available at http://www.icfi.com/Markets/Energy/doc_files/water-markets-china.pdf (accessed 10 October 2011).

Jiang, Yi and Xiaoting Zheng (2010), 'Private sector participation and performance of urban water utilities in the People's Republic of China', ADB Economics Working Paper Series No. 237, Manila: Asian Development Bank.

Lee, S. (2003), 'Expansion of the private sector in the Shanghai water sector', Occasional Paper No. 53, SOAS Water Issues Study Group, available at http://www.soas.ac.uk/water/publications/papers/file38396.pdf (accessed 10 October 2011).

Lee, S. (2006), *Water and Development in China: The Political Economy of Shanghai Water Policy*, Hackensack, NJ: World Scientific.

Lee, S. (2007a), 'Private sector participation in the Shanghai water sector', *Water Policy*, **9**(4), 405–23.

Lee, S. (2007b), 'Public–private partnership in the urban water sector in Shanghai', in Fulong Wu (ed.), *China's Emerging Cities: The Making of New Urbanism*, Abingdon: Routledge, pp. 44–65.

Lee, S. (2010), 'Development of public private partnership (PPP) projects in the Chinese water sector', *Water Resources Management*, 24(9), 1925–45.

Leman, Edward A. (1996), 'Key features of successful BOT: structuring and managing risk', Proceedings of the Seminar on BOT in the Water Supply Sector in the People's Republic of China, Asian Development Bank/Ministry of Water Resources, People's Republic of China.

Li, Q. (2009), 'Administrative rules of Shenzhen for BOT sewage treatment projects: defining ownership of BOT water projects in China', *China Bulletin*,

available at http://www.kingandwood.com/files/20091027/File/2009-10-3-LQ.pdf (accessed 10 October 2011).
Li, S. (2007), *Legal Environment and Risks for Foreign Investment in China*, Berlin: Springer.
Mahony, T.M. and L. Fong (2002), 'Foreign investment opportunities in China's water distribution sector', China Law and Practice.
MIGA (Multilateral Investment Guarantee Agency) (2011), 'MIGA: guaranteeing investments in water projects', available at http://www.miga.org/documents/waterbrief.pdf (accessed 20 February 2012).
PPIAF (2010), 'Private activity in water and sewerage declines for second consecutive year', PPI data update Note 37, available at http://ppi.worldbank.org/features/June2010/PPI-Water-note-2009.pdf (accessed 10 October 2011).
Rozner, S. (ed.) (1998), *Infrastructure Financing Strategies in the PRC, A China Law and Practice Guide,* Hong Kong: Asia Law & Practice Publishing.
Sino French Water (SFW) (2010), 'Corporate brochure – committed to China's sustainable development', available at http://www.sinofrench.com/asset/files/Publication/Corporatepercent20Brochure/51855-30-08-E.pdf (accessed 10 October 2011).
Tjoa, L. and J. Zhou (2005), 'In the mood for investment: new opportunities in China's infrastructure projects?', *China Law and Practice*, April.
Turner III, E.L. and A.D. Seem (1999), 'The new project finance – legal solutions, legal strategies', in B. Sorab and B. Rogers (eds), *Project Finance Models for Greater China*, Hong Kong: Asia Law & Practice, pp. 13–21.
Veolia Water (2011), 'Veolia Water in China', available at http://www.veoliawater.com/about/locations/china.htm (accessed 10 October 2011).
Vernay, S. and S. Gasne (2004), 'China water sector – is a new model for private investment taking shape in the Chinese water sector?', *Water Business Intelligence*, pp. 19–22, available at http://www.gide.com/front/files/GlobalWaterIntelligence_102004.pdf (accessed 10 October 2011).
Wang, H. and X. Zhao (2004), 'PPP case study on China's urban water supply industry – Factory B of Veolia Chengdu No.6 waterworks', Unirule Institute of Economics, China Center for Public Partnerships.
Wang, Shou Q., R. Tiong, S. Ting and D. Ashley (2000), 'Evaluation and management of political risks in China's BOT projects', *Journal of Construction Engineering and Management*, **126**(3), 242–50.
Wang, Z. (2008), 'Urban water sanitation in China', 2nd PECC Water Management Seminar, New Caledonia, 2008, available at http://www.pecc.org/resources/doc_view/1133-urban-water-sanitation-in-china (accessed 12 October 2011).
Woetzel, J., L. Mendonca, J. Devan, S. Negri, Y. Hu, L. Jordan, X. Li, A. Maasry, G. Tsen and F. Yu (2009), *Preparing for China's Urban Billion*, Beijing: McKinsey Global Institute.
Yuan, J. and Y. Qin (2004), 'A look at concession rules for the urban water industry', *China Law and Practice*, April, http://www.chinalawandpractice.com/Article/1692880/Channel/9932/A-Look-at-Concession-Rules-for-the-Urban-Water-Industry.html.
Zhang, Xue-qing and Mohan M. Kumaraswamy (2001), 'BOT-based approaches to infrastructure development in China', *Journal of Infrastructure Systems*, **7**(1), 18–25.
Zhong, L., A.P.J. Mol and T. Fu (2008), 'Public–private partnerships in China's urban water sector', *Environmental Management*, **41**, 863–77.

10. Development of public–private partnerships in the water sector in Singapore

Robert L.K. Tiong, Zheng Sha and Abu Naser Chowdhury

10.1 INTRODUCTION TO WATER MANAGEMENT IN SINGAPORE

Singapore is a small city-state located south of the Malay Peninsula, 137 km north of the equator. With a land area of only 712.4 km^2 and population size of 5.18 million (as of 2011) (Department of Statistics Singapore, 2011a), Singapore has a population density of 7252 per km^2, making it the third-highest among 216 countries in terms of population density in 2010 (World Bank, 2011b). The country has a tropical rainforest climate and in 2010 the annual rainfall was 2075mm (Department of Statistics Singapore, 2011a). However, Singapore does not have enough land to store the rainfall. Singapore has a gross domestic product (GDP) of US$303.7 billion with 14.5 percent real economic growth in 2010 (Department of Statistics Singapore, 2011a) and its per capita GDP ranked thirteenth-highest in 2010 (World Bank, 2011a). Figure 10.1 shows the time-series on population and GDP to demonstrate Singapore's economic and human growth over time (Department of Statistics Singapore, 2011b, 2011c).

The high population density, active economy and critical land constraints impose considerable pressure on water management in Singapore. From the 1980s to the 1990s Singapore made huge efforts to create a comprehensive environmental management system including water supply, control of river pollution, establishment of well-planned industrial estates, and a world-class urban sanitation system for the whole island. More recently, the Singapore Government has made sustainable water supply the main target of water management. The Public Utilities Board (PUB), Singapore's national water agency, is responsible for managing the entire water loop from rainwater collection, water production, water distribution to used water treatment. It is a statutory board under the Ministry

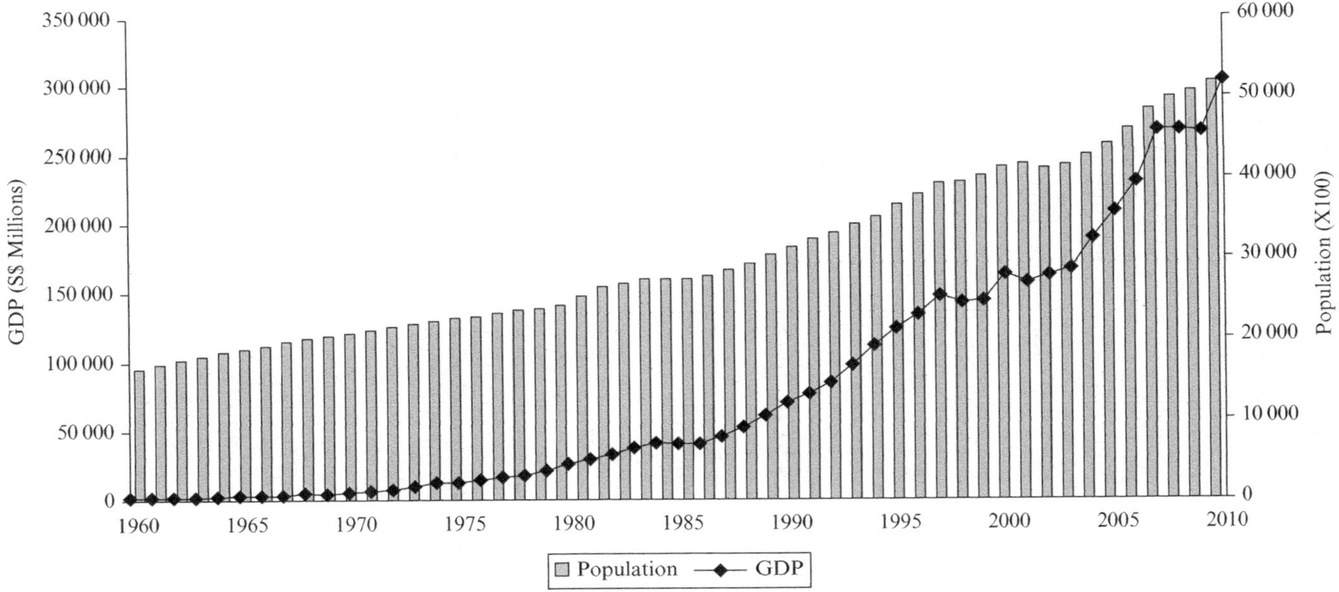

Source: Department of Statistics Singapore (2011b, 2011c).

Figure 10.1 Singapore's GDP and population

of the Environment and Water Resources (MEWR). The straightforward administrative framework is efficient for integrated water management. PUB has been investing steadily in water infrastructure in Singapore. Starting with in-house design and supervision combined with outsourcing of construction (that is, build-to-design approach), PUB has progressed to outsourcing some of its major water infrastructure projects through public–private partnership (PPP) to handle water demand and water resource management in Singapore. Private investment in the form of PPP is expected to increase the water utilities business, which would further improve investment efficiency and promote the capacity of the local water industry.

The key purpose of this chapter is to discuss the development of PPP in Singapore and its procurement in the water sector. It provides an overview of four PPP water projects with particular focus on the Tuas Desalination Plant, delivered based on the design–build–own–operate (DBOO) model.

10.2 STRATEGIES FOR SUSTAINABLE WATER SUPPLY

10.2.1 The Four National Taps

Singapore has implemented the 'Four National Taps' strategy to ensure sustainable water supply. The four National Taps are water supply from local catchment, imported water from Malaysia, high-grade reclaimed water known as NEWater and desalinated water.

After independence in 1965, reservoirs and imported water from Malaysia were the two main water sources for Singapore. Imported water from Malaysia has played an important role in Singapore's water supply. Two water agreements between Singapore and Malaysia were signed in 1960 and 1961 respectively to import water from Johor in Malaysia for 50 years (expired in August 2011) and 99 years (to expire in 2061), respectively (Soon et al., 2009). In order to strengthen local water storage capacity, Singapore is enlarging its local catchment. Currently, there are 17 reservoirs and about two-thirds of Singapore's land is local catchment (Figure 10.2). These two sources (water supply from Malaysia and local sources of Singapore) form Singapore's first two National Taps.

NEWater and desalination water are the third and fourth National Taps of Singapore. NEWater and desalinated water have been introduced more recently as non-conventional water resources to supplement catchment and imported water. Singapore started to supply NEWater in 2002, and currently has four plants at Bedok, Kranji, Ulu Pandan and Changi. The

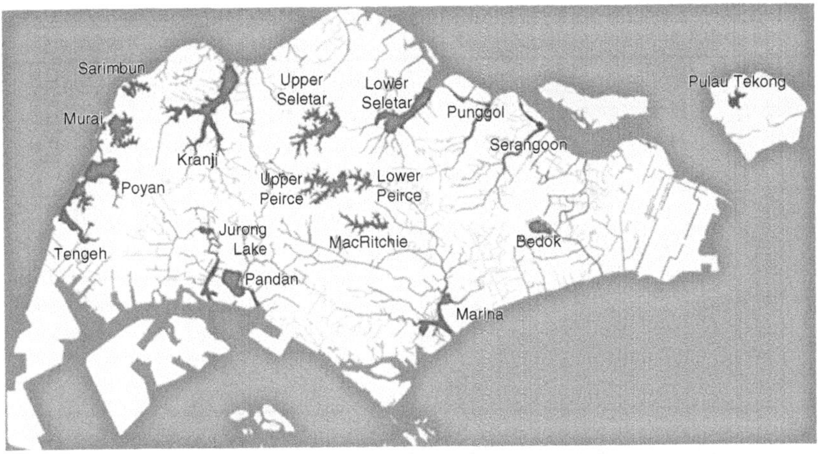

Source: Public Utilities Board (http://www.pub.gov.sg/water/Pages/LocalCatchment.aspx).

Figure 10.2 Reservoirs of Singapore

Ulu Pandan and Changi plants were developed under the PPP approach, and they are operated by private operators. Under this PPP arrangement, a private concession company is engaged to design, build, own and operate (DBOO) the NEWater plants and the desalination plants to supply water to PUB. Distribution of water to the consumers remains under PUB's purview. NEWater is mainly used for non-potable industrial and commercial uses with a small portion added into the reservoirs for indirect potable use (that is, blended with reservoir water before conventional treatment at PUB's waterworks). Desalinated water is a more expensive alternative that requires extensive energy consumption, but considering that seawater is a drought-proof resource and Singapore is surrounded by the sea, seawater desalination is an important 'non-conventional tap' to diversify Singapore's water resources. With advancement in technology and innovation to improve energy efficiency, seawater desalination can potentially become an affordable source of water. Singapore commissioned the first desalination plant in Tuas in 2005, and the second Tuas desalination plant is currently under construction, and it is expected to be completed by 2013. The total supply capacity of desalinated water and NEWater in 2010 met 10 percent and 30 percent of the water demand in Singapore respectively. With new developments planned, NEWater and desalinated water are targeted to supply 80 percent of the water demand (30 percent desalinated water and 50 percent NEWater) in 2060 (Public Utilities Board Singapore,

Table 10.1 Water tariff structure in Singapore

Tariff category	Consumption block (m^3 per month)	Tariff (SGD/m^3) (before GST)	Water conservation tax (% of tariff) (before GST)	Waterborne fee (SGD/m^3) (after GST)	Sanitary appliance fee (after GST)
Domestic	0 to 40	1.17	30	0.30	SGD 3.00 per chargeable fitting per month
	Above 40	1.40	45		
Non-Domestic	All units	1.17	30	0.60	
Industrial Water	All units	0.52	–	–	–

Source: Public Utilities Board Singapore (2011b).

2010b). This is the national strategy to explore unconventional water supply alternatives and reduce reliance on imported water from Malaysia.

10.2.2 Water Tariff Structure

Long-term sustainability in water cannot be achieved by boosting water supply alone. Water demand management is also critical and pricing tools are one of the ways to achieve it. In Singapore, water is priced not only to recover the full costs of its production and supply, but also to reflect the higher cost of additional water supplies such as desalination through a Water Conservation Tax.

The structure of the water tariff is shown in Table 10.1. The increasing block tariff structure is used for domestic use. Waterborne fee and sanitary appliance fee are used to offset the cost of treating used water and for the operation and maintenance of the public sewerage system. Industrial water is mainly non-potable reused water, thus water conservation tax is not applied to encourage water reuse. Tortajada (2006) concluded that the progressive increase in water tariff between 1997 and 2000 has reduced the household water demand in Singapore.

As shown in Table 10.1, an increasing block rate water tariff structure is used. An increase in fees up to SGD 1.4/m^3 is collected when the amount of water used exceeds 40 m^3/month. The government levies a Water Conservation Tax to reinforce the water conservation message. In 2010, the water demand was about 1730000 m^3/day, with 45 percent for domestic use and 55 percent for non-domestic use (Public Utilities Board Singapore, 2010b). The domestic water demand per capita has decreased from 176 liters per day in 1994 to 154 liters per day in 2010 (Ministry of the Environment and Water Resources Singapore, 2008, 2011). In addition to water pricing, Singapore has implemented several incentives, national campaigns, and public education programs for water conservation. These together constitute effective measures for water management.

10.3 DEVELOPMENT OF PPPS IN THE WATER SECTOR IN SINGAPORE

10.3.1 Overview

Traditionally, water-related facilities were constructed under conventional government procurement practices and managed by PUB. The private sector was involved as contractors, equipment suppliers and consultants under traditional construction procurement methods which did not involve capital investment and/or long-term operational input from private sector players. Greater private participation in the water sector has been introduced since 2003 through PPP procurement. PUB is the first public agency in Singapore to adopt a PPP approach. PUB outsourced some of its major water infrastructure projects through PPP where the design, construction, operation and maintenance are undertaken by the private sector. In order to select the best sourcing strategy, PUB has progressed with the DBOO model. Under the DBOO model, a private project company is responsible to design, build, own and operate the relevant water facilities, and deliver water supply services during a contracted period. The reason for PUB to adopt the DBOO scheme is to optimize the life-cycle costs of the projects, optimize the configuration of technologies and thus to achieve a competitive price.

The SingSpring Desalination project in Tuas, awarded in 2003, is the first such PPP project in Singapore. With the experience gained from this project, three other PPP water projects were subsequently implemented: the Ulu Pandan NEWater project, the Changi NEWater project and the Second Tuas Desalination project. As is the case with most PPPs in the utilities sector which had been previously financed, developed and managed by the public sector, PUB is responsible for purchasing water supply services from the private project companies in Singapore. Thus there is no direct engagement between the end-users and the private sector developers. Another feature of the PPP model used in the water sector in Singapore is the strong presence of local companies in each of the four water PPP projects. The Second Tuas Desalination project is the most recent one, and it is currently under construction. The other three projects are currently under operation. A summary of the four water PPP projects in Singapore is listed in Table 10.2.

10.3.2 Procurement of PPP Projects

The Government of Singapore applies PPP to large-value projects where the project value is likely to be more than SG$50 million. A typical PPP procurement process in the water sector is shown in Figure 10.3. PUB conducts pre-tender feasibility study for a water project to ensure whether it will be feasible under PPP procurement before issuing a PPP tender. For the SingSpring Desalination project, bidders were invited by PUB for prequalification exercise, before the Request for Proposal was issued to prequalified bidders. For the Second Tuas Desalination project, the Ulu Pandan NEWater Project and the Changi NEWater project, the bidders did not have to go through the pre-qualification stage in the procurement process because of the learning experience from SingSpring Desalination procurement.

For the SingSpring Desalination project (or the first Tuas Desalination plant), the technology was new, and so was the PPP procurement model. For the NEWater projects, PUB had operated three NEWater plants before the first NEWater PPP project was put up for tender. PUB shared the relevant experiences and NEWater technology with the private sector for the Ulu Pandan NEWater project. This enhanced the confidence of the market in NEWater PPP projects. The Request For Proposal (RFP) for the SingSpring Desalination project was issued in November 2001 and it achieved financial closure in January 2004. It took about 26 months to reach the construction stage. The second desalination project took only 12 months for the tender period from the issuance of the RFP in June 2010 to financial closure in July 2011. The RFP for the first NEWater PPP project, the Ulu Pandan NEWater Project, was issued in December 2004 and it achieved financial closure in March 2005. It took ten months to reach the construction stage. The subsequent NEWater PPP project, the Changi project, took only nine months from the issuance of the RFP in August 2007 to achieve financial closure and to commence construction in May 2008.

Figure 10.4 shows the time span in the procurement of the projects. After the first desalination project, the procurement duration for the subsequent three water PPP projects improved. It took a long time for the first desalination plant to settle its land issue. Due to the flexibility given to the tenderers to propose their own site (besides the site on state land identified by PUB), the due process of leasing the site from the landowner could only kick in after the award of the desalination project. To reduce the lead-time

Table 10.2 Water PPP projects in Singapore

	SingSpring Desalination Plant	Second Tuas Desalination Plant	Ulu Pandan NEWater Plant	Changi NEWater Plant
PPP model	DBOO	DBOO	DBOO	DBOO
Client	PUB	PUB	PUB	PUB
Contracted service	Water supply of 136 000 m^3/day	Water supply of 318 500 m^3/day	NEWater supply of 148 000 m^3/day	NEWater supply of 225 000m^3/day
Tenure of contract (years)	20 (2005–25)	25 (2013–38)	20 (2007–27)	25 (2010–35)
Date of agreement	April 2003	April 2011	January 2005	February 2008
Project company or special purpose vehicle (SPV)	SingSpring	TuasSpring	Keppel Seghers NEWater Development	SembCorp NEWater
Commencement for water supply	September 2005	2013 (expected)	March 2007	May 2010
Current status	In operation	Under construction	In operation	In operation

Source: Ministry of Finance Singapore (2011) and Public Utilities Board Singapore (2010a).

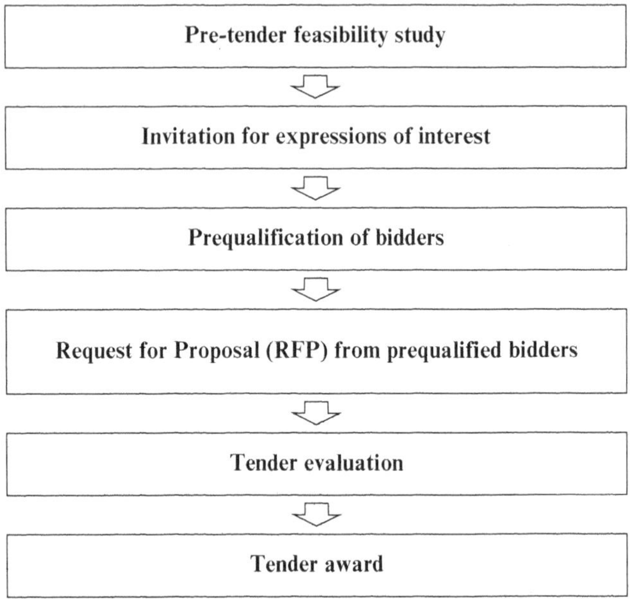

Figure 10.3 PPP procurement process

for financial closure, PUB alienated the land (prior to award of tender) and subleased it to the special-purpose vehicle (SPV) for the subsequent three PPP projects. This is one of the reasons why the tender periods of the subsequent three water PPP projects were reduced significantly.

10.3.3 Water Purchase Agreement

In water PPP and DBOO projects, there are many agreements to define and specify the responsibilities of the project parties and for allocation of risks among them. These include water purchase agreements, concession agreements, step-in agreements, financing agreements, shareholder agreements, land lease agreements, engineering procurement and construction (EPC) contracts, operation and maintenance (O&M) contracts, energy supply contracts and so on. Of these agreements, PUB is a party to the water purchase, step-in and land lease agreements. The water purchase agreement is the key contract that governs the legal and contractual rights of the private project company and PUB. The amount of water that PUB will purchase over the concession period and the payment mechanism are specified in the contract. The step-in agreement allows lenders, who have provided the debt portion of the project development cost, to step in and

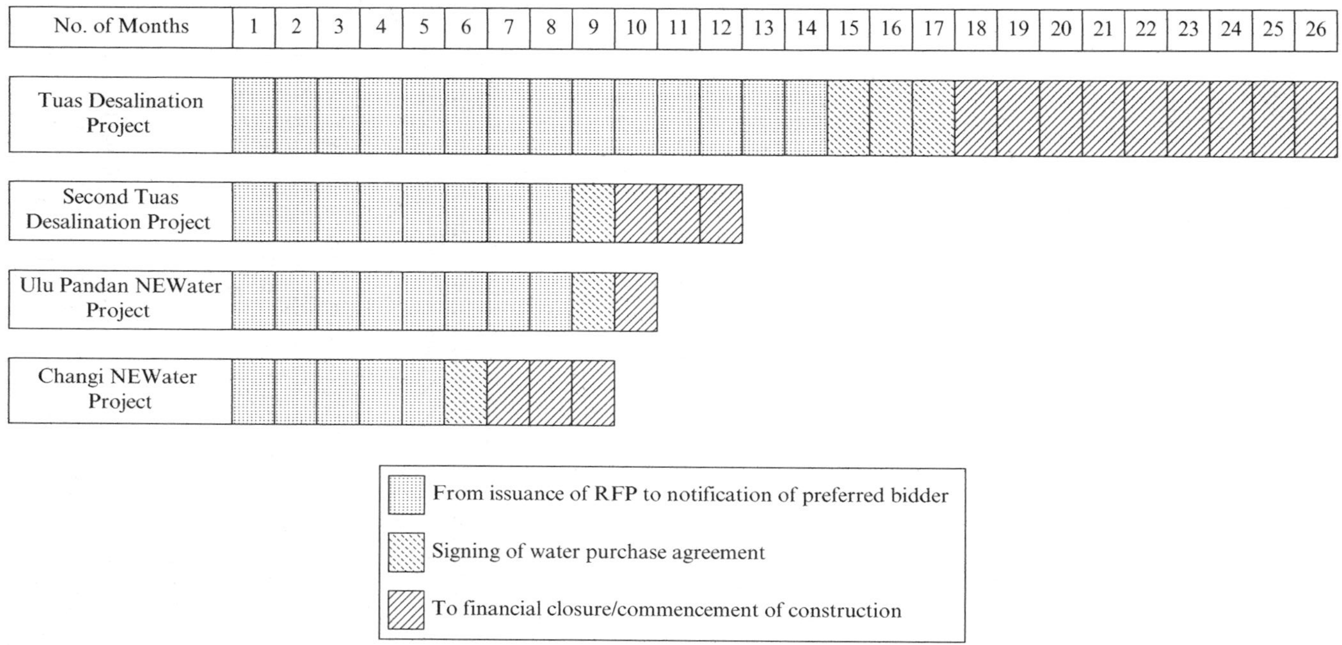

Figure 10.4 Procurement duration of PPP projects in Singapore

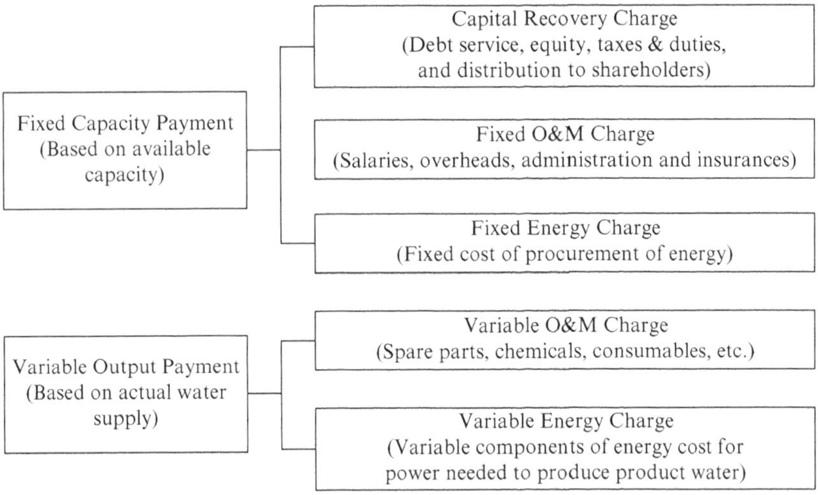

Source: Public Utilities Board Singapore (2010a).

Figure 10.5 Payment structure

rectify project's default under certain conditions defined in the water purchase agreement. By the land lease agreement, the Singapore government provides land access through lease from the Jurong Town Corporation (JTC).

Under the DBOO arrangement, the concession company primarily takes the construction and operational risks. Therefore, the concession company is given significant latitude to determine or select the various design parameters that will achieve the specified performance required under the agreements. Generally, the technical specifications are performance-based, which gives the concession company room to propose innovative and cost-effective means to meet the contractual obligations. In Singapore, an example of innovation is the coupling of desalination with on-site power generation for the second desalination plant in Tuas. Another example is the installation of reverse osmosis membranes on top of the micro-filtration feed tank in the Ulu Pandan NEWater factory.

10.3.4 Tariff Structure

The tariff structure for the two NEWater PPP projects and the two desalination PPP projects is shown in Figure 10.5. It has two components: fixed capacity payment and variable output payment. The fixed capacity payment is based on available capacity of the water plant, including capital recovery charge, fixed O&M charge and fixed energy charge. The capital recovery charge is to pay off the debt service payment, equity investment, taxes and duties, and distribution to shareholders. The fixed O&M charges cover costs such as manpower costs, overheads and administration and insurances. The fixed energy charge covers the fixed component of energy costs. The variable output payment is based on the actual water supply, including variable O&M charge and variable energy charge. The variable O&M charge covers the variable O&M costs such as spare parts, chemicals and consumables. The variable energy charge covers the variable components of energy cost for power needed to produce water. To share the long-term risks in future changes in cost due to inflation and fluctuations in fuel price, the base rates offered by the concession company are subject to indexation based on the Singapore Consumer Price Index and the HSFO 180 CST index.

10.3.5 Performance Monitoring of the Concession Company

Performance monitoring and measurements (such as key performance indicators, value for money) are critical to PPP contracts because they form the basis of the payments to the concession company. Monitoring includes day-to-day monitoring of performance, assessing whether the contracted services are delivered to the contracted standards, and evaluating the remedial actions taken by the concession company when the performance standards are not met. For the desalination and NEWater projects, a comprehensive monitoring and audit system has been put in place to allow PUB to regularly check the water quality and operation and maintenance of the plants. This includes linking the plants' key online water quality monitoring system to PUB's monitoring systems. The water is also sampled and analysed regularly by an accredited laboratory. In addition, audits are conducted regularly by PUB to ensure that the plants' operations meet the required quality standards.

10.4 CASE STUDY: SINGSPRING DESALINATION DBOO PROJECT

10.4.1 Project Features and Structure

The SingSpring Desalination DBOO project is the first PPP and the first desalination project in Singapore. It is valuable to study the detailed features and structure of the project. This section compares the SingSpring Desalination DBOO project and the Second Tuas Desalination DBOO project. The information on the two projects is taken from various sources including Chan (2011), Chowdhury and Tiong (2011), Hyflux (2011), Ministry of Finance Singapore (2011), Public Utilities Board Singapore (2010a, 2011a) and Water Technology (2011).

Hyflux was awarded the first project tender to design, build, own and operate a desalination plant with a capacity of 136 000 m^3/day. The concession period is 20 years from 2005 to 2025 and the capital cost of the plant was SG$200 million. It was the largest desalination plant in Asia at the time, and can meet about 10 percent of the water demand in Singapore. The capacity of the Second Tuas Desalination plant is even larger (318 000 m^3/day) and the project cost is SG$890 million with a 25-year concession from 2013 to 2038. The Second Tuas Desalination project was also awarded to Hyflux. Both desalination plants use the sea water reverse osmosis (SWRO) process to remove salts and other minerals from seawater.

The project structure of the first SingSpring Desalination project is shown in Figure 10.6. Hyflux is one of the world's leading fully integrated water solutions companies listed on the Singapore Stock Exchange (SGX). At the time when the contract was awarded in January 2003, Hyflux partnered with Ondeo Degremont, a unit of the French environmental company Suez, with a 70:30 stake. However, Ondeo withdrew from the partnership in June 2003 due to its restructuring. The project company, SingSpring, thus became fully owned by Hyflux. Later, in 2007, SingSpring was restructured into a trust structure, which is owned by CitySpring Infrastructure Trust (70 percent) and Hyflux (30 percent). The engineering–procurement–construction (EPC) contractor and the O&M contractor are Hydrochem and Hyflux Engineering, respectively, which are subsidiaries of Hyflux. In the Second Tuas Desalination project, the same project structure was applied. The EPC and O&M contracts were again undertaken by the fully owned subsidiaries of Hyflux. This structure has helped Hyflux to access more business opportunities, and develop its service capacities as an integrated package. In terms of efficiency, it is useful to combine the interests of different subsidiaries under the same parent company.

The SingSpring desalination plant was built in Tuas, which is located in the western part of Singapore. SingSpring leased the land from Jurong Town Corporation (JTC) (Singpaore's principal developer and manager of industrial estates and their related facilities) for the first desalination plant. As tenderers were given the flexibility to propose an alternative site, the due process of leasing the site from the landowner could only kick in after the award of the desalination project. Experience from the first desalination project has shown that the land-lease process is a protracted

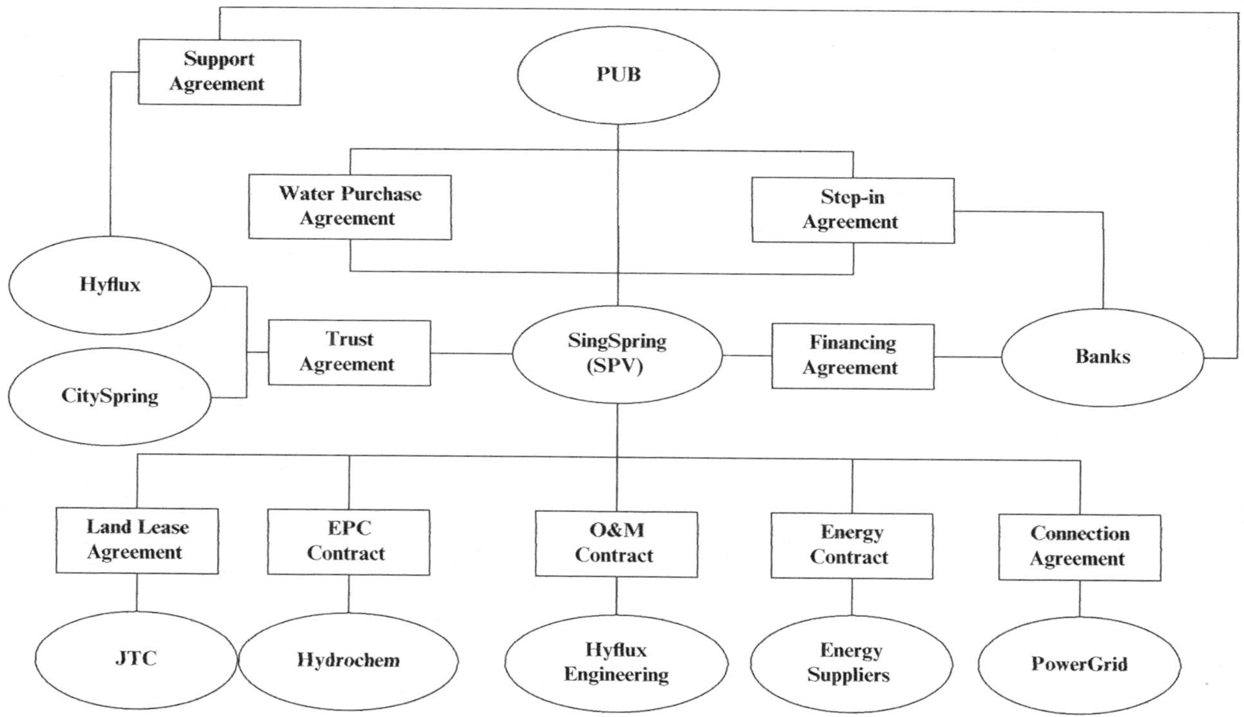

Figure 10.6 Project structure of SingSpring Desalination Plant

one which could potentially lead to project delays. As such, for the subsequent PPP projects where the choice of land was fixed, PUB purchased the land before tender award and subleased it to the concession company.

The project was financed by a SG$165 million loan and a SG$35 million equity investment from Hyflux. A senior debt facility of SG$158.5 million was arranged to fund 80 percent of the project cost, and a contingency facility of SG$6.5 million was to cover 70 percent of the cost overrun. The loan syndication was led by the Development Bank of Singapore (DBS), and KBC Bank, ING Bank and Standard Chartered Bank were the participants.

The tender price for the first year tariff was SG$0.78/$m^3$. With the good track record in the SingSpring desalination project and extended experience overseas in China and Africa, Hyflux had the confidence to reduce the first-year desalinated water price to SG$0.45/$m^3$ in the Second Tuas Desalination project. Hyflux has proposed innovative plant design and energy management for the Second Tuas Desalination project. A combined cycle gas turbine (CCGT) power plant will be built in the desalination plant to provide the energy required for the desalination process. This proactive energy management enables a secure power supply, lower energy cost, and optimal efficiency. The concession period of the Second Tuas Desalination project is 25 years, which is five years longer than the SingSpring Desalination plant. The tariff rate of the Second Tuas plant is thus lower than that of the SingSpring plant. This reflects that the longer the concession period, the lower is the tariff. Given Hyflux's innovation, technological and management expertise in desalination plant, and the low price quoted, it is not surprising that PUB perceived that Hyflux's proposal provided the best value to Singapore's water users.

10.4.2 Risk Management

There are various risks in a typical PPP project. It is important to identify those risks and manage them in a proper way. The basic principle is that risks should be allocated to the party best able to manage them. The allocation of some major risks in the SingSpring desalination project is summarized in Table 10.3.

PUB and SingSpring share the *force majeure* risk due to acts of God, war and natural disasters. In case of government *force majeure*, PUB still needs to make the capacity payment to SingSpring. PUB retains the risk of change in law and it is mitigated by tariff adjustment. The construction and operation risks are retained with SingSpring. SingSpring appointed Black & Veatch as the technology partner, which is very experienced in

Table 10.3 Risk allocation for SingSpring desalination project

Risks	Risk allocation	
	PUB	SingSpring
Force majeure risk	x	x
Change in law	x	
Construction risk		x
Operation risk		x
Demand risk	x	
Price risk	x	x
Energy supply risk	x	x

water technology and solution (that is, standard delivery of services in terms of both cost and quality). Black and Veatch designed the plant and selected the membrane technology and reliable pre-treatment processes. The construction risk and operation risk could be transferred to the EPC contractor, Hydrochem, and the O&M contractor, Hyflux Engineering; however, both of them are the wholly owned subsidiaries of Hyflux, thus both construction and operation risks are still retained with Hyflux. Demand risk is allocated to PUB as it is the only purchaser for the water supply. PUB and SingSpring share the price risk. The Singapore Consumer Price Index (local portion) and the US Producer Price Index (foreign portion) index the fixed and variable O&M charges, and HSFO CST 180 indexes the variable energy charge. The variable energy charge is completely passed through to PUB when the energy supply agreement is renewed on an annual basis after the third anniversary of the project's date of commercial operation. The energy supply risk is shared between the parties as well. In the case of grid failure, the capacity payment is still made by PUB. In the SingSpring project, PUB showed the positive attitude to share the risks and to work with the private party to achieve the best value for money of the project.

10.4.3 Lessons Learnt

Some valuable lessons can be drawn from the experience of Singapore's water PPP projects:

1. Singapore has a clear policy framework and strong institutional capability to ensure a transparent and fair tender process. As a PPP project has a complex structure and relationship, it is important to have a procurement process to ensure that the selected PPP project company

has the capacity to deliver good service of water supply under the PPP structure.

2. PUB is a public sector entity in charge of integrated water management. Its institutional simplification removes the administrative barriers to private sector participation, and enhances the efficiency of communication between the public sector and the private sector.
3. The Singapore Government has a clear water policy, and strategy for planning and implementation of water PPP projects. PUB cooperates with the private sector and provides support in various forms, and thus is an active project participant. Further, it is willing to share its experience and technical knowledge with the private sector. For instance, before the two NEWater PPP projects – the Ulu Pandan NEWater Plant and the Changi NEWater Plant – PUB had prior experience in managing NEWater plants. Hence, PUB provided training to staff from the private companies from its earlier NEWater project.
4. Proper risk allocation and management are keys to ensure value-for-money outcome and efficiency in delivering water supply under the PPP model. This significantly depends on whether the government or the public sector is willing to share risks and have reasonable risk allocation instead of pushing all the risks towards the private sector. In Singapore, it is clear that PUB is willing to share risks and the government advocates fair risk-sharing between partners.
5. In Singapore's PPPs in the water sector, it is common to see that the EPC and O&M contractors are the subsidiaries of the same parent company, which is also the equity provider and major sponsor of the PPP project company. This arrangement can allow the parent company to leverage more business opportunities, and it is easier to unite the interests of different subsidiaries under the same parent company.
6. All the four PPP water projects in Singapore have the local companies sponsoring and investing in the project company. PUB and the project company involve companies with international reputation and experience as consultants and advisers for PPP projects. This project structure can allow the PUB and the local companies to benefit from the experience and expertise of international players, and at the same time help Singapore to build up local competence in PPP water projects. This local competence can be exported overseas eventually, and provide more business opportunities for Singapore companies. For instance, Hyflux has achieved a good track record and experience in

the SingSpring Desalination project, and subsequently has expanded its desalination business in China and Africa.

10.5 CONCLUSION

The use of PPP models in delivery of water services is beneficial for both the public and private sectors. It segregates the government from the construction and operation of the water facilities, and the government can focus more on its main function as a policy maker and regulator. With stringent monitoring and checks by the government, water quality will not be compromised. With greater private participation, long-term water supply can be secured for the public. In addition, the PPP model adopts the life-cycle costing approach, and it can also bring the private sector's expertise, efficiency and innovation to the provision of water services. This can result in cost savings and lower the water price. For instance, PUB reduced the NEWater prices from SG\$1.30/m^3 to SG\$1/m^3 in 2007 due to the overall productivity gains, improved operational efficiency and increased economies of scale after the Ulu Pandan NEWater PPP project (Wong, 2011). On the other hand, the private sector can benefit from long-term clients and predictable cash flows, project experience and access to new business opportunities. However, the PPP model is complex and multiple parties are involved. Effective communication is important to achieve meaningful outcomes. The PPP contract is normally a long-term contract, and flexibility should be built in to allow necessary changes for unforeseen circumstances. As the project contracts are the major measures to define the responsibilities and rights of different parties in a project, the terms and conditions of the contract must be clearly defined to avoid potential disputes. Attention to the financial market and financial exposure is needed as these will affect the project and financing cost and eventually govern the water price. Proper risk allocation and mitigation measures must be in place to ensure value for money from PPP water projects.

The future market for water PPP projects in Singapore is promising. According to PUB, NEWater and desalinated water are targeted to supply 50 percent and 30 percent, respectively, of Singapore's water demand in 2060. More than five sites along the coastline have been identified for future desalination plants (Public Utilities Board Singapore, 2010b). With technological innovation and advances in energy efficiency in the future, desalination plants would become much cheaper. This indicates the huge potential for water projects. With the consistent service quality delivered by the PPP model, greater private participation in the water sector can be encouraged in Singapore.

ACKNOWLEDGEMENT

The authors would like to thank the Public Utilities Board, Singapore, especially Mr Koh Siong Teck, General Manager (Water Reclamation Department), for providing relevant information on which some of the analyses are based.

REFERENCES

Chan, F. (2011), 'Hyflux to build and run second desalination plant', *Straits Times*, Singapore: Singapore Press Holdings.

Chowdhury, A.N. and R.L.K. Tiong (2011), 'Financing and case study of Singapore PPP desalination project', *CECI Engineering Technology*, **90**, 114–25.

Department of Statistics Singapore (2011a), 'Singapore in figures', Singapore: Department of Statistics Singapore.

Department of Statistics Singapore (2011b), 'Time series on annual GDP at current market prices', available at http://www.singstat.gov.sg/stats/themes/economy/hist/gdp2.html (accessed 28 September 2011).

Department of Statistics Singapore (2011c), 'Time series on population (mid-year estimates)', available at http://www.singstat.gov.sg/stats/themes/people/hist/popn.html (accessed 28 September 2011).

Hyflux (2011), 'Hyflux named preferred bidder for Tuas II Desalination Plant, Singapore', Singapore: Hyflux.

Ministry of Finance Singapore (2011), 'Public private partnership', available at http://app.mof.gov.sg/ppp.aspx (accessed 1 October 2011).

Ministry of the Environment and Water Resources Singapore (2008), 'State of the environment 2008 report Singapore'.

Ministry of the Environment and Water Resources Singapore (2011), 'Water resource management', available at http://app.mewr.gov.sg/web/Contents/Contents.aspx?ContId=682 (accessed 1 October 2011).

Public Utilities Board Singapore (2010a), 'PUB Singapore's experience in public–private partnership (PPP) projects', in *Singapore–Commonwealth Leadership Programme on Public Private Partnership*, Singapore: Nanyang Technological University.

Public Utilities Board Singapore (2010b), 'Water for all: conserve, value, enjoy', available at http://www.pub.gov.sg/LongTermWaterPlans/index.html.

Public Utilities Board Singapore (2011a), 'PUB selects Hyflux as the preferred bidder for the second and largest desalination plant in Singapore', available at http://www.pub.gov.sg/mpublications/Pages/PressReleases.aspx?ItemId=280 (accessed 1 October 2011).

Public Utilities Board Singapore (2011b), 'Water tariff', available at http://www.pub.gov.sg/GENERAL/Pages/WaterTariff.aspx (accessed 1 October 2011).

Soon, T.Y., L.T. Jean and K. Tan (2009), *Clean, Green and Blue: Singapore's Journey Towards Environmental and Water Sustainability*, Singapore: Institute of Southeast Asian Studies.

Tortajada, C. (2006), 'Water management in Singapore', *Water Resources Development*, **22**(2), 227–40.

Water Technology (2011), 'Tuas seawater desalination plant, Singapore', available at http://www.water-technology.net/projects/tuas-seawater-desalination/ (accessed 1 October 2011).
Wong, M. (2011), 'NEWater to supply 30% of Singapore's water needs by 2011', available at http://www.channelnewsasia.com/stories/singaporelocalnews/view/264240/1/.html (accessed 1 October 2011).
World Bank (2011a), 'GDP per capita (current USGD)', *Indicators*, available at http://data.worldbank.org/indicator/NY.GDP.PCAP.CD (accessed 28 September 2011).
World Bank (2011b), 'Population density', *Indicators*, available at http://data.worldbank.org/indicator/EN.POP.DNST (accessed 28 September 2011).

11. Urban water governance in Australia: the private sector at the margins

Michael Paddon

11.1 INTRODUCTION

In 2011, a comprehensive review of urban water in Australia by the national Productivity Commission, which generally promotes the more extensive use of market mechanisms in government activities and an increased role for the private sector, concluded that governments play the dominant role in the provision of water services and will continue to do so (Productivity Commission, 2011). While the governance arrangements for urban water vary considerably between the states and territories in Australia, the private sector has not taken a major role in the delivery of water or wastewater services. Involvement of the private sector in Australian water and wastewater services has been limited to specific activities or points in the supply chain, and/or in specific (atypical) locations.

This chapter provides an overview of the governance arrangements for urban water across the jurisdictions of Australia and a review of the involvement of the private sector in these arrangements in which government provision remains dominant. As defined in Chapter 1 in this volume, this chapter also adopts the conventional definition of water governance which refers to 'the range of political, economic and administrative systems that are in place to develop and deliver water resources and the delivery of water services at different levels of society' (Global Water Partnership, quoted in OECD, 2011). While the major responsibilities for water service delivery have not been privatized in Australia, contracting-out of specific services by urban water utilities is widespread, particularly in the metropolitan areas, and there has been extensive engagement between public and private sectors in schemes to build infrastructure at certain points in the supply chain. The chapter also discusses the one major privatization of urban water provision for the city of Adelaide, through concession arrangements, which has been in operation since the mid-1990s. Although public policy has largely rejected a model of extensive private provision of water or funding of water services, the public utilities providing water and wastewater services have been affected profoundly by a range of reforms since the 1990s intended to refocus the ways in which public utilities in all sectors operate through a 'corporatization model' and policies aimed at introducing greater competition (National Competition Policy, 1995–2005). The chapter suggests that there has now been a major shift in the drivers for change in water governance, with less emphasis on investment in supply (a source of one imperative for engagement with the private sector), and more attention being given to broader policy issues of integrated water planning, with more extensive demand-side considerations.

11.2 URBAN WATER IN AUSTRALIA DELIVERED BY PUBLIC BODIES

Australia is overwhelmingly an urban society. While its land area from east to west is broadly equivalent to that of the whole of North America, and from north to south would extend from the northern tip of Denmark to the southernmost point of mainland Greece on a map of Europe, the population is concentrated in cities and urban areas. As of 2011, over three-quarters of Australians live in 18 major cities with populations over 100000. The majority of urban Australians live in the five largest cities: Sydney, Melbourne, Brisbane, Perth and Adelaide (Infrastructure Australia, 2010, 2011a), which are the capitals, respectively, of the states of New South Wales, Victoria, Queensland, Western Australia and South Australia. They are the centres of the 'metropolitan areas' described by the larger circles in Figure 11.1.

The overall governance of the water sector is usually directed by government departments in Australia's six states and two territories while water and wastewater services to the urban areas are delivered, with only few notable exceptions (discussed below), by government-owned and -controlled utilities (Productivity Commission, 2011). The Organisation for Economic Co-operation and Development (OECD)'s comprehensive review of water governance arrangements across developed countries concluded that though there is considerable diversity of arrangements, most OECD countries have decentralized water policy. In its characterization of these countries, based on the actual involvement of subnational entities in water policy design and implementation, Australia was placed in the group along with the United States of America (USA), Canada and Belgium in which the 'main actors' are subnational rather than central governments (OECD, 2011).

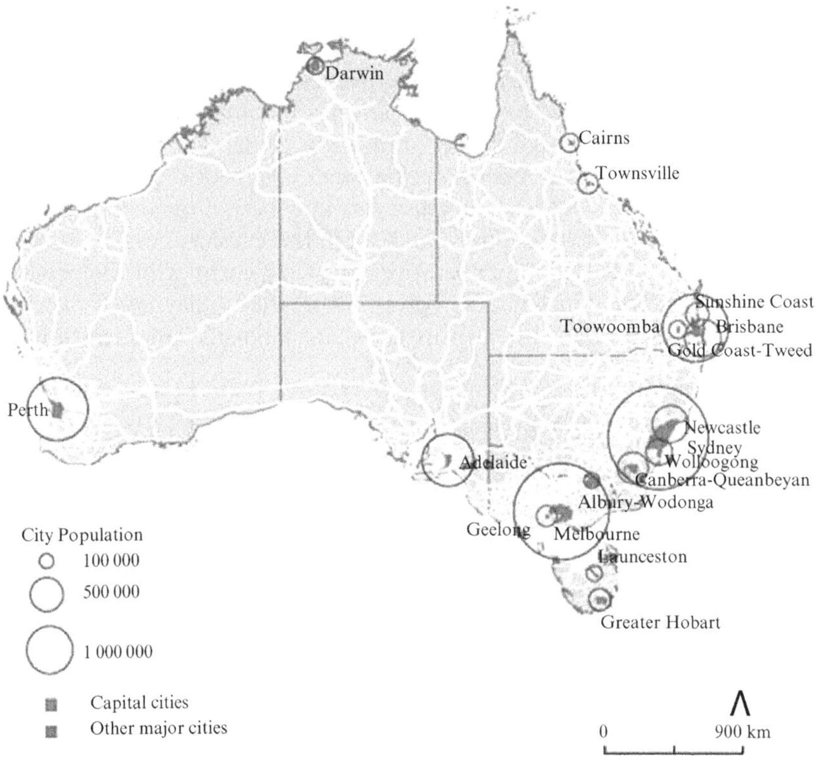

Source: Infrastructure Australia (2011a, p. 11).

Figure 11.1 Major cities in Australia

The arrangements within Australia would provoke similar conclusions to those made by the OECD about the international scene: the diversity of provision is striking but there are obvious general characteristics. In 2007, as part of an assessment of the challenges in pursuing national water and sustainability initiatives, McKay undertook a broad survey of the governance arrangements across Australia and charted the main responsibilities of the different levels of government and the range of corporate structures across the jurisdictions, which are reproduced in Table 11.1 (McKay, 2007). Three things stand out. Firstly, subnational state governments and local governments are the suppliers of water services. Secondly, there are significant differences between the jurisdictions; for instance, water services are mainly delivered by the local government utilities in New South Wales (NSW) and Queensland, while the role of the local government utilities is less important in other states, and local governments

Table 11.1 Corporate governance of major water supply businesses in Australia's states and territories, 2007

	ACT	NSW	NT	QLD	SA	TAS	VIC	WA	TOTAL
Local government regional council	0	2	0	0	0	0	0	0	2
Shire council	0	46	0	92	0	17	0	14	169
City/town council	0	0	0	15	0	5	0	0	20
Local government-owned corporation	0	14	0	4	0	0	0	0	18
Joint local government organization	0	5	0	1	0	3	0	0	9
TOTAL LOCAL GOVERNMENT	0	67	0	112	0	25	0	14	218
Government department licenser	0	0	0	1	0	2	0	0	3
Government-owned corporation	1	5	1	1	1	0	6	1	16
Statutory body	0	0	0	0	0	0	18	0	18
Corporations law company	0	3	0	1	2	0	0	1	7
TOTAL STATE GOVERNMENT	1	8	0	3	3	2	24	2	44
Water boards	0	0	0	0	0	0	0	2	2
Irrigation trusts	0	2	0	0	4	0	0	0	6
Hybrid	0	2	0	0	0	0	0	0	2
TOTAL OTHER ARRANGEMENTS	0	4	0	0	4	0	0	2	10
Undetermined	0	0	0	0	0	2	0	4	6
TOTAL	1	79	1	115	7	29	24	22	278

Source: Summarized from McKay (2007).

have no role in the provision of water services in the two territories, the Australian Capital Territory (ACT) and the Northern Territory. Thirdly, there is diversity in the formal administrative and legal arrangements for government entities; in local government, some utilities operate under the general auspices of shire councils, city or town councils, while others are established as local government-owned corporations. State government bodies include corporations, statutory bodies and companies. This means that different legislation governs the institutional or corporate activities of the utilities, depending on whether they are operating under a general Corporations Act, generic local government legislation or specific corporations legislation (which is usually the case with state government-owned corporations; see McKay, 2007). This has implications for significant aspects of their operations including corporate responsibilities, funding and investment opportunities.

These summary observations still hold. However in the five years since 2007, there have been significant changes in several jurisdictions, arising either directly from water service reforms, or indirectly from changes in the scale, responsibilities and structures of local government. As discussed later on in the chapter, reforms of local government have been a major driver for changes in water service delivery. In each case the reforms have been instigated by the respective state governments – a reminder that in Australia local government is a creation of the state and operates under the legal authority of the state government. The current arrangements and recent changes are summarized below by jurisdiction.[1]

11.2.1 Australian Capital Territory (ACT)

Water and wastewater services in the ACT are supplied by a government-owned vertically integrated utility, ACTEW. ACTEW contracts Actew-AGL Distribution, a joint venture with the private company AGL, to operate and maintain its water and sewerage networks including billing, sales, planning, design and maintenance of the network.

11.2.2 New South Wales (NSW)

Metropolitan areas

In Sydney, the provision of water and wastewater is vertically separated. The state government-owned corporation, Sydney Water, is responsible for water distribution and retail functions. The Sydney Catchment Authority is responsible for bulk water services (except the desalination plant) and water treatment. Water treatment plants are generally owned and operated by private companies. Sydney's desalination plant is owned by a subsidiary of Sydney Water, Sydney Desalination Pty Ltd, but maintained and operated privately (see below). In late 2011, the NSW state government established a taskforce of Sydney Water and government departments to start a process to privatize the Kurnell desalination plant (*Sydney Morning Herald*, 2011).

In Newcastle and other local government areas in the Hunter region (an area 200 km to the north of Sydney), bulk and retail water, wastewater and storm water services are provided by Hunter Water, a state government-owned, vertically integrated utility. In the Gosford and Wyong Council areas (which are on the coast between Sydney and the Hunter region), the Gosford–Wyong Councils' Water Authority provides bulk and retail water, wastewater and storm water services.

Urban arrangements in regional NSW

Water and wastewater services in regional NSW are provided by local government-owned water utilities. In mid-2008, there were 106 local water utilities (including the Gosford–Wyong Joint Water Authority discussed above); a reduction from 126 local water utilities in the 1980s resulting from amalgamations of councils (Armstrong and Gellatly, 2008).

There is a range of service delivery models across these 106 local water utilities (Armstrong and Gellatly, 2008). In 96 of the councils, water utilities were operated by councils as part of their general activities and were not established as separate legal entities. In five cases, the water and/or sewerage operations were run through a single-purpose county council operation, operating independently but with boards of management made up of councillors from the constituent councils. There were five water supply authorities established under a specific piece of state legislation relating to water management (the Water Management Act 2000). These included the Cobar Water Board, which was a joint establishment between the shire council and three local mining companies supplying water to each of the four constituent parties.

11.2.3 Northern Territory

Urban water and wastewater services in metropolitan and regional areas of the Northern Territory are provided by the Power and Water Corporation, a state government-owned vertically integrated utility. In remote indigenous communities, a subsidiary of the Power and Water Corporation, Indigenous Essential Services Pty Ltd, provides water and wastewater services.

11.2.4 Queensland

Southeast Queensland

In southeast Queensland, the most heavily urbanized part of the state, there have been several major phases of structural reform in the water sector since 2008. Prior to that, councils were responsible for retail water and wastewater services and a single state government entity, Seqwater, provided bulk water services. In 2008, the bulk water activities were divided between four separate entities: bulk water supply, bulk transport, manufactured water (desalination and recycling) and grid management. The bulk water and manufactured water authorities were merged in 2011. In 2010, the council operations were merged into three local government-owned distributor-retailers, each covering a defined geographical area. However, the merger was not deemed to be totally successful. Hence, in

2011, councils were given a once-only opportunity to withdraw from the three distributor-retailers and return to operating the water and sewerage business within the council (by mid-2012).

Rest of Queensland

Outside the southeastern part of Queensland, water and wastewater services in the state are provided by vertically integrated local government-owned utilities. This sector has also undergone recent reforms with local council amalgamations in 2008, resulting in a reduction in the number of regional urban water utilities to 71. The majority of these are local council utilities, although there are two commercialized water boards (the Gladstone Area Water Board and the Mount Isa Water Board), and a corporatized utility (the local government-owned Wide Bay Water).

11.2.5 Victoria

Metropolitan areas

The present arrangements in the Melbourne metropolitan area are a product of structural reforms in 1995. Prior to that, Melbourne Water was the vertically integrated state government-owned utility providing water and wastewater services. In 1995, bulk and retail functions were separated vertically, with Melbourne Water taking on the provision of bulk water and stormwater services and the establishment of three retailer-distributors providing retail water and wastewater services in their own geographic areas: Yarra Valley Water, South East Water and City West Water.

Urban arrangements in regional Victoria

Up to 1994, water, wastewater and storm water services in regional Victoria were provided by vertically integrated local government-owned utilities. In 1994, in a major structural reform as part of wider local government reforms, these were amalgamated to create 15 catchment-based state government-owned vertically integrated water and wastewater utilities providing water and wastewater services. Catchment Management Authorities were also established to coordinate the management of catchments in their area. In 2005, the utilities were merged into 13 utilities (Armstrong and Gellatly, 2008).

11.2.6 South Australia

South Australia (SA) Water, a state government-owned, vertically integrated utility, is the main provider of water and wastewater services in South Australia. From 1996, the provision of services to Adelaide, the state's capital city, and the surrounding areas has been contracted out in the only major privatization of water service delivery in Australia (discussed in more detail below). In addition to SA Water, there are some small local government service providers, such as Coober Pedy Council.

11.2.7 Tasmania

Until 2009, most of Tasmania's water sector was vertically separated. Twenty-eight of the 29 local government councils of the state provided retail and sometimes bulk water services, and 27 provided wastewater services. Many also sourced bulk water from three bulk water utilities: Hobart Water, Esk Water and Cradle Coast Water. The sector was reformed fundamentally in 2009 when the responsibility for water and wastewater services was restructured into three vertically integrated local government-owned trading enterprises – Southern Water, Ben Lomond Water and Cradle Mountain Water – each servicing a defined geographic area. Services such as information technology, finance and billing (which are often outsourced by water utilities in metropolitan areas; see below) are provided by Onstream, which is owned by the three corporations.

11.2.8 Western Australia

The Water Corporation, a state government-owned vertically integrated utility, provides water and wastewater services to most of urban Western Australia. Exceptions are in three small areas in the west and the south-west of the state (Bunbury, Busselton and Rottnest Island), where government statutory authorities provide water and wastewater services, and three areas in the north-west (Dampier, Paraburdoo and Tom Price), which are serviced by a private company, Hamersley Iron Pty Ltd.

11.3 ROLE OF THE PRIVATE SECTOR IN PUBLIC WATER SERVICES

With very few local exceptions, it is subnational governments which deliver water services to urban areas in Australia and own the assets and infrastructure. However, the picture is less straightforward beyond the questions of formal responsibilities and ownership of infrastructure. Figure 11.2 provides a schematic diagram showing the respective roles of public and private sectors in delivering urban water services by separating different aspects of the management roles. Moving from left to right, the figure depicts an increasing involvement by the private sector in the types

Management roles	Publicly built, owned and operated	Private construction	Private construction and operation	Private construction, operation and maintenance
Ownership				
Finance	PUBLIC	SECTOR		
Planning				
Design				
Construction			PRIVATE	SECTOR
Operation				
Renewal				
Examples	Most pre-1990 water utilities	Most regional and rural areas	Water treatment, recycling and desalination plants in major conurbations	Adelaide, ACTEW

Source: Amended from Department of Prime Minister and Cabinet (2006).

Figure 11.2 Schema of the respective roles of public and private sector in Australian water provision

and extent of management roles. However, as one moves from left to right, the number of illustrations or cases gets smaller. So, while it is relatively commonplace for subnational government water utilities to contract with private companies to design and build infrastructure, with instances in virtually every jurisdiction of state or local governments contracting with private (usually transnational) companies for construction and operation of particular facilities within the supply chain, there are only two major urban areas (Adelaide and the Australian Capital Territory) in which the delivery of services themselves has been contracted out to entities other than publicly owned utilities.

11.3.1 Widespread Contracting for Services

Every state and local water utility will be contracting, to some extent, with the private sector for delivering services, as is the case with nearly all publicly owned and publicly run entities in Australia at different jurisdictional levels. Among the services most commonly outsourced by water utilities are legal services; design services for capital works; project management and construction of capital works projects; information technology (IT) services; water meter reading and replacement; land management; and in some cases, billing services (Department of Prime Minister and Cabinet, 2006).

For larger metropolitan utilities, contracting for capital works and construction services is particularly significant. In 2004–05, an estimated 80 percent of a total of AU$1.15 billion capital expenditure on water and sewerage services by seven of the largest state-owned utilities in metropolitan NSW, Victoria, Western Australia and Queensland was contracted with third parties (data provided by the Water Services Association of Australia, quoted in Department of Prime Minister and Cabinet, 2006). In the same financial year, these utilities outsourced 35 percent of their operating expenditure (AU$1.5 billion) to third parties. However, this scale of contracting or outsourcing is unlikely to be matched in non-metropolitan areas, particularly smaller regional and local government utilities, which in some instances are the major employers and are essential to the sustainability of the local economy.

11.3.2 Building and Operating Infrastructure

Governments across jurisdictions and not limited to metropolitan areas have also entered into longer-term contractual arrangements, either as direct contractual arrangements or as part of more complex joint ventures, to finance, build and operate specific plants or facilities which contribute to the overall supply chain for water and wastewater. Figure 11.3 isolates the main stages in the supply chain: these types of contractual and/or partnership arrangements are most frequently found in bulk water sourcing (desalination) and water and wastewater treatment plants.

These arrangements have provided the avenue for the engagement in Australia by the major multinational companies which have been operating in water services internationally and have also been actively engaged in various forms of water privatization globally. Two French companies dominated the expansion of the private sector's role in water and wastewater services globally in the 1990s, and between them shared around 60 percent of the market served by multinational companies in the early part

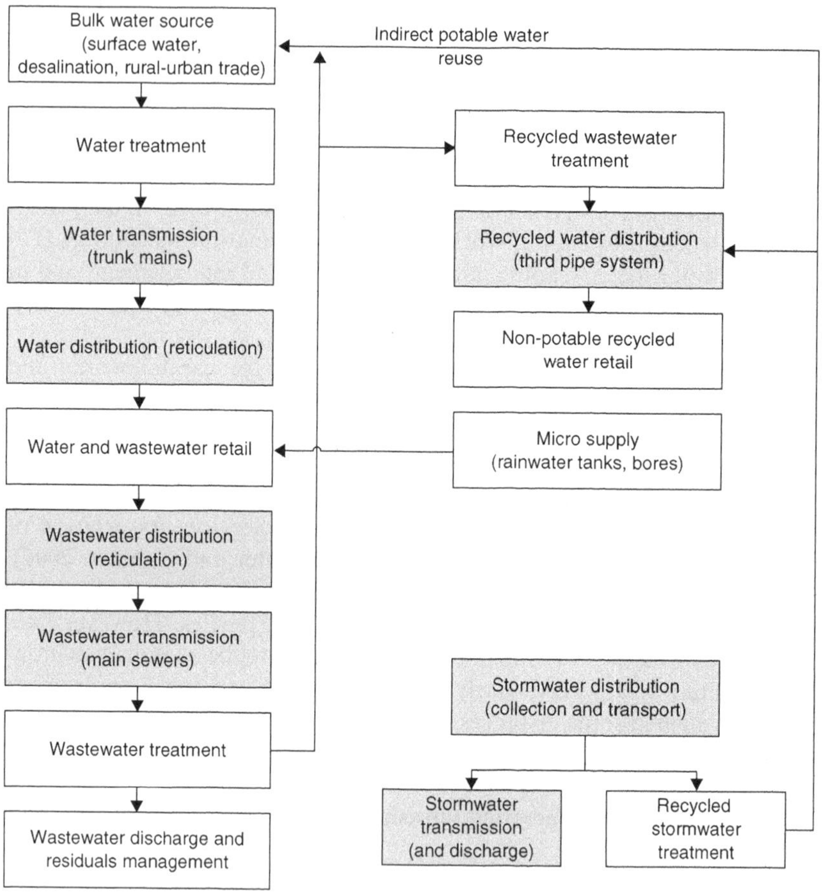

Source: Productivity Commission (2011, Chapter 2, p. 13).

Figure 11.3 Supply chain for water, wastewater and stormwater

of this century (Hall and Lobina, 2007, 2008). Table 11.2 lists the major water and wastewater projects in which one of these multinational companies, Veolia (previously Compagnie Générale des Eaux and Vivendi) has been involved in Australia. Full contractual details for these projects are rarely in the public domain but involvement is invariably for relatively long periods of 16 to 20 years or more and the scale of projects ranges from those with a total value in excess of AU$1 billion (Sydney desalination plant) to those with values around AU$50–55m (Ballarat North Water Reclamation Plant and Gerringong Gerroa Sewerge Scheme). Table 11.3

Table 11.2 Water and wastewater projects in Australia involving Veolia, 2011

Project	Client	Nature of involvement	Duration
Ballarat North Water Reclamation Plant	Central Highlands Region Water Corporation	DBO	15 years from 2016
Bendigo, Castlemaine and Kyneton Water Treatment Plants	Coliban Region Water Authority	BOOT	25 years from 2002
Gerringong Gerroa Sewerage Scheme	Sydney Water	exact details unknown	20 year from 2002
Illawarra and Woronora Water Treatment Plants	Sydney Water	BOO	Commenced in 1996–97; still operating
Noosa Water Treatment Plant	Noosa Council Queensland (now Sunshine Coast Regional Council)	DBO	15 years from 1998
Rosehill Recycled Water Scheme	Sydney Water	DBO	20 years from 2008
Gold Coast Desalination Plant	Gold Coast Council through Gold Coast Desalination Alliance – the Alliance comprises the Council, the state government (through Sure Smart Water) and four private companies	DBO Alliance Contract	Commenced operation in 2009; mothballed in 2010
Sydney's Desalination Plant	Sydney Water	Construction joint venture between Veolia and John Holland; O&M by Veolia	Operating contract 20 years; plant operating from 2010
Western Corridor Recycled Water Project	Seqwater	O&M	Online 2007, construction complete 2009

Sources: Partnerships Victoria http://www.partnerships.vic.gov.au; Infrastructure (Ballarat North); Partnerships Australia http://www.infrastructure.org.au/Content/CaseStudies.aspx; Company Website (Bendigo, Castlemaine, Kynerton); Boake (2006) (Gerringong-Gerroa); Chapman and Cuthbertson (1996) (Illawarra and Woronora); Treasury, NSW Government http://www.wwg.nsw.gov.au; (Rosehill) Veolia company website http://www.veoliawater.com.au; (Gold Coast Desalination Plant); Queensland Government (2010) (Western Corridor); Sydney Water http://www.sydneywater.com.au (Sydney Desalination Plant).

Table 11.3 Water and wastewater projects in Australia involving Degremont (Suez Environment, previously Lyonnaise des Eaux), 2011

Project	Client	Nature of involvement	Duration
Victorian Desalination Plant	Victoria state government Department of Sustainability and Environment	Degremont is a partner in AquaSure (the consortium also includes Thiess and Macquare Capital Group)	Construction started in September 2009 to deliver water by end of 2011. Duration of contract 30 years from 2009
Prospect Water Filtration Plant	Sydney Water	BOOT	25 years from 1996
Perth Seawater Desalination Plant	Water Corporation of Western Australia	DBO	25 years from 2007
Noosa Waste Water Treatment Plant Queensland	Noosa Council, Queensland	DBO	25 years from 1997
Cronulla Waste Water Treatment Plant, Cronulla, New South Wales	Sydney Water Corporation	O&M	2001–2003

Sources: Degremont company website, http://www.degremont.com.au/; Partnerships Victoria http://partnerships.vic.gov.au (Victorian Desalination Plant).

covers the other multinational company, Degremont (Suez Environment, previously Lyonnaise des Eaux).

Since the early twenty-first century, there have been significant adjustments in the global operations of these multinational companies, including shifts in the patterns of industrial domination. In response to difficulties in earning expected financial rates of return and an increasing opposition to water privatization internationally linked to an apparent failure by the private sector to actually deliver anticipated investments, multinational groups have been curtailing their activities in water and wastewater (Hall and Lobina, 2008, pp. 7–10). There appear to have been two consequences. Firstly, the patterns of ownership of the private companies involved in water and wastewater services have changed, with regional and national actors, including financial and equity investors, taking an increasing role. Secondly, the anticipated drivers for greater 'finance privatization' (diverse financial arrangements involving private capital and equity), that they would be generated by requirements for investment in infrastructure on the one hand and restrictions on public sector borrowing on the other, have not been as significant as expected (Paddon, 1998).

The company now operating in Australia as TRILITY is illustrative of the first. TRILITY, formerly known as United Utilities Australia Pty Ltd, is a subsidiary of United Water, the British company which operates the privatized North West Water in England. An international consortium of Mitsubishi Corp, INCJ, Manila Water and JGC Corp took over United Utilities in 2010, and rebranded it TRILITY. TRILITY has since bought out joint-venture partners for long-term build–own–operate–transfer (BOOT) schemes with Coliban Water (Victoria), McCarthur (Sydney) and Yan Yean Water (Melbourne) water treatment plants (Global Water Intelligence, 2011). The current spread of its water activities is indicated in Table 11.4.

It is difficult to be precise about the financial agreements underpinning BOOT and similar arrangements, since they are generally governed by considerations of commercial in-confidence. For this reason, the 'official' estimate in Australia of the extent of public–private partnerships (PPPs) in water has given an inaccurate picture of the extent of private sector engagement. The federal government's reporting of the total number of PPP arrangements involving federal or state governments and agencies across all sectors in Australia identifies only six out of a total of 51 PPPs relating to water or water recycling schemes, and only one out of the 12 PPPs currently 'in the market' (Infrastructure Australia, 2011b). Notwithstanding the absence of reliable and comprehensive information on financing of water and wastewater services, it is clear that one stage of the supply chain, in which there has been investment in each major metropolitan area, is 'climate-independent supply augmentation projects', that is, desalination plants (Productivity Commission, 2011, Chapter 2, p. 11). Table 11.5 indicates the six desalination plants completed since 2006–07 or due for completion in 2012. However, there is now evidence that the investment in desalination plants, as an adjunct or addition to water supply to meet projected demand, was both unnecessary and disproportionately expensive (Productivity Commission, 2011). In late

Table 11.4 Water and wastewater projects in Australia with public water utilities involving TRILITY (previously United Utilities Australia), 2011

Project	Client	Nature of involvement	Duration
Adelaide Desalination Project	South Australian Water Corporation	DBO (as a Member of Adelaide Aqua consortium, which includes three other companies)	20-year O&M contract; plant due to be fully operational 2012
Agnes Water/ Seventeen Seventy Integrated Water Project	Gladstone Regional Council	Construction of a desalination plant, sewerage treatment plant upgrade, water and sewerage reticulation and a reservoir	Operational August 2012
Berri Barmera Reuse Scheme	Berri Barmera Council	BOO: partnership between Council, Hardy Wine and TRILITY	25 years
Campaspe Asset Management Services	Coliban Region Water Authority	O&M of 16 water treatment plants, 13 waste reclamation plants and 227 water and wastewater pumping stations	10 years from 2003
LinkWater; Field Operations and Maintenance	LinkWater	O&M undertaken in conjunction with another private company Transfield Services	5 years from 2010
Riverland Water Project	South Australian Water Corporation	Design, finance, build operate and maintain 10 water treatment plants	25 years from 1996–97
Macarthur Water Treatment Plant	Sydney Water	Design, finance, build operate and maintain	25 years from 1995
Townsville City Council Water Supply Upgrade Project	Townsville City Council	Design, finance, build and operate	25 years from 2008
Victor Harbor Wastewater Treatment Plant	South Australian Water Corporation	Design, finance, build, operate and maintain	20 years from 2005
Yan Yean Water	Melbourne Water	Design, build, operate and transfer	25 years from 1994

Sources: South Australian Water Corporation, http://www.sawater.com.au/; Gladstone Regional Council, http://www.gladstone.qld.gov.au/; Berri Barmera, http://www.berribarmera.sa.gov.au; TRILITY website, http://trility.com.au/projects/.

Table 11.5 Desalination plants built in Australia, to 2012

Desalination plant	How funded and built	Initial capacity GL/year	Initial investment AUD (millions)	Full completion
Perth Seawater Desalination Plant, Western Australia[a]	25-year design, build and operate (Alliance) contract awarded to Degremont with State Water Corporation.	50	387	2006–07
Gold Coast Desalination Plant, Tugun, South East Queensland[b]	Southeast Queensland Water awarded a design, construct and maintain (Alliance) contract to an alliance of Veolia Water, John Holland Group and SureSmart Water, representing the state of Queensland.	49	1200	2009
Sydney Desalination Plant, Kurnell, New South Wales[c]	Sydney Water, Sydney's state-owned water supply corporation, awarded a 20-year design, build, operate and maintain contract to Blue Water Joint Venture, which consists of the construction company John Holland and the operator Veolia Water.	90	1890	2010
Southern Seawater Desalination Plant, Binninyup, Western Australia[d]	25-year design, construct and operate contract with Southern SeaWater Alliance (SSWA), comprising Técnicas Reunidas, Valoriza Agua, AJ Lucas, WorleyParsons and the Water Corporation.	50	955	2012
Victoria Desalination Plant, Wonthaggi[e]	PPP contract awarded to Aquasure, a consoritum of Degremont, Theiss and Macquarie Capital, for the construction and operation of the desalination plant.	150	3500	2011–12

Table 11.5 (continued)

Desalination plant	How funded and built	Initial capacity GL/year	Initial investment AUD (millions)	Full completion
Adelaide Desalination Plant, South Australia[f]	South Australian Government and SA Water awarded a 20-year design, build, operate and maintain contract to Adelaide Aqua, a multinational consortium of four companies.	100	1830	Late 2012
TOTAL		484	9762	

Sources:
a. Perth Seawater Desalination Plant, available at http://www.water-technology.net/projects/perth/ (accessed 3 May 2012).
b. Gold Coast Desalination Plant, available at http://www.belvessels.com/sites/default/files/projects/case-study.pdf (accessed 3 May 2012).
c. Sydney Desalination Plant, New South Wales, available at http://www.veoliawater.com/ressources/documents/1051,Sydney-desal.pdf (accessed 3 May 2012).
d. Southern Seawater Desalination Plant, available at http://www.watercorporation.com.au/D/desalination_plant2.cfm (accessed 3 May 2012).
e. Victorian Desalination Plant, available at http://www.partnerships.vic.gov.au/CA25708500035EB6/0/8ACBA1C56F57CF23CA25736E0001DCE8 (accessed 3 May 2012).
f. Adelaide Desalination Plant, available at http://www.sawater.com.au/SAWater/WhatsNew/MajorProjects/ADP_ProjectInfo.htm (accessed 3 May 2012).

2010 the state government in Queensland was considering mothballing the Gold Coast desalination plant, and in 2011 the plant was operating at below capacity or was in what was termed 'hot standby' mode (World Desalination Report 2011–12). By the end of 2011, the NSW government had initiated a process to sell off the Kurnell (Sydney) desalination plant.

11.3.3 Private Operation of Water Services

In Australia, there are two significant exceptions to the generalization that urban water services are delivered by state or local governments. In the Australian Capital Territory, the location of the national capital Canberra, water supply and sewerage are provided by a joint venture between the state-owned utility and a private company, Australian Gas and Light (AGL). In Adelaide, the capital city of South Australia (and known as the driest state in the driest continent in the world), water services have been delivered under contract since 1996.

Adelaide

Adelaide is the only major city in Australia where water services have been contracted out completely to a private sector consortium. In 1996, SA Water, the state-owned corporation, awarded a 15-year concession contract for the city's water and wastewater services. In 2011, the concession was retendered and a second contract was awarded to a different consortium.

There was a common set of drivers to the various forms of privatization which took place in Australia in the mid- to late 1990s (Fairbrother et al., 2002). In this instance, the process was initiated by an incoming state government committed to introducing 'efficiencies' into publicly run organizations and agencies, with the first step being the commissioning of an audit review which identified the need for such efficiencies and the areas in which they could potentially be obtained (though the claims that the efficiency gains were necessary or that the claimed savings would be achieved were contested at the time; Ranald and Black, 2000). The outsourcing was preceded by 'corporatization' of what had been a state government department focused primarily on funding and preparing infrastructure, to a more commercially and service-focused state-owned corporation, SA Water (Sheil, 2000). From 1995, the South Australian government implemented the intergovernmental agreements on the National Competition Policy (NCP), which are discussed in more detail in section 11.4. Though privatization and outsourcing were not obligations under or strict requirements of the NCP, the structural reforms of public enterprises, which were part of the package, gave a further impetus for these initiatives in those states which had already embarked on these types of public sector reforms.

A restricted tendering process resulted in the award of a 15-year contract to United Water, a consortium of British-based Thames Water and the French transnational corporation Compagnie General des Eaux. The French company has been through a number of transformations over the contract period, renamed as Vivendi, and now operating as Veolia. While the outsourced arrangements are generally presented in current, conventional accounts as a success story, the initial tendering process was controversial and the subject of both audit and parliamentary enquiries (Ranald and Black, 2000). One issue of concern was the restriction of tendering to a select list of transnational corporations following from the state government's stated intention that the outsourcing would contribute to regional and state economic development by providing a base for the companies' Asian operations from Australia. Sixteen months into the contract, there was a major operating failure at a wastewater treatment plant, resulting in complaints to the state's environment agency and a technical audit report which revealed a 'story of persistent breaches of the public trust in the management, operation and maintenance of the plant' (Sheil, 2000, pp. 57–66). In 2009, SA Water filed an action against Veolia in the state's Supreme Court on the company's pricing policy (*Australian*, 2009).

At the completion of the 15 years of the original contract, the contract was retendered in 2011 and awarded to the Allwater Consortium, a joint venture of Transfield Services with Degremont and Suez Environnement (the current operating name of the other French transnational corporation which shared the domination of private involvement with Vivendi/Compagnie Generale des Eaux in the early 2000s). In an unusual development, the contract was let as an 'alliance' contract. This is a form of collaborative joint venture contracting, promoted by a number of large infrastructure firms operating in Australia, and used in the construction of a number of the desalination plants listed in Table 11.5. It is promoted on the ground that it provides for a more balanced distribution of risk between the contracting parties than is the case with more conventional forms of contracting. However, it has been used previously in new construction arrangements rather than maintenance and service arrangements.

11.4 DRIVERS FOR REFORM AND REFASHIONING THE ROLE OF THE STATE IN URBAN WATER SUPPLY

In focusing on the predominance of government provision of water services and the limited role of private organizations, so far this review has not given attention to the extent to which these public organizations and utilities have been reformed significantly over the past two decades. These reform initiatives were ostensibly driven by the same policy objectives as those which, internationally and in Australia, provided a rationale for privatization in the form in which it is most conventionally understood (that is, change in ownership). In Australia, this approach to policy reform is often referred to as a narrow 'economic rationalist' agenda of focusing predominantly on achieving greater economic efficiencies (Johnson and Rix, 1993). The outcome of the reforms, if a broader definition of privatization is taken, is that there has been extensive 'management privatization' (through corporatization) in the water utilities (Paddon, 2001; Fairbrother et al., 2002). State-owned water enterprises across all jurisdictions have been corporatized in processes starting in the 1990s and carried forward under a series of intergovernmental agreements through the Council of Australian Governments (COAG, covering the federal and state governments), which were codified and extended in the NCP. In a series of agreements intended to open up sections of the Australian economy to increased competition, particularly those in which governments operated, the NCP was a driving force for governmental reforms for ten years from the mid-1990s. The reforms were given impetus by the linking of significant payments from the federal to state governments for the implementation of the various reforms. The core elements of the NCP package, which had direct or indirect impacts on the state and local government bodies providing water services, are summarized below (Paddon, 2001):

1. Infrastructure reforms. The NCP contained specific obligations for reforms in each of the areas of infrastructure involving utilities (electricity, gas and water). For water, the NCP reforms included a governance/structural obligation to separate water management and delivery from regulation and other requirements in terms of pricing, investment and water allocation.
2. Reform of public monopolies. Regulatory activities were required to be disaggregated from commercial ones in cases where the government intended to introduce competition in areas in which public enterprises were operating under the NCP. Governments were also required to review the commercial objectives of government businesses and their financial relationship with the government owner. There was also a commitment for governments to review the provision for other users and competitors to have access to infrastructure used by government business enterprises.
3. Competitive neutrality arrangements. The principle of 'competitive neutrality' was that significant government business activities should not have competitive advantage purely as a result of being publicly owned. Under the NCP, governments agreed to corporatize significant government-owned enterprises and move towards full cost recovery pricing in designated government-owned businesses.
4. Reforms in local government. Each jurisdiction also agreed to introduce a policy framework for application of the NCP to local governments, focused on application of competitive neutrality to 'significant businesses' operated by councils. There were significant variations between the states in the frameworks they developed and applied. However, for most councils, particularly those outside the main conurbations, water and/or wastewater services were the most 'significant businesses' they operated with the result that 'in all cases, [the NCP] resulted in some reforms: either corporatization, greater commercialization, or fundamental changes in pricing structures so that prices charged are more reflective of the full costs of provision' (Paddon, 2001). The interpretation of the application of the NCP to local governments in Queensland, for example, was that the 18

> councils which were deemed to have the most significant business activities were required to review the 'net public benefit' of applying corporatization, commercialization or full-cost pricing. Of the 25 'significant business activities' identified by these councils, 17 were water or wastewater services for which the councils resolved to apply full commercialization and removal of cross-subsidies. In addition to these large facilities, another 75 local councils across the state elected to consider application of competitive neutrality policies and NCP water reforms to their water and sewerage activities even though they were not legally mandated to do so (Paddon, 2001).

The other significant drivers for reform, which have an impact on water services and have resulted in many of the changes identified in section 11.2 of the chapter, relate to the roles, scale and administrative arrangements for local governments. There has been a long-standing and recurrent policy and research debate in Australia about the appropriate size and scale of local councils. The terms in which the debate has been conducted echo the NCP: it focuses predominantly on the 'effectiveness' and sustainability of local government operations, and in particular, a discussion of the potential, if any, for economies of scale to be derived from consolidating local councils into a small number of councils which are larger in area, population and resources (for example, see Byrnes and Dollery, 2002; Dollery et al., 2006). Over the past two decades, there have been rounds of mandated amalgamations of local councils in Victoria and South Australia. Most recently, the amalgamations of councils in Queensland in 2008 outside the metropolitan south-east has reduced the number of local government-owned and -run water utilities considerably.

At around the same time that these reforms were implemented in Queensland, the NSW state government commissioned an independent inquiry into 'secure and sustainable' urban water supply and sewerage services for the non-metropolitan parts of that state. The inquiry reviewed the performance of all the local government water utilities against 23 indicators covering economic (financial and efficiency), environmental and social (cost and public health) dimensions (Armstrong and Gellatly, 2008). Based on the variations in performances across councils and assumptions about the increasing challenges from demographic changes, an ageing infrastructure and the effects of climate change, the inquiry concluded that there was a 'need for structural change in most instances, to enable local water utilities to better plan for and manage their responsibilities in the face of major challenges such as climate change, skills shortages, demographic changes and increasing human health and environmental standards' (Armstrong and Gellatly, 2008). The inquiry commented in particular on the relatively poor performance of some of the smaller utilities and expressed particular concern about water quality issues since 17 local water utilities had failed to meet the microbiological water quality requirements of the Australian Drinking Water Guidelines. Moreover, there had been two 'boil water alerts' in the two years between 2006 and 2008.

However, the 'structural change' recommended was not change in ownership of the water utilities: 'The recommendations contained within this report . . . seek to maintain council ownership and management of water supply and sewerage services' (Armstrong and Gellatly, 2008). All council submissions to the inquiry argued that local governments should continue to own water supply and wastewater assets. The majority of council submissions also argued for the continuation of direct responsibility for the operation and management of water supply and sewerage services by local government on the ground that the removal of direct responsibility would result in a substantial reduction in council revenues (30–40 percent); have an adverse impact on employment; and due to the importance of local government and local council employment in towns, it would have an overall detrimental socio-economic impact in their areas.

The 2007–08 NSW Inquiry recommended aggregation of the 104 local water utilities into regional groupings (either into 15 groups, the preferred option, or 32). It also reviewed potential models for the organizational structure of the aggregated entities and, while rejecting the notion that one organizational model would be appropriate for the range of circumstances in the state, proposed that three models should be considered for the future delivery of water supply and sewerage services. These include:

1. Binding alliance. This is a new organizational concept for local water utilities under which individual councils retain their assets and responsibility for service delivery but establish a separate and distinct legal entity (the Alliance Entity) for strategic and financial planning and for providing specialist technical skills to the member councils.
2. Council-owned Regional Water Corporations. The assets and operating responsibility are transferred to the new entities, acknowledging that there would be potential for significant detrimental socio-economic impacts and consequences for council revenues, which would need to be considered and mitigated.
3. Retaining current structure arrangements. These arrangements are recommended only for certain existing large water utilities.

The NCP, the related reforms and the local government reorganizations (in some jurisdictions linked to the perceived obligations of the NCP) have, therefore, been major drivers for reform. However, the 2007–08 Report of Inquiry in NSW, whilst being conducted within the same broad terms of reference which informed the NCP, is illustrative of what, in the Australian context, are generally taken to be the limits to acceptable reform. It set the parameters that whatever other changes and amendments are made to the scale and administrative arrangements of water utilities, they would remain publicly owned and that therefore governments should retain direct responsibility for services.

The recent Productivity Commission report recognizes that 'water is perceived to be different from other utility services (electricity, gas, telecommunications and mail) because water is "essential for life" and/or exhibits common property characteristics' (Productivity Commission, 2011, p. XVIII). When compared with the other areas in which public utilities have operated, public policy debate is arguably influenced significantly by an apparent community anxiety about water supply quantity (after years of prolonged drought in Australia) and quality. Though the number of events in Australia involving poor water quality is ostensibly quite small, when they do occur they attract considerable public scrutiny, which has made both governments and potential providers risk-averse with an acceptance that the ultimate responsibility for water quality remains with governments. There were two high-profile events relating to water and wastewater quality at the end of the last century. In Adelaide, the 'Big Pong', an event in which bacteria generated widespread and unpleasant odors, prompted investigations by the state auditor and by the environmental protection agency (Sheil, 2000). In the second and even higher-profile national event, residents in Sydney experienced three serious alarms concerning parasites in their water supply as a result of which they were instructed to boil water before using it for cooking, drinking, washing up and so on between July and September 1998. The immediate source of the apparent contamination was water from a privately operated water treatment plant at Prospect. The official inquiry after the event revealed interesting and significant absences from the contractual arrangements with the operators of the treatment plant, and confusion about the roles and responsibilities within Sydney Water, which resulted in a re-emphasis on its social and environmental obligations and the establishment of a new and separate catchment authority. In a broader context, the crisis of water quality could be traced back to the manner in which Sydney Water was corporatized and its operations commercialized (Sheil, 2000; Chapman and Cuthbertson, 1996).

Indirectly, the extent to which these parameters are established and widely accepted in public policy debate was acknowledged in an explicit initiative in 2006 by the then federal government to increase the role of the private sector in the supply of water and wastewater services. The fundamental premise of a discussion paper released by the Prime Minister's Department, and hence indicative of a significant intention to develop national policy, was that there would be an increasing need for investment in the sector, partly driven by the requirements of meeting future water demand (Department of Prime Minister and Cabinet, 2006). The discussion paper identified three barriers to more extensive private sector involvement:

1. Water pricing arrangements. Since the government regulates water prices, the concern in the report was whether the regulated monopoly prices can provide for a commercial return on investment.
2. Third-party access arrangements. Since the fixed infrastructure requirements for water and sanitation make generation of competition through provision of alternative infrastructure and hardware difficult if not impossible, the concern from those advocating more competition through the involvement of additional potential private providers was that they should have access to the existing infrastructure.
3. Risk management. The allocation of risk between a public or government contracting agency and a private provider is seen as a barrier where potential providers have the perception that a disproportionate degree of risks resides with them.

The discussion paper made two proposals to promote greater private sector involvement, neither of which amounted to a significant change in the fundamental structural arrangements for water and sewerage services. The first was for a national code for 'third-party access', to produce some standardization of arrangements between the different states and territories and to minimize 'complexities over regulations'. Third-party access in relation to public infrastructure was a component of the NCP. In 2005, the Australian Competition Tribunal, the ultimate national appeal body for cases relating to fundamental legislation underpinning the NCP, handed down what was its 'only reported judgement . . . relating to water' (Department of Prime Minister and Cabinet, 2006, p. 14). The Tribunal, ruling on an application by a private company, Services Sydney Pty Ltd, gave the company a legal right to use the sewer reticulation network of the state of NSW for 'sewer mining': treating and recycling wastewater to sell for industrial, agricultural and other uses. The discussion paper attributed considerable significance to the judgment in opening out the potential for private use of publicly owned infrastructure. However, there is no evidence that the company actually made use of this right following the ruling or that others have done so since.

The second proposal was for national guidelines for contracting with the private sector, which arrangement is extensively used in water and wastewater services by the large utilities (but is likely to be less significant in non-metropolitan urban areas). In identifying areas with an apparent potential for greater private sector involvement, the discussion paper also reveals the limits of that involvement in the Australian context. Since the paper was released in the mid-2000s, the national government has changed. The 'potential' has also been severely curtailed by a greater questioning of the initial premise of the report: that there is an extensive need for increased investment. The assumption that the need for increased investment would drive financial arrangements involving private institutions and finances (including public–private partnerships), and that hence 'finance privatization' would become more prevalent, would have seemed entirely plausible in the late 1990s in a context of thinking of 'supply-side' solutions for water and wastewater. Indeed where Australian jurisdictions have continued with supply-side initiatives, notably commissioning desalination plants, they have done so through ventures with extensive private sector involvement. However, in Australia, as with water policy internationally, there was a discernable 'shift in the focus on the sector during the 1990s. The political agenda has gradually moved from "technical" supply-driven, infrastructure-led solutions towards demand-based approaches emphasising the role of institutions and economic and social instruments' (OECD, 2011).

In the water policy initiatives, which have followed the NCP and the related COAG initiatives, it would appear that competition (as a major driving force) has a more limited role. Issues of ownership, where they appear as major concerns, are in relation to water rights (rather than what types of organizations deliver water services). Given the topography and hydrology of Australia, the management and distribution of water rights in the dominant Murray–Darling river system has been a major policy focus in recent years. In these discussions, major governance issues have been the division of responsibilities between state and federal authorities with attention focused on the interrelationship between state and federal regulatory priorities and practices (or lack thereof) and the consequences of trying to meet several competing if not conflicting demands (for irrigation; for providing water for urban areas; and maintaining a sustainable environment).

The broad issues identified by the OECD summary as themes across the developed economies are also brought to the fore in Australian water governance discussions with the more recent emphasis on demand management and integrated planning for urban water systems. These are the broader issues articulated in Australia's current intergovernmental National Water Initiative, the guiding principles for the current formal approach to water governance (National Water Commission, 2011b).

11.5 CONCLUSION

The basic character of urban water and wastewater provision in Australia – that it is a responsibility of subnational state, territory and local governments, which own the infrastructure – is therefore unlikely to change. Reforms to the public sector and to public utilities over more than a decade have certainly changed the operating and to some degree the financial model, but urban water services remain essentially public services. The two variations from this governance and operational model, in South Australia and the Australian Capital Territory, are well known and are always quoted in discussions about the potential role for the private sector but are unlikely to be replicated more widely. The most extensive engagement with the private sector by most water utilities is through contracting for services; this is financially significant but works within the fundamental operating model. In the late 1990s and early 2000s the predictions were that the requirements for investment in water infrastructure in conditions of public sector financial stringency would operate as drivers for a greater involvement by the private sector through funding arrangements including PPPs. To some degree this has been borne out in the widespread engagement of transnational water companies in building and operating water treatment plants and desalination plants around many cities and urban areas. However in Australia, as elsewhere in the OECD, the agenda has arguably shifted from investing in supply-side solutions to a greater emphasis on planning and other instruments focused on the demand side. This suggests a future governance agenda with more emphasis on administrative, political and social components and less on technical, financial or ownership elements.

NOTE

1. These summaries are based on Productivity Commission (2011), which in turn draws heavily from details made available by the National Water Commission for which the current source is NWC (2011a).

REFERENCES

Armstrong, I. and C. Gellatly (2008), 'Report of the Independent Inquiry into Secure and Sustainable Urban Water Supply and Sewerage Services for Non-Metropolitan NSW', Vol. 1, State of New South Wales.

The Australian (2009), 'Rann sues French firm Veolia for water "overcharging"', 31 August, available at http://www.theaustralian.com.au/news/rann-sues-french-firm-veolia-for-water-overcharging/story-e6frg6n6-1225767978086 (accessed 28 November 2011).

Boake, M.J. (2006), 'Recycled water: case study Gerringong-Gerroa', *Desalination*, **188**, 89–96.

Byrnes, J.D. and B.E. Dollery (2002), 'Do economies of scale exist in Australian local government? A review of the research evidence', *Urban Policy and Research*, **20**(4), 391–414.

Chapman, R. and C. Cuthbertson (1996), 'Sydney's water: a suitable case for private treatment', *Agenda*, **3**(1), 45–58.

Department of Prime Minister and Cabinet (2006), 'A discussion paper on the role of the private sector in the supply of water and wastewater services', Canberra: Australian Government.

Dollery, B.E., L. Crase and A. Johnson (2006), *Australian Local Government Economics*, Sydney: UNSW Press.

Fairbrother P., M. Paddon and J. Teicher (2002), *Privatization, Globalisation and Labour: Studies from Australia*, Sydney: Federation Press.

Global Water Intelligence (2011), 'Trility pushes through a trio of transactions', *Global Water Intelligence*, **12**(7), available at http://www.globalwaterintel.com/archive/12/7/general/trility-pushes-through-trio-transactions.html (accessed 10 December 2011).

Hall, D. and E. Lobina (2007), 'International actors and multinational company strategies in Europe 1990 Global Water Intelligence 2003', *Utilities Policy*, **15**(2), 64–77.

Hall, D. and E. Lobina (2008), 'Water privatization', Public Services International Research Unit, Business School, University of Greenwich, London.

Infrastructure Australia Major Cities Unit (2010), 'State of Australian cities 2010', Canberra: Commonwealth of Australia.

Infrastructure Australia Major Cities Unit (2011a), 'State of Australian cities 2011', Canberra: Commonwealth of Australia.

Infrastructure Australia (2011b), 'Australia's public private partnership pipeline', available at http://www.infrastructureaustralia.gov.au/public_private/ (accessed 10 December 2011).

Johnson, M. and S. Rix (eds) (1993), *Water in Australia: Managing Economic, Environmental and Community Reform*, Sydney: Pluto Press.

McKay, J.M. (2007), 'Water governance regimes in Australia: implementing the National Water Initiative', *Water: Journal of the Australian Water Association*, February, 150–56.

National Water Commission (2011a), 'Water governance', available at http://www.nwc.gov.au/home/water-governancearrangements-in-australia (accessed 29 November 2011).

National Water Commission (NWC) (2011b), 'National Water Initiative', available at http://www.nwc.gov.au/water-reform/national-water-initiative (accessed 10 December 2011).

OECD (2011), 'Water governance in OECD countries: a multi-level approach', OECD Studies on Water.
Paddon, M. (1998), 'Restructuring and privatization of utilities in the Asia Pacific region', in L. de Luca (ed.), *Labour and Social Dimensions of Privatization and Restructuring: Public Utilities*, Geneva: International Labour Office, pp. 43–108.
Paddon, M. (2001), 'Social and labour consequences of the decentralization and privatization of municipal services: the cases of Australia and New Zealand', WP.177, Geneva: International Labour Office.
Productivity Commission (2011), 'Australia's urban water sector', Report No. 55, Final Inquiry Report, Canberra.
Queensland Government Department of Infrastructure and Planning (2010), 'Final progress report: Western Corridor Recycled Water Project State 2', available at http://www.environment.gov.au/water/publications/urban/pubs/wsa-qld-08-final-report.pdf (accessed 10 December 2011).
Ranald, P. and B. Black (2000), 'Privatising water in the driest state', *Labour and Industry*, **11**(2), 17–38.
Sheil, C. (2000), *Water's Fall: Running the Risks with Economic Rationalism*, Sydney: Pluto Press.
Sydney Morning Herald (2011), 'Desalination leaseback offer flushes out French bidder', 11 September, available at http://www.smh.com.au/business/desalination-leaseback-offer-flushes-out-french-bidder-20110918-1kfwj.html (accessed 13 January 2012).
World Desalination Report (2011–12), available at http://www.desalyearbook.com/current-issues-in-desalination/plant-mothballing-be-or-not-bepercentE2percent80percentA6 (accessed 13 January 2012).

12. Risks in urban water reforms: a challenge to public–private partnerships[1]

Claude Ménard

12.1 INTRODUCTION

Despite their slow pace, reforms based on public–private partnerships (PPPs) are gradually making their way into the urban water sector. In an extensive study of reforms in water and electricity distribution over the period 1973 to 2005, Gassner et al. (2009) showed the significant impact of private sector participation on 977 utilities in the water and sanitation sector, with 141 reforms, mostly concentrated in Latin America and the Caribbean over that period, involving PPPs and the remaining 836 being operated as state-owned enterprises. More recent data from the Public–Private Infrastructure Advisory Facility (PPIAF) confirm the trend, although there is a clear shift in PPP from Latin America to Asia, particularly China.

Forces pushing public authorities to reform their water sector are relatively well identified (Ménard and Shirley, 2002, pp. 4–8). Reform of utilities in the urban water sector is attempted due to a combination of three factors: (1) unmet demand and service problems in the water system, so that a significant number of actual as well as potential users would benefit from reform; (2) macroeconomic crisis characterized by growing public deficits, public debt and inflation creating major financial pressures on public authorities; and (3) a regime change or coalition shift bringing to power a governing group that perceives net political benefits from reform. Excluding cases in which water utilities remain directly managed by public authorities, these reforms can take essentially three forms, in theory: (1) corporatization, which transfers major decision rights exercised by the public body to a relatively independent board, which is publicly accountable for its decisions while property rights remain in the hands of public authorities; (2) full divestiture, which remains marginal in the water sector (it represented only 5 percent in the sample of Gassner et al., 2009); and (3) various forms of private sector participation. Among the reforms involving private sector participation, the same study showed that 66 percent were concessions and 24 percent lease and management contracts.

However, such reforms also face risks. Risks are defined here as any factor, event or influence that threatens the successful completion of a transaction in terms of time, cost or quality (adapted from EC, 2003). Although several risks are threatening all types of reform, PPPs particularly face challenging ones because of the difficulties in risk allocation among the key partners and because many of these risks are deeply embedded in institutional factors, thus remaining exogenous to decision variables on which private partners (or even public ones) have control.

These 'institutional risks' are amplified by the very characteristics of urban water sector. On the one hand, this sector is a typical network industry, that is, an industry characterized by: (1) the 'club effect', with the increasing rate of connection generating important economies of scale; (2) large economies of density, with concentration of users reducing the network's average costs; (3) important sunk costs needed to establish and maintain the network; and (4) the existence of critical functions (for example, interconnections, interoperability) that are essential for a network to remain operational. As such, the urban water sector is exposed to standard risks in network industries, coming out of the large fixed costs involved and the impact of potential disruptions in the system. However, the urban water sector also displays specificities when compared to other network industries, such as electricity, railroad, urban public transportation, and so on. Being scarce and having limited capacity of regeneration, water resources have economic specificities such as low price elasticity, since there are no substitutes for water; and strong environmental and public health externalities make it a quasi-public good. There are also specificities stemming from the nature of technologies involved, for example: building dams, thus deviating water; pumping important quantities of underground water, with consequences on alternative usages; and meeting high transport costs, which make the localization of the resource a key issue.

These specificities have determined the way the urban water sector has traditionally been structured as a public monopoly and contribute to explain why the pace of reforms in the urban water sector has been slow compared to other network industries. Nonetheless, the water sector has undergone important changes and remains under strong pressure for further reforms, triggered by inefficiency, underinvestment, and new environmental challenges such as climate change and the threat to water security. Hence, reforms are considered and canvassed, in search of innovative arrangements that intend to introduce some competition in the market, to turn towards participation of the private sector, and to increase autonomy for those water utilities remaining within public sector management.

Apart from the traditional risks faced by any water sector operator in the execution of its functions, there are also new risks created by reforms intending to introduce or develop private sector participation. For instance, the contractual nature of this participation, with the accompanying protective clauses, the asymmetries of a very special nature among partners, and the regulations that accompany the transfer of rights to private partners might produce new sources of uncertainties and vulnerabilities for the management of water utilities. The chapter argues that these uncertainties and vulnerabilities result from 'unbundling' functions which were traditionally integrated in the publicly owned utilities, from the introduction of operators with preference functions that differ in many respects from those of public authorities, as well as from the new institutional designs imposed by these changes.

This chapter mainly focuses on the 'institutional risks' that accompany reforms in the water sector, with an emphasis on risks associated with changes involving the private sector. Section 12.2 identifies the many different types of risk involved in the water sector with particular attention paid to the specific risks faced by private water operators in relation to the institutional environment and the resulting factors of vulnerability plaguing contractual agreements. Section 12.3 emphasizes the difficulties in evaluating these risks *ex ante*, although such evaluation is essential for legitimizing the choice by public authorities of PPPs over alternative arrangements. Section 12.4 examines the sensitive issue of the allocation of risks between public authorities and private participants in a context in which one player can determine and even unilaterally change the rules of the game, and the key role of enforcement in determining if PPP was the right choice. In section 12.5, in conclusion, the chapter provides some insights on the ambiguous role of regulation as an institutional response to the vulnerability of parties facing these risks.

12.2 FACTORS OF VULNERABILITY: THE KEY ROLE OF INSTITUTIONAL RISKS

In general, reforms are about the reallocation of decision rights and possible changes in the distribution of property rights. In corporatizing a public utility, public authorities transfer a significant part of decision rights to the board of the corporation while retaining the ownership of property rights. At the other end of the spectrum, full privatization involves the transfer of all decision and property rights to a private actor, although the exercise of these rights is often regulated by primary and subsidiary legislation enacted by the state, particularly in the water sector. In between are the different forms of partial transfer of rights and responsibilities associated with various degrees of private sector participation, which can be grouped under the general umbrella of PPPs. Transfer of rights among technologically separable units, transfer thereafter labeled as 'transaction', might concern the provision of specific tasks, as with management contracts; the operation and maintenance of the system, as in lease contracts; or the building and/or extension of the system and its management, as with concession arrangements. These variations in the allocation of rights, and the blurred areas they delineate in comparison with the polar cases of full divestiture or corporatization, make PPP particularly vulnerable to different types of risks.

In order to better grasp the risks at stake, it is useful to distinguish between endogenous risks, inherent to the transaction itself or generated by institutional factors directly determining the environment within which a transaction is processed; and exogenous risks, due to factors independent from the transaction at stake. The different categories of risks are summarized in Figure 12.1 and this chapter mainly deals with endogenous risks.

12.2.1 Business Risks

Business risks have attracted the attention of policymakers and operators as well as financial analysts. These risks are associated with the normal running of a business that all operators of utilities face, whether public or private. Although they are not specific to water systems, the importance of sunk costs in urban water systems and the duration of investment make them part of what has been identified in the literature as 'critical infrastructures' (Moteff, 2003; Kunneke et al., 2010), in which transactions are exposed to amplified business risks along four dimensions: construction, demand, operation and finance. These risks are well established, with detailed protocols for their evaluation, and will not be developed here.[2]

12.2.2 Institutional Risks

Political risks

Although institutional risks have challenged investors and traders for centuries, political risks have recently attracted more attention in the economics and managerial literature. A leading factor that pushes towards the reform of public utilities, including water systems, is the idea that divestiture or at least some participation of private partners in association with

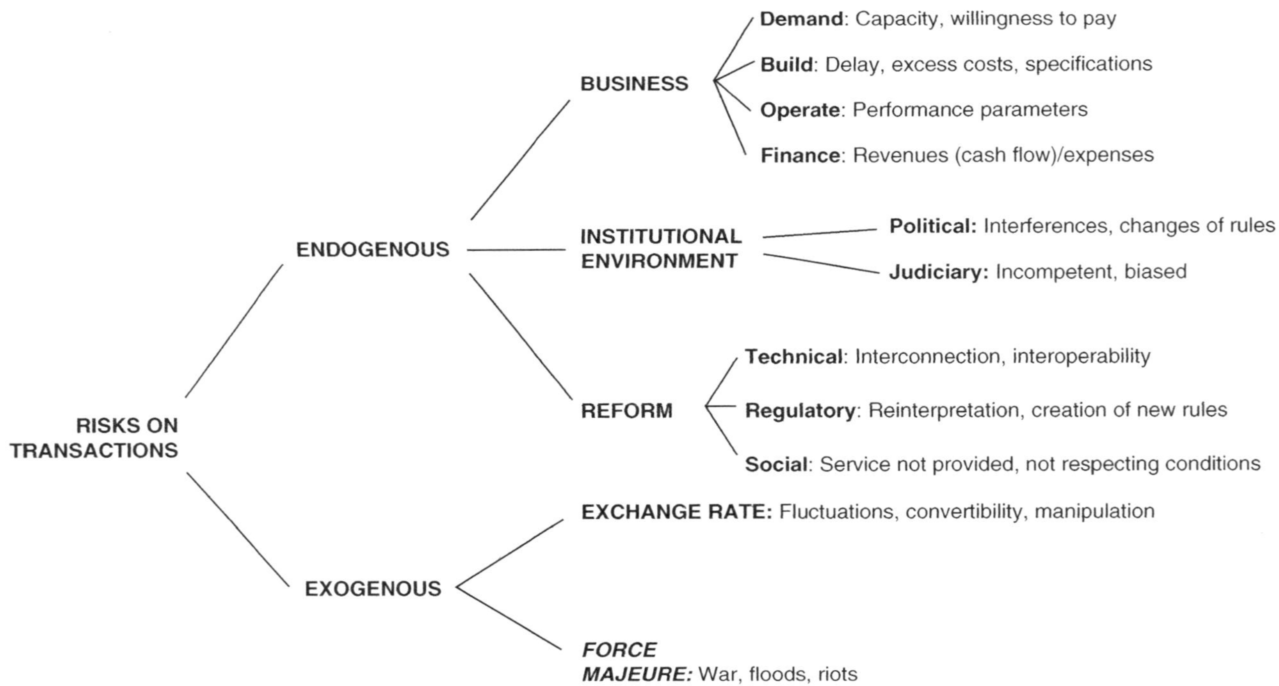

Figure 12.1 Risks faced by operator(s) in the urban water sector

public authorities in the provision of goods or services could eliminate or relax the risk of political interference in the decision process of these utilities. However, several recent contributions have shown that there might also be political risks coming out of the very nature of the contractual approach favored by adepts of PPP (Guasch, 2004; Medda, 2007; Spiller, 2009).

Indeed, partnerships of this type are confronted with two major problems. Firstly, there are the classical risks associated with one party to the arrangement (public authorities) changing the rules of the game, as when they introduce new environmental requirements during the contract and impose these changes on the other party (the private operator). There are risks associated with the temptation for public authorities to capture the rent either through anticipated renegotiation of the contract (Guasch, 2004) or even more radically through expropriation once sunk investments have been made by the private operator(s). Symmetrically, there is the risk of capture by private operators, for example through corruption of the decision-makers. Contractual guarantees and neutral dispute resolution mechanisms such as arbitration could only partially mitigate these risks. Secondly, there is another type of political risk associated to PPP, what Spiller (2009) has identified as third-party opportunism. Indeed, PPPs introduce an element of transparency, even if a limited one, since they are based on formal agreements. As a result, they can be challenged by political opposition, special interest groups, environmental associations, and so on, on a variety of grounds ranging from violation of public policy to infringement of human rights. In order to protect themselves from these political risks, particularly acute in a democratic system, policymakers endorsing PPP have a strong incentive to make contracts as explicit and constraining as possible. This strategy makes contracts much more rigid than, say, contracts between typical business entities, since the latter are less exposed to public outcry and more prone to renegotiation. With PPPs, contracts are easily looked at with suspicious eyes by opponents, and even more when renegotiation occurs, thus generating a different type of political risk that makes adaptation to unexpected circumstances (for example, changes in regulation or in the natural environment) particularly difficult.

Judicial risks

Because of their very nature as contractual arrangements, PPPs are also confronted with problems relating to fulfillment of obligations. When conflicts relating to interpretation of the contractual undertakings emerge, they can sometimes be solved by way of direct negotiation and/or mediation. However, not all disputes can be solved through negotiation or mediation. Hence, most contracts make provision for formal dispute resolution through courts, which submits parties to another type of risk: depending on which courts have the jurisdiction to hear disputes and depending on the judges in charge, such mechanisms are not always efficient in producing satisfactory results. Firstly, judges can be incompetent or corrupt. Secondly, even honest and competent judges might be biased: they may have their own agenda (for example, getting promoted), they have their own ideology and beliefs, and so on. Thirdly, almost by definition a judge will have to make decisions that remain unpredictable *ex ante*, since they have some discretionary power in interpreting contractual clauses, determining damages, and so on (otherwise no judge would be needed). Fourthly, judicial procedures are costly, these costs are often difficult to evaluate *ex ante* by the parties, and they can be quite substantial. All of this concurs to create judicial risks as part of the risks coming out of the institutional environment.

12.2.3 Risks Inherent to Reforms

There is another category of risks that has increasingly attracted attention in the economic literature, which are the risks coming out of the reform process itself. Most contributions so far have focused on the role of regulation and regulators, with risk of capture at the forefront of the analysis. However, there are other dimensions to be taken into account.

Technical risks

These risks are essentially relevant when some decision rights as well as property rights initially under the control of public utilities are transferred to private actors, and they tend to be particularly significant when this transfer accompanies the unbundling of functions. Technical unbundling might not be as spectacular in water systems as it is in electricity or railways. Yet, there are many examples where responsibilities are split or shared between the parties, making the governance mechanism ineffective at times. For example, this can happen when construction of the physical infrastructures for water remains in the hands of a publicly owned entity, while the private sector is in charge of distribution and connection of water to households and firms (Ménard and Clarke, 2002). In such situations, problems of interoperability, interconnection and capacity management might surface (Newbery, 2001; Finger et al., 2005; Kunneke et al., 2010). At the same time, maintaining the integrity of the system is essential for guaranteeing safety and the quality of the service provided, like fixing leaks rapidly to avoid spreading pollution in the system. This requires regular and coordinated assessment of the infrastructure as well as a decision process to fix problems that command cooperative behavior among partners. Difficulties in building confidence might transform technical risks into a major threat.

Regulatory risks

Another type of risk inherent to reforms of urban water systems involving private participation comes out of regulation and its implementation. Whatever the type of reform adopted, urban water and sanitation systems remain highly regulated everywhere. Regulation plays a specific role in PPP in that one partner, namely the public entity(ies) involved, determines the rules of the game, either directly through legislation, or through the exercise of regulation. Risks then come out of the possibility that policymakers create new regulations or that regulators reinterpret existing regulations, thus affecting costs and revenues for the operator(s) (Bakovic et al., 2003). Another problem can emerge when regulation is defined on a sector basis: although the contract is monitored by the sector regulator, there might be actions by regulators from other sectors and/or by distinct public agencies that deeply impact the reform, for example, when changes in environmental requirements or in public health standards involve substantial investments and even changes in techniques (for example, filtering process, purification techniques).

A distinctive difficulty comes out of the very nature of urban water systems, which are decentralized and most of the time monitored at the local or regional level. In most countries, general laws or rules frame the possibility to introduce PPP in the provision of urban water, but the implementation is transferred to local or regional level. As a result, micro-institutions develop locally to monitor the arrangements. Municipalities or regional authorities with little experience and/or poor expertise become very active in the definition, adaptation or implementation of rules, thus adding to the number of risks that are usually faced by the private sector operator(s).[3]

Social risks

One more category of risks associated with the reforms of public utilities and particularly relevant when it comes to choosing PPP as an alternative solution is what can be identified as social risks. Water is a very specific good in that it is essential to human survival. Its use is universal and it is expected that it will be available to everyone. At the same time, all users are also political constituencies, making water reforms highly sensitive to political changes (Savedoff and Spiller, 1999, Chapter 1; Shirley and Ménard, 2002, Chapter 1). This is particularly significant in democratic regimes, with political cycles largely disconnected from investment cycles in the water system, with the active role of lobbies, and so on. Moreover, users share beliefs, for example viewing water as a gift from Mother Nature or God, and they exert some control over policymakers through their vote, through lobbying, through public protest and demonstrations, and so on, which may influence the way these policymakers may turn to PPP as the means to satisfy the essential need: that is, having drinkable water available. In that respect, they can apply pressure to benefit from as low tariffs as possible.

On the other hand, private water providers might have an incentive to raise tariffs, because they need to finance major investments in the system; because they have to absorb major shocks (for example, financial crisis); or simply because they are greedy and wish to take advantage of their monopolistic position in the delivery of the good. However, such tariff policies might reach socially unacceptable levels, opening the way to breach of contract (see the case of Buenos Aires). Private operators may also behave opportunistically in other ways, for example in giving priority to the expansion or improvement of their activities in the more profitable urban areas, ignoring the areas populated with low-income groups or groups opposed to dealing with water as an economic good. At the same time most reforms involving private participation have a major target, which is the end or at least limitation of subsidies, since alleviating public finance is usually an essential motivation to endorse PPP. Ultimately, risks should then be borne by consumers, through higher tariffs or sometimes having no access to the service. Hence, equity may no longer be guaranteed, or perception of tariffs policies implemented by private operators may be perceived as inequitable, even when backed by public authorities. The resulting social tensions might provide further incentives to policymakers and/ or regulators to transfer these risks to the operator(s), for example by imposing a minimum service or interfering in price determination beyond what was in the contract.

Although we lack a good theory and a sound methodology to capture these institutional risks (institutional since they are rooted either in the political regime or in the design of the reform and the conditions of its implementation), private sector water operators are well aware of their existence. This might explain their reluctance to bear investments that require durable commitments. Indeed, urban water systems require heavy investments that are specific and translate into long-term sunk costs. These investments correspond quite closely to what Williamson (1985, Chapter 3) identified as specific assets. Risks described above might challenge the sustainability of such fixed investments by private operator(s) (Luis-Manso, 2007). Hence there is the need for a careful examination of these risks *ex ante*, before deciding to go for PPP.

12.3 CHOOSING TO GO FOR PPP: LEGITIMACY AT STAKE

Organization theory, and transaction cost economics, which provides its modern backbone, suggests that a mode of organization will perform better if it is aligned with the type of assets at stake (Williamson, 1996, Chapter 12). From a normative point of view, it means that choosing an organizational solution for the provision of a public good such as urban water should be assessed comparatively, which requires the assessment of risks at stake. More precisely, if the choice of an organizational solution is to take all costs into account, deciding to adopt PPP should be based on an initial evaluation of the advantage and costs of the different risks identified above, comparatively to what they would be under alternative solutions (for example, keeping the water system public or fully divesting it to the private sector) (Shirley, 2002; Ménard and Oudot, 2009). In some countries, it is even required by the law to make such an evaluation in order to legitimize the choice of PPP.[4]

Recent developments in organization theory, parallel to the actual diversification of arrangements in the provision of public utilities, help in understanding better the various forms among which public authorities can choose to provide urban water. A useful distinction to capture this diversity is between decision rights and property rights. While the latter deal with the use, abuse and transfer of rights on goods and services, the former concern only the delegation of rights to use. Figure 12.2 provides a simplified version of the different modes of organization available for delivering water.

The area between public corporation (the 'corporatization' solution) and full divestiture delineates the domain within which contracts between public authorities and private operators are implemented. Differences among these arrangements depend on: (1) the duration of the contracts at stake; (2) the definition and degree of specification of the goals; (3) incentives embedded in the contract; (4) rules for the implementation and monitoring of the contract; and (5) mechanisms of adaptation and (possibly) renegotiation. An important contribution of transaction cost economics is that these clauses not only shape relationships among partners but also intend to limit contractual hazards. These 'contractual hazards' may come out of:

1. The specificity of investments required by the transaction at stake. The higher the specificity of these investments, the more incentives partners have to adopt clauses protecting them against contractual hazards. As a result, contracts tend to become more complex (*ex ante*)

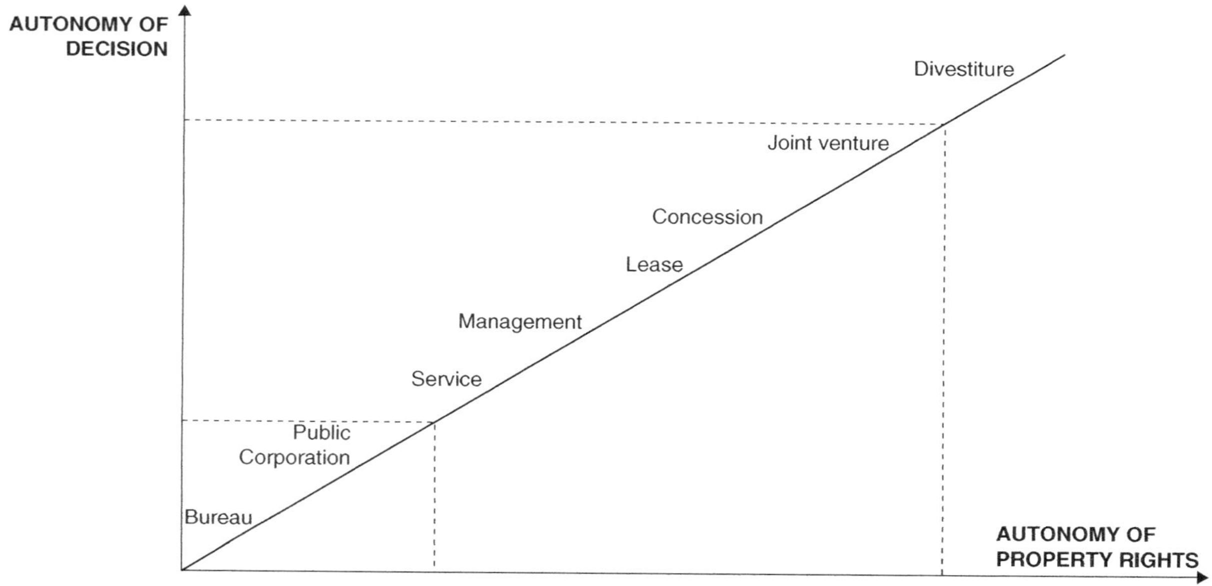

Source: Adapted from Ménard and Peeroo (2011).

Figure 12.2 *Modes of organization for delivering water*

and more constraining (*ex post*), so that their transaction costs tend to increase when sunk costs increase.

2. The uncertainty surrounding the transaction. Uncertainties may arise due to a changing or even unstable environment, asymmetries of information among partners, and/or risks described above becoming increasingly difficult to assess. The higher the uncertainty, the more significant become clauses regarding adaptation rules and/or risk-sharing. As a result, transaction costs of contracting tend to increase.
3. The frequency of transaction. The higher the frequency of a transaction in which one partner is engaged, the more informational advantages he gets. As a result, his transaction costs tend to go down. This is clearly an advantage for a multinational corporation over public authorities, particularly those public authorities representing relatively small communities with infrequent opportunities to negotiate contracts.

The theory goes a step further in emphasizing that the efficacy of an arrangement in organizing transactions depends on its appropriate 'alignment' with the transaction at stake, that is, its capacity to reduce contractual hazards and therefore to minimize transaction costs. This is the now well-known 'Williamsonian principle of discrete alignment', defined as 'The assignment of least-cost governance structure to manage transactions' (Williamson, 1996, p. 378). An important consequence of this principle for our topic is that when institutional risks increase, so do transaction costs, to the point where costs exceed advantages, making recourse to PPP an economic mistake.

The implementation of this 'discrete alignment principle' requires a comparative approach. The efficacy of one mode of organization cannot be assessed without assessing what it would be if another mode would be chosen. This is precisely the task of 'preliminary assessments'. However, this *ex ante* evaluation is confronted by the difficulty of determining the specificity of investments required (but business people are usually not bad at doing this) and, above all, the difficulty of evaluating risks. Notwithstanding these difficulties, such an assessment is essential to arbitrate the trade-offs between stability and flexibility (for example, long-term versus short-term contract, or detailed contract versus framework) and between incentives and guarantees (for example, high profitability associated to high risks versus low rate of return guaranteed over a long period).

An important consequence is that a sound *ex ante* comparison of alternatives available among the different forms identified in Figure 12.2 should take into account all costs, including costs associated to the various risks identified in Figure 12.1. Of course, practitioners on both sides of contracts have developed methodologies, with the help of financial analysts, to evaluate 'business risks'. Most financial models estimate the feasibility of a project by its bankability, that is, its capacity to get financial support from banks and other lenders. Classical methods mobilize ratios such as: (1) debt over funds ratio; (2) debt service cover ratio (DSCR); and (3) loan life cover ratio (LLCR). Alternative sources of financing might also be particularly significant because of the very nature of water projects, for example, the existence of public subsidies or public guarantees that reduce the risk premium. Evaluating the other types of endogenous risks, those associated with the institutional environment or coming out of the reform itself, is much more difficult and may translate into high transaction costs.

Taking these transaction costs into account, even if one focuses only on direct costs associated with the elaboration of a contract with clauses anticipating some of these risks, is much more significant than usually acknowledged (if acknowledged at all) and tends to be underestimated. We know from transaction cost economics that these costs are those necessary to find a partner (for example, writing and managing a 'call for projects'), to identify the best partner, to estimate the realism of its proposal in terms of costs, to compensate competitors who have been discarded, to negotiate and write the contract. In a study of the costs of consultancy needed to prepare contracts in the British private finance initiative (PFI), Valila (2005) has estimated that these costs can represent as much as 10 percent of the value of a project and are rarely less than 4 percent.[5] This is particularly significant in situations characterized by a stable and well-known environment, with significant cumulated experience in this type of contract. Since PPP as a solution must be assessed comparatively to what it would cost to make 'in-house' (that is, building the infrastructure and/or providing the service directly through a public entity), there is also a need to estimate the administrative costs in case of direct provision through a public bureau or a public corporation; for example, the cost of the staff involved, the market value of the funds needed, and so on. This estimation tends to be neglected or ignored in most decisions to go public rather than introducing PPP.

A preliminary evaluation comparing alternative solutions should also take into account the possible revenues of activities that are complementary to the core activity. For example, if a basin in which clean water is stored gives the opportunity to organize side activities such as camping, swimming and boating, for which users will have to pay, revenues should be estimated with an underlying question: can it be assumed that these revenues are 'neutral', that is, independent from the organizational solution chosen?

Lastly, there is the difficult issue of social costs at stake. Because urban drinkable water satisfies survival needs that are universal, the introduction of private sector participation, which is usually accompanied by extensive metering and pricing, tends to face sensitive issues regarding the provision of water to the poorest segment of the population, the quality of water delivered and its monitoring, or the impact of for-profit organizations with respect to environmental concerns and the sustainability of the resource. This is not to suggest that social costs delegitimize resorting to PPP. Very often, PPP is precisely implemented to allow investments that public authorities could not make, thus connecting households previously deprived of drinkable water of quality; or to introduce managerial practices that were missing in publicly controlled utilities, thus making access and quality much more reliable at a lower cost. But there is a major problem of perception in the delivery of water by for-profit organizations that raises the issue of the legitimacy of this choice.[6]

Therefore, it is essential to keep in mind this sensitivity with respect to preliminary assessments. The lack of a rigorous *ex ante* examination of the costs and benefits of a PPP solution comparatively to a public solution (or a full divestiture in that respect), the legitimacy of this evaluation when it is done, and the legitimacy of decisions made by policymakers based on such evaluation can be and has repeatedly been challenged. Firstly, the choice of PPP is often questioned by opposing political parties, which is not a surprise. What can be more surprising is that this challenge does not necessarily stem from doctrinal positions: it can result from strategic choices by conservative parties as well as by leftists. Secondly, challenges can come out of vested interests that the reform threatens, as when unions protest against the impact of private sector participation on employment and/or wages. But it can also result from protests by groups of interest concerned by aspects of the contract that they perceive as putting at risk specific issues, for example environmental ones. This is what has been called 'third-party opportunism' (Spiller, 2009, pp. 52). Public authorities may be tempted to reduce such risks and build the legitimacy of their choice by making contracts that are extremely detailed and extremely costly to establish and to adapt, thus creating rigidities that might make the contract look even less legitimate in comparison with, say, provision through a public corporation.

To sum up, the absence of *ex ante* estimation of the comparative costs and benefits of PPP with respect to alternative solutions, or weak hypotheses and lack of rigor in such estimations when they are done, or lack of transparency about hypotheses that were used, can challenge the legitimacy of the choice of PPP as a solution to the problems facing delivery of water in an urban environment. An extensive review of surveys on PPP and privatization in Latin America over the period 1995–2005 (Bonnet et al., 2006) shows the progressive erosion of what was initially positively perceived by populations as a solution to the problems of lack of service and/or bad quality in water systems. Another negative impact of deficiencies in *ex ante* evaluation can be that public authorities might become reluctant to endorse modes of organization that could feed their opposition and that could also generate costly (and sometime devastating) actions in courts. On the other hand, detailed evaluation *ex ante* might have exorbitant costs and might carry the risks of major errors that could also sabotage the legitimacy of the entire reform. For example, potential operators might be confronted with classical observability problems: that is, problems in assessing the actual quality of the system for which it is bidding (this was the case for the consortia competing in the Buenos Aires case; see Shirley, 2002, Chapter 3); or unexpected risks, as with the Asian financial crisis that plagued one consortium in the Manila reform (Xun and Malaluan, 2008). There is an extremely delicate equilibrium between rigorous evaluations that can make them credible and support the decisions made, and evaluations that become so complex and costly that they represent a challenge to the *raison d'être* of the exercise.

12.4 ALLOCATING RISKS

Once risks have been estimated, whatever the rigor of the method, two problems remain to be solved: (1) how to allocate these risks between the contracting parties; and (2) how effective are the institutional rules in monitoring relationships between these parties? The first problem comes out of the blurred allocation of rights between parties. The second problem comes out of the difficult alignment of the regulatory regime with the mode of organization chosen.

12.4.1 Allocating Risk when Rights are Poorly Designed

One important difficulty with reforms involving PPP is that by definition there are two or more contracting parties with different preference functions although sharing some key decision rights and, in many contractual arrangements such as leases and concessions, some property rights as well. However, this is not enough to identify the specificity of the problem at stake: after all there are also business-to-business contracts in the private sector with poorly defined rights. What makes PPP more at risk is that one party to the agreement, the public authorities, is the initiator of the agreement, the propeller which engineers the rules within which the agreement shall operate, and the supervisor in charge of guaranteeing the implementation of the contract. As a result, there are risks inherent to the very nature of PPP.

An extensive literature suggests that when it comes to endogenous risks, the party which is at the origin of the actual materialization of a risk should bear its resulting costs.[7] In other terms, the party that did not respect its commitments or that did not control adverse events that it could have controlled should pay. This is the so-called 'responsibility principle' (for its economic interpretation, see Holmström and Milgrom, 1991, p. 27). As for risks of exogenous origin, for example an unanticipated change in environmental regulation that has an impact on the costs of processing water or a brutal shock on exchange rates as with the Asian crisis from 1998, the existing literature tends to suggest that the burden should be on the shoulders of the public partner because it could take advantage of favorable financial conditions on financial markets; and/or it could more easily transfer the burden to the entire collectivity (for example, through taxes), operating as a universal mutual insurance.

However, the actual implementation of these two principles – responsibility, which implicitly refers to incentives to behave properly; and mutuality, which refers to assurance needed for parties to get involved in a contractual relationship – is not that simple. Even in very detailed and carefully crafted contracts, there remains a potentially important subset of risks associated with decisions that have to be made jointly in a context in which some rights are almost inevitably poorly designed, badly allocated or derailed by opportunistic behavior.[8]

Some empirical studies suggest that in the absence of clear responsibilities in the materialization of a risk and of a clear allocation of rights, parties tend to converge towards a 50:50 sharing rule: it is the fair rule exhibited in several experiments (Fehr et al., 2007), and empirically observed in contracts between public authorities and private partners, for example in the defense industry (Oudot and Ménard, 2010). One possible explanation is that in having to share risks in a context of poorly defined rights, parties adjust decisions through a 'dialogue' that allows them to converge rapidly towards an equilibrium, thus reducing the transaction costs of finding solutions that could otherwise be better aligned to specific responsibilities. However, a central characteristic of PPP, namely the asymmetry among parties with respect to the rules framing the exercise of rights, exposes private partners to a different variety of risks, of an institutional nature. In endorsing this strategy of mutual adjustments, contracting parties take the risk of facing political challenges and third-party opportunism, as described above. Public authorities might be in the line of fire of political opposition or groups of interests exploiting this 'dialogue' as a source of collusion, even corruption; and private operators face the risk of being questioned as manipulators of public authorities motivated by their greediness. Finding the right institutional setting to monitor the relationships among parties then becomes essential to the success of reforms, particularly those relying on PPP.

12.4.2 Building Adequate Micro-institutions

Indeed, beside the internal characteristics of the contract negotiated between public authorities and private operators, the success of PPP depends largely on the institutions in which they are embedded. Practitioners as well as scholars are increasingly aware of the importance of this dimension. The issue is often encapsulated in topics about 'good governance' (see OECD, 2011). Following the path-breaking paper by Levy and Spiller (1994), attention focused on leading institutions at the aggregate level, typically the political regime and the legal system. For example, there are numerous papers about the role of the judiciary in securing contracts and making commitments of parties credible.

These 'general' institutions are important, particularly when it comes to the development of network infrastructures with a wide national or even supra-national coverage, for example electricity or telecommunications. However, the neglected role of micro-institutions in the success or failure of so many PPPs should be emphasized. In this context, micro-institutions refer to the subsidiary institutional arrangements that translate the general rules of the game into sector and/or regional mechanisms of implementation and control of these rules. In other terms, micro-institutions are essential to make operational those rules that are broadly defined at the level of the macro-institutions, mainly the political and/or the judicial system. Primarily, micro-institutions provide a means of coordination to public authorities in charge of regulating a sector, organizing infrastructures, or delivering specific goods and services at the local or regional level. It also means that these micro-institutions operate under a subsidiarity principle, holding their decision rights from what is determined by the institutions in which they are embedded (their 'institutional environment'). To illustrate, policymakers might decide, by law or decree, to open the water sector to competition, to unbundle a previously integrated monopoly, and so on. This defines the general institutional rules. Policymakers may also create 'micro-institutions' such as regulatory authorities in charge of monitoring the openness to competition, monitoring the market thus created, or deciding the specifics of the unbundling. Another example would be basin agencies, which have been created in several countries along river basins to monitor the allocation of water among different usages (drinking, industrial and irrigation) within the general rules defined by a legislative body or by the judiciary.

Substantial alterations might of course occur in the 'translation' of the general rules into specific ones embedded into these micro-institutions. Moreover, these devices may well have enough power or capabilities to develop rules of their own. If they are well adapted, micro-institutions should help in maintaining coherence in the concerned industry and guaranteeing the adequate alignment of goals between public authorities and private operator(s) in the case of PPP. They can do so by providing formal as well as informal dispute resolution devices and coordination mechanisms among the parties involved, while keeping transaction costs at a sustainable level. Symmetrically, maladaptation of micro-institutions can occur, generating a new category of risks for the parties involved. For example, there might be a misfit between the micro-institutions implemented and the general institutions or the mode of organization (for example, the type of unbundling in the sector at stake), or between the rules they defined and technological changes (for example, the development of smart grids in electricity). Maladaptation might generate incoherence and tensions among partners to the arrangement, pushing towards more reform or even capsizing reforms. The many layers of micro-institutions involved in the introduction of PPP in Manila and the resulting lack of coordination that brought the reform to the brink of collapse illustrate the point (Xun and Malaluan, 2008).

12.4.3 The Case of Regulatory Regimes[9]

In order to better understand how micro-institutions can generate risks of their own, the issue can be substantiated with the case of regulatory regimes. Regulation is almost universal when it comes to the provision of network infrastructures, even in the case of full privatization (Finger and Kunneke, 2011). Regulation is both a set of rules and a body in charge of their implementation. It is embedded in the general institutions, that is, its role and power are usually defined by a parliament (whatever the status of this parliament, either democratic or submitted to an authoritarian party) and/or complemented by the judiciary. At the same time, regulators translate the general rules of the game into operational guidelines that may alter the general principle or that may change over time. As such they can also generate risks that are distinct from risks associated with the flaws of the general regime (such as non-credible or predatory central governments).

Things become more complex when differences in regulatory regimes are taken into account. These regimes are more or less adequate for the development and credibility of PPP. Indeed, their characteristics and the power devolved to their jurisdiction can affect quite substantially the risks that private operators will face through the more or less intense control different regulatory regimes have over key variables such as the duration of the contract, the credibility of the commitments of the regulator, budget constraints on a regulatory agency and the degree of independence it signals, the potential a regulator has to interfere with the management of the operator, and so on. The varying capacity of regulatory regimes to manipulate these variables makes decisions of private partners more or less vulnerable, with a potential impact that may deeply alter investments and, more generally, the risks operators are willing to assume.

There are basically three types of regulatory regimes with respect to their capacity to interfere with the allocation of property rights as well as decision rights (Figure 12.3): (1) command-and-control, which means that public authorities have the power to interfere directly, through bureaus or politically dependent administrators, with the exercise of these rights; (2) regulation by contract, which means that public authorities commit to delegate rights defined *ex ante* to private or semi-public operators and transfer power to monitor in respect of the commitment to external agencies (ideally, independent ones) or to the judiciary; and (3) regulation by general laws when full divestiture goes hand-in-hand with the introduction of some competition and translates into full transfer of rights to private operators (although most of the time these operators remain constrained by obligations defined by the law).

Command-and-control regime

The command-and-control type of regulatory regime corresponds to a situation in which ownership as well as decision rights are in the hands of public authorities that usually build specific micro-institutions to monitor these rights (think about 'Department of Transport' or 'Bureau of Water and Sanitation'); or situations in which property rights are transferred to a private monopoly while decision rights remain tightly controlled by public authorities, which was often the case with classic 'public utilities' (think about the telecom industry before the so-called deregulation of the sector). In both cases, the operator is responsible for the running and management of the system. Risks are mostly coming out of: (1) political changes and/or political interference in the decision-making process; (2) incentive problems because of the statute of employees; or (3) in case of public ownership, lack of funds due to budget constraints or mismanagement.

Regulation by general laws

At the other end of the axis, regulation by general laws tends to prevail in situations in which property rights as well as decision rights are transferred

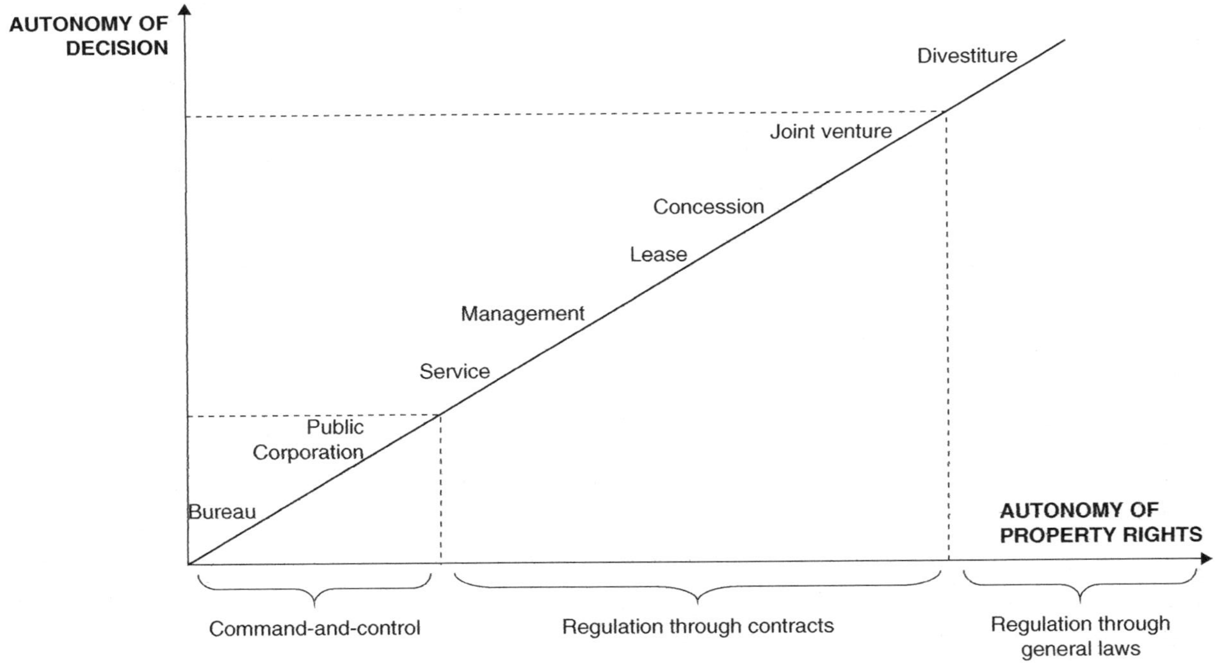

Source: Ménard and Peeroo (2011).

Figure 12.3 Forms of regulation in relation to modes of organization

to private operator(s) with the goal of introducing not only competition for the market (which might be the case when the provision of the infrastructure or of goods and services remain in the hand of a private monopoly, falling under the regulatory regime above) but also competition in the market. Regulation by general laws can be exercised through the judiciary, as was initially the case with the full privatization of telecommunications in New Zealand; or through more or less independent administrative bodies, for example the authorities in charge of competition or specialized agencies in charge of implementing a market and monitoring its development (the model adapted by most European countries in the privatization and liberalization[10] of telecoms).

The underlying rationale here is to introduce a clear separation between policymaking and the exercise of regulation by specific institutions: in principle, the extension of regulatory power and the responsibilities of regulators are defined *ex ante*, usually by law with no further interference of policymakers. This 'independence' is critical to the credibility of the relevant institutions, which in turn commands the willingness of private operators to assume risks in making highly specific and durable investments. Hence the risk of underinvestment, which has been observed in many of the most radical reforms of this type, even in contexts with well-developed general institutions (for example, railroads or electricity in the United Kingdom). These difficulties pinpoint the importance of designing adequate micro-institutions, but also the problem of keeping critical infrastructures or essential goods and services out of interference from public authorities. One can even argue that this is impossible, even illegitimate in a democratic regime with elected authorities concerned with the general well-being of their constituencies (whether they do that efficiently or not is another issue).

Regulation by contract

PPPs are mostly concerned with the third type of regulation, regulation by contract. Under this regime, most rules and regulations should in principle be specified in the contract among parties, including conditions under which adjustment and/or renegotiation can happen. Nevertheless, actual contracts are confronted with sources of risks that impose some flexibility, with a thin line between flexibility and discretionary power. The intensity of risks that PPPs face varies according to the nature of the contract, which itself depends on the degree of transfer of property rights as well as decision rights. These transfers are quite limited in the case of management or services contracts and much more significant with lease or concession contracts.

Among the risks that regulation by contract should help to overcome but might as well contribute to feed under some circumstances, the following are particularly noticeable. Firstly, there is the risk of inappropriate alignment between contract duration and the durability of investments required from the parties. From the operators' viewpoint, there is a trade-off between the advantages of long-term contracts, which can secure returns on specific investments, and short-term contracts that introduce some flexibility at the cost of repeated renegotiation. Duration is also often an issue on the side of public authorities, particularly in those contracts in which they commit to invest. They may do so for certain types of investment and according to a certain timing that determines the possibility for the private operator to meet its own obligation. For example, this is so when public authorities are in charge of developing the primary network in water supply, or more generally network infrastructures, while private operators are in charge of the secondary network or of services to be provided through the network. Private operators might then be exposed to delays for which users will hold them responsible. In this situation, what is the institution in charge of adjusting the contract and/or penalizing public authorities, and at what costs for both sides of the agreement?

Secondly, public authorities might mis-evaluate the actual capacities of the operator(s) with which they intend to contract. In the water and sanitation sector, technical and managerial expertise as well as financial credibility has long given a comparative advantage to a handful of multinational firms. However, opposing forces have weakened this dominance. On the side of operators, financial as well as political risks have pushed several of them to retreat from areas once considered as particularly promising. On the side of public authorities, opposition to the transfer of rights over essential services to foreign companies, and disappointments relating to the performance of several utilities operated by multinationals under PPPs, have provided reasons to look at local enterprises as possible participants in water governance.[11] However, risks of default, technical as well as financial, tend to combine with increased risks of favoritism and corruption to sabotage the credibility of PPPs as solutions to the provision of critical infrastructures and essential services. Moreover, once committed by contract, public authorities may get trapped in a vicious spiral: once a project has been started and significant rights transferred to a private operator, there might be powerful incentives for public authorities to renegotiate a contract with terms more favorable to the private operator rather than admitting failure. But such a strategy also exposes parties to the contract to high political risks.

So far it has been assumed that contract clauses are relatively exhaustive, which raised problems of adaptation and/or renegotiation. However, and this is a third source of risk, contracts are often relatively open, defining a general blueprint rather than detailed compulsory clauses, leaving plenty of room for adaptation and interpretation. This is particularly so in sectors with long-life assets and important sunk costs due to the very nature of infrastructure at stake, as with pipes in urban water systems, rail tracks in railroad, and high-voltage transmission lines in electricity. Changes in the technology, the demography, incomes, and so on can deeply alter the conditions under which long-term contracts can be implemented. Again, risks of dispute and renegotiation tend to increase (Shirley and Ménard, 2002, Chapter 1; Guasch, 2004, Chapters 3 to 6). Credibility of partners, the existence of efficient conflict resolution mechanisms (competent courts, qualified arbitrators), and adequate devices for implementing solutions become critical to the success of contracts. Contracts are very rarely self-fulfilling with respect to these conditions. Regulation by contract then exhibits its dependence on the quality of its institutional environment.

This is where institutional risks come once more into the picture. Indeed, these risks are not exclusively associated to a command-and-control type of regulation or to regulation by general laws. They are also very significant in regulation by contract, threatening the capacity of PPPs to deliver what was agreed upon, with a strongly negative impact on expectations that are often too high from the start. A contractual approach faces trade-offs about the desirable degree of flexibility, which opposes precision of contractual clauses to regulatory discretion for making required adjustments; about the optimal degree of transparency, which opposes open access to information with the rights of operators to keep their business practice private; and about the intensity of institutional safeguards, which opposes institutions powerful enough to make agreements credible with rules preserving operators from predatory practices by public authorities.

An important lesson from numerous empirical cases, from Buenos Aires to Manila and Paris, is that institutional risks should be contained, when it comes to PPP, by reducing discretionary power in proportion to the specificity of investments at stake. This idea can be encapsulated in the following central proposition: the more specific investments are, the less discretionary power should be left in the hands of regulators.[12]

However, this approach faces a difficult problem, that of acceptability. Indeed, it involves almost complete transfer of responsibilities to private operators when major investments are at stake, which deprives policy-makers of their capacity to orient investments and might also provoke reactions from users who might view this as loss of control in favor of technocrats in a democratic regime.

12.4.4 Strategies of Actors

The issues discussed above lead naturally to another factor of risk, ultimately the main one: the strategies of the actors involved. The four main sets of actors that might have strategies putting PPPs in water at risk are: public authorities, regulatory entity, private operator(s) and the so-called 'third party.'

Public authorities

Because of the nature of infrastructure or goods and services in which PPPs develop, public authorities remain responsible in the last resort for their provision, for delineating the rules that shape the agreements, determining the type and role of regulation to be adopted, and remaining accountable for the protection of the 'general interest'. They do so by defining the legal characteristics within which private operators will intervene; in proposing and negotiating a contract; and ultimately in deciding the degree of autonomy and decentralization that they will accept with respect to the exercise of decision rights. In that process, public authorities develop strategies with two different goals: (1) they must give some satisfaction to their constituencies, particularly when submitted to periodic elections; and (2) they must overcome a problem of legitimacy, which is the acceptability by users who are also citizens of the transfer to private operators looking for profits of rights over goods and services that are perceived as essential. The resulting strategies, which do not necessarily coincide, are already a source of risk. It is complicated by the existence of the multiple layers composing public authorities. For example, in the failure of the water concession in Buenos Aires, conflicts among levels of government with different political orientation were a major issue (Alcazar et al., 2002, p. 65). Similarly, the multiple layers of policymakers involved in the reform of the water system in Manila have been a continuing source of tensions (Xun and Malaluan, 2008).

Regulators

Regulators in charge of supervising PPPs might also develop strategies which create risks. Under a command-and-control regime, as when contracts are monitored by a public bureau, risks result from interference by public authorities motivated by strategies of the same nature as the one described above; or they may result from specific interests of bureaucrats in charge, including bribery and other forms of corruption. With regulation through general laws, risks can result from strategies developed by those involved in the design of the law, particularly policymakers or their subordinates, or in the actual implementation of the law, for example by the judges; from strategies by those in charge of checking the conformity of contracts to the rules who are biased due to their incompetence, or asymmetries in the information available; or from strategies permitted by too much discretionary power in the interpretation of the rules of the game. Lastly, with regulation by contract, which prevails in PPP, several clauses might open the way to strategic behavior. Some clauses might allow interference of public authorities in access to privately monitored infrastructures or in the private management of services. Other contractual clauses, or the lack of adequate clauses, might increase risks of capture by private operator(s) or predation by public authorities. And risk might also be inherent to strategies coming out of the very mechanisms defined by the contract. For example, price determination might be subject to discretionary interference that introduces biases. Rate-of-return regulation includes a 'fair' profit on capital investments that has been criticized, rightly so, for the risk of arbitrariness involved. Price cap regulation, which has been introduced to reduce these risks by making profit a function of performance, also confronts strategic behavior since measuring performance in a way acceptable to both parties to a PPP and with adequate information is not trivial.

Private operators

Private operators also have their own strategies, which might hamper the success of PPP. Firstly, there is the now well-known risk of capture of public authorities or regulatory agencies, making conditions in the delivery of services vulnerable to the specific interests of an operator, as well as to political changes or to the appointment of new regulators. Secondly, decision to invest can always be revisited even when specific clauses of contracts make explicit commitments. For example, private operators can argue that unexpected exogenous factors have introduced extra costs that can hardly be assessed by public authorities or regulatory entities. Such opportunistic strategies might find favorable terrain in unpredictable natural factors, for example geology or hydrology; or environmental factors, for example volatility in the price of energy or changes in demography. Thirdly, there is the risk of collusion among competitors when competition is introduced, translating either to a limited number of competitors or to incentives to develop joint strategies. Under some circumstances, such strategies might have a rationale, as when private operators have an incentive to coordinate in order to face a powerful and possibly predatory government.

Third party

Lastly, strategic behavior that is a source of risks may come out of what recent literature has identified as 'third-party opportunism' (Spiller, 2009). As already mentioned, PPP is often chosen as a mode of organization to provide network infrastructures and/or to deliver specific goods and services because public authorities do not have the financial resources to do it directly or because existing public entities show very poor performance. However, asking private operators to overcome these problems might confront different oppositions that put PPP at risk. Firstly, vested interests are likely going to offer resistance, as when unionized employees face reductions in staffing. Secondly, political opponents may challenge the choice of PPP, arguing about its cumulated costs in the long run and of profits picked up by private operators. Thirdly, ideological opponents might also introduce risks, for example pressures from ecologists pushing towards changes in norms and standards that affect costs while the contract is still going on. Fourthly, other groups of interests may also develop strategies that put efficiency at risk, for example when a private operator faces local lobbyists opposing their decisions, as in the so-called NIMBY (not in my back yard) attitude.

There is a paradox in PPP: the more efficient it is, the more likely that it will confront risks coming out of the strategies of the different parties. Making the operation more efficient by increasing productivity and lowering costs might involve reduced staffing and pressures on providers of inputs to reduce their prices. Developing efficient transmission lines or adequate public transportation might involve confrontation with landowners opposing the implementation of these lines on (or near) their properties, or with car drivers opposing the implementation of dedicated sites for buses. Providing clean water on a regular basis to the urban population might require debasing the course of a river.

12.5 CONCLUSION

The challenges that PPP faces to be successful are numerous. Looking at the impressive list of risks identified in this chapter might suggest that it is not worth the effort. However, this is not the message that the chapter intends to convey. The central message is that PPPs in the delivery of water and sanitation services are here to remain. This is so because PPP offers a credible alternative 'architecture' to other solutions, such as entirely public provision or full divestiture. However, to avoid undue expectations with potentially counterproductive effects in the long run, it is essential to assess properly the risks that a PPP might face. Beside financial risks, which often focus the attention of negotiators on both sides because sunk investments are so significant in the water (and sanitation) sector, there is the need to take into account risks associated with contract design, something that is progressively permeating the community of agents involved in the negotiation and implementation of PPP; and risks of an institutional nature, which are most of the time neglected or considered at a very general level and mostly boiled down to political risks. There is an urgent need to go beyond general institutions such as political regimes and to dig deeper in order to understand at the theoretical level, and assess at the operational level, those 'micro-institutions' that translate general rules into operational ones, and implement them at the sector and/or local level.

The second message is that risks, particularly organizational and institutional ones, should be assessed comparatively. We have learned from organization theory that all organizational solutions involve transaction costs, which are the costs of organizing and monitoring transfers of rights among technologically separable entities (Williamson, 1996, p. 379). Choosing a mode of organization in the hope that it will be efficient in the delivery of water is bidimensional: (1) it must meet requirements regarding the specificity of delivering drinkable water (continuity, safety, and so on); and (2) it must fit requirements of the transactions at stake while minimizing costs, which requires assessing comparatively the economic costs of different solutions, for example delivering water through PPP versus delivering water through a public entity such as a public waterworks or public corporation. When it comes to the provision of critical infrastructures such as urban water systems and/or the delivery of goods and services considered as essential by a community, as water is, there is an extra dimension to take into account: the political transaction costs involved. PPPs are endorsed in the context of reforms. These reforms generate their own risks, which should be properly assessed *ex ante*.

The third and last message is that PPP raises problems of political legitimacy that are particularly acute in the water sector. Because they are about the transfer of decision rights, and in many cases of property rights as well, from public authorities to private operators, they involve very specific types of contracts. On the one hand, if public authorities transfer most rights to private operators and transfer their monitoring to independent specialized agencies or to courts, they might well face the opposition of citizens who consider that it is the role of policymakers to control and be accountable for the delivery of essential goods and services and critical infrastructures. On the other hand, if public authorities interfere systematically with the management of private operators to monitor their actions and respond to changing wishes of citizens (or groups of interest), they are at risk of losing all the benefits of PPP.

Therefore, there is a dual equilibrium required for PPP to sustain: (1) an economic equilibrium based on the comparative costs and benefits of this mode of organizing for the delivery of water and, more generally, of 'public' goods and services and the provision of critical infrastructures; and (2) a political equilibrium based on the assessment of the political transaction costs of reaching a coalition supportive to PPP as a durable solution for the activities at stake.

NOTES

1. This chapter draws partially from a research project developed with Patricia Luis-Manso and Matthias Finger while I was a visiting professor at the Ecole Polytechnique Fédérale de Lausanne (EPFL), and from a report to the Centre d'Expertise Français pour l'Observation des Partenariats Public-Privé (CEF-O-PPP) and the French Ministry of Finance (Ménard et al., 2009). However, the usual disclaimer fully applies.
2. For a summary, see Ménard et al. (2009) and for more details Oudot (2011).
3. This contradicts risk management theories, according to which the party at the origin of risk factors or having more control over these factors should bear the burden.
4. For example, notwithstanding its long tradition of private participation in the urban water sector, France has adopted only recently an ordinance (2004–559, June 17, 2004) reinforced by a Law (2008–735, July 28, 2008) making preliminary comparative assessment compulsory. This law and its following jurisprudence have established that beside emergencies or complexity (so that public authorities do not have the required expertise), search for economic efficiency can justify choosing PPP rather than relying on traditional public entities or public procurement, but it must be based on an *évaluation préalable*. The theoretical problems of assessing comparative costs of alternative arrangements *ex ante* have already been pointed out in Masten et al. (1991).
5. See also Marty and Voisin (2008).
6. Of course the issue could be different if private participation would be through non-profit organization. But: (a) there is no significant experience of that type for large-scale cities; (b) a non-profit organization may face many problems similar to those confronting public utilities, for example issues of incentives for their employees, or of penalties for users who are bad payers because they can hardly afford the costs, and so on.
7. For a detailed review, see Oudot (2011, Chapter 7).
8. For an empirical analysis, see Clarke and Ménard (2002) about the overlapping rights between the public entity and a private operator in the reform of water supply in Guinea in the 1990s.
9. Elements in this subsection and the next one draw freely from Ménard and Peeroo (2011) and Luis-Manso et al. (2008).
10. There is a need to make a distinction between privatization, which concerns the transfer of property rights to private operators, and liberalization, which has to do with forms of competition. The two concepts are too often confused: there can be privatization without liberalization, as when there is transfer from a public to a private monopoly; and there is also the possibility of liberalization without privatization, as when publicly owned banks are put into competition. Details are provided in Ménard and Peeroo (2011).
11. Recent data from the PPIAF about PPP in China illustrates this trend. See Marin (2009, pp. 31sq.), and Chapter 2 in this volume.
12. For several examples in this direction, see Shirley (2002), Marin (2009, particularly Chapter 3, p. 39), and Xun and Malaluan (2008). Note that we often observe the reverse

trend in the real world, that is, increasing discretionary intervention when long-term contracts are involved (see Guasch, 2004; Spiller, 2009).

REFERENCES

Alcazar, L., M.A. Abdala and M.M. Shirley (2002), 'The Buenos Aires water concession', in M. Shirley (ed.), *Thirsting for Efficiency: The Economics and Politics of Urban Water Reforms*, Amsterdam: Elsevier-Pergamon, pp. 65–102.

Bakovic, T., B. Tenenbaum and F. Woolf (2003), 'Regulation by contract: a new way to privatize electricity distribution?', Energy and Mining Sector Board Discussion Paper, Washington, DC: World Bank.

Bonnet, C., P. Dubois, D. Martimort and S. Straub (2006), 'Empirical evidence on satisfaction with privatization in Latin America: welfare effects and beliefs', Working Paper, Toulouse: IDEI.

Clarke, G. and C. Ménard (2002), 'A transitory regime: water supply in Conakry, Guinea', in M. Shirley (ed.), *Thirsting for Efficiency: The Economics and Politics of Urban Water Reforms*. Amsterdam: Elsevier-Pergamon, pp. 273–316.

European Commission (EC) (2003), 'Guidelines for successful public–private partnerships', Brussels.

Fehr, E., A. Klein and K.M. Schmidt (2007), 'Fairness and contract design', *Econometrica*, **75**(1), 121–54.

Finger, M., J.P. Groenewegen and R.W. Künneke (2005), 'The quest for coherence between technology and institutions in infrastructures', *Journal of Network Industries*, **6**, 227–59.

Finger, M. and R. Kunneke (2011), *International Handbook of Network Industries: The Liberalization of Infrastructure*, Cheltenham, UK and Northampton, MA, USA: Edward Elgar Publishing.

Gassner, K., A. Popov and N. Pushak (2009), *Does Private Sector Participation Improve Performance in Electricity and Water Distribution?* Washington, DC: Public–Private Infrastructure Advisory Facility, World Bank.

Guasch, J.L. (2004), *Granting and Renegotiating Infrastructure Concessions: Doing it Right*, Washington, DC: World Bank.

Holmström, B. and P. Milgrom (1991), 'Multitask principal–agent analyses: incentive contracts, asset ownership and job design', *Journal of Law, Economics and Organization*, **7**, 24–52.

Kunneke, R.W., J.P. Groenewegen and C. Ménard (2010), 'Aligning modes of organization with technology: critical transactions in the reform of infrastructures', *Journal of Economic Behavior and Organization*, **75**(3), 494–505.

Levy, B. and P. Spiller (1994), 'The institutional foundations of regulatory commitment', *Journal of Law, Economics and Organization*, **9**(Fall), 201–46.

Luis-Manso, P. (2007), 'Reform and risk management in the urban water sector: the role of regulation', PhD thesis, Ecole Polytechnique Fédérale de Lausanne, Switzerland.

Luis-Manso, P., M. Finger and C. Ménard (2008), 'Institutional responses to the risks created by the urban water sector reform', Working Paper, Lausanne: Ecole Polytechnique Fédérale de Lausanne.

Marin, P. (2009), *Public–Private Partnership for Urban Water Utilities. A Review*

of Experiences in Developing Countries, Washington, DC: Public–Private Infrastructure Advisory Facility, World Bank.
Marty, F. and A. Voisin (2008), 'Partnership contracts, project finance and information asymmetries: from competition for the contract to competition within the contract?' Working Paper, Observatoire Français des Conjonctures Economiques, no. 2008-06.
Masten, S.E., J.W. Meehan and E.E. Snyder (1991), 'The costs of organization', *Journal of Law, Economics and Organization*, **7**(1), 1–25.
Medda, F. (2007), 'A game theory approach for the allocation of risks in transport public private partnerships', *International Journal of Project Management*, **25**, 213–18.
Ménard, C. and G. Clarke (2002), 'A transitory regime: water supply in Conakry, Guinea', in M. Shirley (ed.), *Thirsting for Efficiency: The Economics and Politics of Urban Water System Reform*, Amsterdam: Elsevier-Pergamon, pp. 273–316.
Ménard, C. and J.M. Oudot (2009), 'L'évaluation préalable dans les contrats de partenariat', *Revue Française d'Administration Publique*, **130**(Eté), 349–64.
Ménard, C., J.M. Oudot, S. Ben Ali, J. De Brux, J. Gatignol, L. Garcin and M.J. Kodjovi (2009), 'L'évaluation préalable dans les contrats de partenariat. Contribution à l'analyse de l'évolution des pratiques d'évaluation', Rapport au CEF-O-PPP, Institut de la Gestion Déléguée.
Ménard, C. and A. Peeroo (2011), 'Liberalization in the water sector: three leading models', in R. Kunneke and M. Finger (eds), *Handbook of Liberalization*, London: Routledge, pp. 310–27.
Ménard, C. and M. Shirley (2002), 'Cities awash: a synthesis of the country cases', in M. Shirley (ed.), *Thirsting for Efficiency: The Economics and Politics of Urban Water System Reform*, Amsterdam: Elsevier-Pergamon, pp. 1–42.
Moteff, J., C. Copeland and J. Fischer (2003), 'Critical infrastructure: what makes an infrastructure critical?', Report to the Congress, Congressional Research Service.
Newbery, D. (2001), 'Regulating unbundling network utilities', European Association for Research in Industrial Economics, Trinity College, Dublin, Ireland.
OECD (2011), *Water Governance in OECD Countries: From Theory to Implementation*, Paris: OECD.
Oudot, J.M. (2011), 'Choix contractuels et performances. Le cas des contrats d'approvisionnement de défense', Editions Universitaires Européennes.
Oudot, J.M. and C. Ménard (2010), 'Opportunisme ou équité ? Le cas des contrats d'approvisionnement de défense', *Revue Française d'Economie*, **24**(3), 196–226.
Savedoff, W. and P. Spiller (1999), *Spilled Water. Institutional Commitment in the Provision of Water Services*, Washington, DC: Inter-American Development Bank.
Shirley, M. (ed.) (2002), *Thirsting for Efficiency: The Economics and Politics of Urban Water Reforms*, Amsterdam: Elsevier-Pergamon.
Shirley, M.M. and C. Ménard (2002), 'Cities awash: a synthesis of the country cases', in M.M. Shirley (ed.), *Thirsting for Efficiency: The Economics and Politics of Urban Water System Reform*, Amsterdam: Elsevier-Pergamon, pp. 1–42.
Spiller, P. (2009), 'An institutional theory of public contracts: regulatory implications', in C. Ménard and M. Ghertman (eds), *Regulation, Deregulation, Re-regulation*, Cheltenham, UK and Northampton, MA, USA: Edward Elgar Publishing, pp. 45–66.

Valila, T. (2005), 'How expensive are cost savings? On the economics of public–private partnerships', *European Investment Bank Papers*, **10**(1), 94–119.
Williamson, O.E. (1985), *The Economic Institutions of Capitalism: Firms, Markets, Relational Contracting*, New York: Macmillan–Free Press.
Williamson, O.E. (1996), *The Mechanisms of Governance*, Oxford: Oxford University Press.
Xun, W. and N.A. Malaluan (2008), 'A tale of two concessionaires: a natural experiment of water privatization in Metro Manila', *Urban Studies*, **45**(1), 207–29.

13. Public and private interfaces: changing international perceptions of public–private partnerships in water services

Håkan Tropp and Camelia Dewan

13.1 INTRODUCTION

For those more than 900 million and 2.6 billion people that lack access to adequate water supply and basic sanitation (WaterAid, 2011), respectively, the question remains: from where will the big push come to significantly improve water services for the poor? In regions with the greatest need to develop water services, such as in sub-Saharan Africa, a huge investment gap remains. WaterAid (2011) estimates an annual funding gap in sub-Saharan Africa of US$15 billion and assesses that on average there is a need for spending 3.5 percent of gross domestic product (GDP) on water and sanitation to reach the Millennium Development Goals on water. Most countries in the region spend in fact well below 1 percent of GDP on water services.

The Millennium Development Goals (MDGs) created by the United Nations and committed to by almost all the governments of the world in 2000 focused on ending poverty by targeting a number of development areas. The MDG on water and sanitation states to halve by 2015 the proportion of people without sustainable access to safe drinking water and basic sanitation. However, the rapidly growing populations and international financial crises make the challenges of reaching the MDG on water and sanitation services formidable. Some countries are on track to reach the water supply target (South-East Asia) but most countries are off-track to reach the sanitation target (most low-income countries in sub-Saharan Africa) (WaterAid, 2011).

Typically water supply and sanitation provision has been heavily government-subsidized public services to the urban middle class and higher-income groups. Poor people suffer mostly as a result of lack of adequate infrastructure providing them with access to safe water at an affordable cost. As a result of government failures to cater for their populations, bilateral and multilateral organizations started in the early 1990s to emphasize on increasing the role of the private sector, market-oriented solutions and the formation of public–private partnerships (PPPs) as alternative means of delivering services to poor people. This was done by promoting procurement practices designed for engaging the private sector for infrastructure development and making available development funds for projects developed with private sector participation. In addition, targets set were to extend the number of people with connections to the municipal water grid, reduce unaccounted-for water, improve cost recovery and increase investments.

The engagement of the private sector and the formation of PPPs was seen as a new development panacea and led to high expectations that the private sector could succeed where the public sector had failed. Many developing countries opened up their water services sector for increased private sector participation and potential competition. In several cases, this also led to national controversy and opposition. However, the results of private sector engagement and PPPs have fallen short of expectations and, as a result, top–down and long-term PPPs, such as water services concessions, are increasingly questioned as means to reduce poverty.

Chapter 1 in this volume has dealt with the evolution of water governance architectures over the last two centuries and linked them with the global events and dominant international perceptions during each period. This chapter digs deeper into the topic and particularly focuses on the changing role and perception of PPPs during the past decade from the perspective of international development agencies, and points to some emerging trends of redefining PPPs. The chapter argues that the enthusiasm for PPPs, characterized by long-term concessions and leases in major urban centers in middle- and lower-middle-income countries, by bilateral and multilateral agencies as seen in the 1990s, seems to have worn off significantly due to very mixed results. Further, it explores whether PPPs in water can find new entry points for making contributions to improved access to water supply and sanitation, particularly among poorer sections of society, or if PPPs now are considered as something that may work under certain conditions but only make marginal impacts in developing countries. This study will approach these issues by analyzing the shifting perceptions of the private sector and the state on changing trends in water services in urban areas and the rise of new actors and partnerships.

13.2 THE WASHINGTON CONSENSUS

13.2.1 The Washington Consensus: PPPs in Water Services Take Off

The post-World War II period saw the push for modernization and a proactive state that guided development in the country, known as the 'developmental state' (Polanyi, 1944; Johnson, 1982). This assigned a strong role for the state and its active involvement in all sectors of the economy. After the economic crises of the 1970s, many developing countries had problems with their national finances and large government deficits. This was in part attributed to the size of the public sector and the perceived inefficiency of state-led economic development. In order to ensure repayment and a healthy fiscal status, the International Monetary Fund (IMF) and World Bank began to promote structural adjustment policies (SAPs) in the 1980s and 1990s. The SAPs were based on a premise that excessive state intervention was inefficient. Instead, markets were seen as more effective in promoting economic growth. SAPs consisted of a set of policies known as the Washington Consensus, which focused on liberalization, privatization and deregulation. One of the policy imperatives was to start to deregulate and privatize public services such as energy, telecommunication and water. Concessions and divestitures were key outcomes of privatization, and were extensively used as contractual arrangements during this structural adjustment period. After the end of the Cold War, this trend accelerated exponentially. According to Rodrik (2006), there was more privatization, deregulation and trade liberalization in Latin America and Eastern Europe than perhaps anywhere else at any point in economic history.

The rationales provided by the Washington Consensus were:

1. The private sector was seen as more efficient in delivering services than the public sector.
2. Markets were superior compared to bureaucracies.
3. The private sector was able to generate much-needed investments.
4. The private sector can expand access and improve service quality to poor people at lower prices due to competition.

Private sector involvement became almost an object in itself and was seen as a panacea for the low development of water services. In the early 1990s, PPPs in the water sector in developing countries started to take off. From 1991 to 2010, there were 731 PPPs in water services (World Bank, Private Participation in Infrastructure Database, PPID). In the water sector, as explained in detail in Chapters 4 and 12 in this volume, England and

France stood as models for the privatization of water services. In England, the water services were privatized under Margaret Thatcher through full divestiture in 1989. In France, concessions were the preferred option, where the municipalities were allowed to delegate service provisioning to the private sector but maintain ownership of assets (Moreau-Le Golvan and Bréant, 2007). The French model of concessions was also a much-preferred model that was promoted by the World Bank in developing countries, given the reservations in most developing countries about full-fledged privatization of public assets (Moreau-Le Golvan and Bréant, 2007).

13.2.2 Responses to the Washington Consensus

From the early 1990s and onwards, there was mounting criticism of neoliberal market reforms from aid agencies, non-governmental organizations (NGOs) and academics. It was argued that market reforms had not promoted growth or reduced poverty (Meagher, 2010). The negative impacts of structural adjustments, particularly in many African societies, were increasingly noted (Mkandawire, 2004). Privatization of water services also emerged as a highly politically contentious issue. After several popular protests in Latin America against privatization of water services and high tariff rises, PPPs in the form of concessions diminished considerably in Latin America and the market approach was seen as a failure in many countries in the region. In addition, bilateral and multilateral organizations have not had a unified view on the Washington Consensus. On the contrary, many of them have criticized it for being too narrow and neglecting important aspects of income redistribution, governance and poverty reduction, and they have thus attempted to act as a counterweight to the World Bank and IMF – the chief proponents of liberalization, privatization and deregulation. For example, during the mid-1980s, the International Labour Organization (ILO) and the United Nations Economic Commission for Africa promoted adjustment policies to the agenda of liberalization, deregulation and privatization to protect employment and to make it work for targeted poverty reduction. Around this time, the United Nations Children's (Emergency) Fund (UNICEF) also argued for policy adjustments with a 'human face' to include health, education and social protection. In 1990, the United Nations Development Programme launched the Human Development Report (HDR). It challenged the priorities of structural adjustment programmes and put growth with equity at the centre for accelerated human development (Jolly et al., 2004, 2009).

However, much of the criticism of the Washington Consensus was more about what was lacking in it rather than criticizing the approach as such. For example, the United Nations Economic Commission for Latin America and the Caribbean (ECLAC) argued for 'reform of the reforms' since privatization and deregulation also created social clashes and imbalances (ECLAC, 1995). However, external as well as internal criticisms (such as through the previous chief economist of the World Bank, Joseph Stiglitz) have shaped the Post-Washington Consensus, which has made a number of additions to the original Washington Consensus. If the original consensus was to 'get prices right', the new emerging consensus was about 'getting institutions right' to ultimately get prices right and to make private sector engagement work. Hence, from the mid-1990s onwards, there has been an increasing focus on governance and institutional reform linked to financial systems as well as for particular development sectors such as water. The World Water Forum, held in The Hague in 2000, was one of the first major international meetings where governments and other water sector stakeholders explicitly mentioned governance as critical for water development. The Post-Washington Consensus also brought targeted poverty reduction and the importance of developing social safety nets to the forefront (see, for example, Saad-Filbo, 2010).

The next section will show how major international development meetings were instrumental in promoting private sector engagement and changing the role of public entities in charge of water services to a development co-driver by enabling and regulating private sector activities in the spirit of the Post-Washington Consensus.

13.3 POST-WASHINGTON CONSENSUS: EMERGING PPP PRAGMATISM

13.3.1 Changing Discourse and Priorities among Bilateral and Multilateral Agencies

As a response to the criticisms of structural adjustment policies, rapid privatization, and demonization of state interference, a more nuanced view of the state and more realistic expectations of the private sector emerged in the Post-Washington Consensus, manifested in Agenda 21 and the MDGs for example. The United Nations Conference on Environment and Development (UNCED) in Rio in 1992 indicated the start of the sustainable development era. The conference adopted Agenda 21, a very comprehensive reform program that called for sustainable and balanced development of social, economic and environmental aspects, not only for present generations but also for future ones.

A clear outcome of the Rio meeting was that governments alone cannot realize the vast sustainable development agenda. New partnerships had to be forged among governments, civil society and private sector. For the private sector, it meant that its active involvement was now increasingly sought by governments and by multilateral and bilateral organizations to implement the sustainable development agenda.

Agenda 21 drew attention to new roles for governments. It was seen as critical that governments established procedures that allowed for an expanded private sector role in development cooperation (Faulkner, 1997). Agenda 21 urged governments and multilateral and bilateral organizations to overcome their suspicion and distrust of the private sector and view it as a new development partner to implement the sustainable development agenda. This new way of looking at private sector permeates most parts of Agenda 21. For example, in Chapter 18 on Freshwater it is stated: 'Enhancement of involvement of the private sector, where appropriate, in human resource development and provision of infrastructure.' Agenda 21 also reaffirmed water as an economic good, a notion that was first acknowledged by the Dublin Principles of 1992. Similarly, it was understood that private sector actors could not act alone and needed the state as well. A former director of the Business Council for Sustainable Development (BCSD) stated:

> UNCED's urgent insistence that the private sector should have an expanded role in moving Agenda 21 forward, in collaboration with the public sector, provided the key to unlocking the door to a completely different approach: the concept of public–private partnerships (PPP). The immediate post-Rio challenge was moving PPP from concept to action . . . This represented a sea change in attitude: recognition, at last, that the private sector has a powerful contribution to offer. (Faulkner, 1997)

For institutions such as the World Bank and the ADB, this provided an incentive to put the state back on the agenda through an emphasis of 'good governance' consisting of stable property rights, reduction of political control of rents (corruption) and reducing transaction costs through legal institutions (Meagher, 2010; Khan, 2004). This view of the state assumes that markets would be able to succeed where states failed, and that they would do so given that the state creates institutions that help facilitate the market and reduce transaction costs. The essence of the Post-Washington Consensus view on state and private sector roles and relations was captured by the former Secretary General of the United Nations Kofi Annan (1998), who stated that governments in the developing world had to accept market forces and focus on creating institutions and laws that provided an 'enabling environment for sustainable development, an environment in which private enterprises can develop and flourish'.

Hence, internationally, there has been a strong push for developing different types of partnerships between the private and public sectors, which paved the way for PPPs under the new paradigm of the Post-Washington Consensus where governments take on the role as development co-drivers to enable and regulate private sector activities. The Post-Washington Consensus was a response to both market and state failures and was based on the premise that institutions were necessary for markets to function. A strong institutional framework could also help 'rein in' the state through a strong judiciary and anti-corruption laws. This trend is evident in World Bank studies that confirm the existence of the good governance agenda in the current day through its emphasis on the importance of clear contracts and a comprehensive regulatory framework (Marin, 2009).

From a PPP perspective, privatization can be seen as a way of changing the role of the state, where governments are increasingly becoming catalysts, promoters, regulators and redistributors (of wealth), and the responsibility for production of goods and services is shifting to those who have a comparative advantage in this area, namely the private sector (Guislain, 1997). Meagher (2010) argues that the Post-Washington Consensus idea of a 'good governance state' illustrates the discursive shift from the 'developmental state' to one in which the state co-drives the development initiatives in partnership with the private sector.

Since Rio 1992, many UN meetings have recognized the important development role of the private sector and the need to develop various types of PPPs at all levels and in all sectors. Similarly, there has been a strong reiteration of the role of governance and of the enabling and facilitating role of governments and the public sector, as well as a pro-poor rationale for PPPs. While PPP was a new concept in the early 1990s it gained ground and was, for example, extensively and directly referred to in the ten-year follow-up summit in Johannesburg 2002. The following quote encapsulates the spirit and reaffirmation of PPPs. The Johannesburg Plan of Implementation 2002 states that:

> It is critical to facilitate the establishment of public–private partnerships and other forms of partnership that give priority to the needs of the poor, within stable and transparent national regulatory frameworks provided by Governments, while respecting local conditions, involving all concerned stakeholders, and monitoring the performance and improving accountability of public institutions and private companies. (United Nations, 2002)

Bilateral support to the private sector is seen as the interplay between the state as a formulator of the 'rules of the game', the 'players in the private sector', and also of civil society. One could argue that while the initial structural reforms propagated rolling back of the state to the most basic functions of law and order, the contemporary view of the Post-Washington Consensus stresses the importance of the state as both a nurturer and a regulator of the private sector and markets. A well-functioning private sector requires a well-functioning state, just as the state needs a well-functioning private sector (Sida, 2004; Amjad, 2003).

13.3.2 PPPs in the Water Sector from the 1990s

As explained in this volume, prior to the 1990s there was very little privatization of piped water services in developing countries. In the few cases where private sector services existed, it was mainly linked to Francophone ex-colonies, such as Cote d'Ivoire, that had inherited a system relying on the private sector for water services, as exists in France (Kirkpatrick et al., 2004). Hence, by and large, water services provisioning was perceived as a public good and something governments alone should provide. The marginal involvement of private sector is illustrated by the fact that between 1984 and 1990 only eight contracts for water and sewerage projects were awarded to the private sector globally and the new aggregated investments in private water services totaled less than US$1 billion (Kirkpatrick et al., 2004).

Various international development agencies such as the World Bank, IMF, Asian Development Bank (ADB) and bilateral donors such as the United States Agency for International Development (USAID), the United Kingdom (UK)'s Department for International Development (DFID) and the Swedish International Development Cooperation Agency (SIDA), to name a few (Zhong et al., 2008), argued in the 1990s that the public sector had not been able to manage the technical, administrative and financial costs of water by itself, leading to highly inefficient and uncompetitive utilities (Rees, 1998; Zhong et al., 2008). Privatization in water did not take off entirely until 1993 and reached its investment level peak around 2000 (World Bank, 2007; Pinsent Masons, 2011a). Privatization, or PPPs in the form of concessions and divestitures, was seen as a way of attracting much-needed investment, while the sales of utilities would also boost government income (Gwynne and Kay, 2000). It was expected that PPPs would simultaneously improve service coverage, quality and efficiency by replacing conventional public sector systems suffering from underinvestment and inefficiencies (Zhong et al., 2008; Hall et al., 2005; Annan, 1998; Labonne, 1998). The hope was that PPPs would turn around poorly performing public utilities by bringing new expertise, financial resources and a more commercial orientation (Marin, 2009). As a result, a variety of PPP constructions in water took place in developing countries and especially in Latin America and parts of Asia (Zhong et al., 2008; Checchi et al., 2009). Chapter 2 in this volume provides a detailed statistical overview of PPPs in the water sector in high-, middle- and low-income countries across the world.

13.4 PPP EXPERIENCES: POVERTY ALLEVIATION, PUBLIC OPPOSITION, RISK AND PERFORMANCE

13.4.1 Problem of Reaching the Poor

One of the main arguments for many bilateral and multilateral organizations to promote private sector investments and to advocate PPPs was the high expectation that they would lead to improvements in water and sanitation for poor people. However, the success of the PPP approach in reaching poor people has been minimal. Targeting improved water services for poor people has not been perceived as a viable business operation and developing PPPs has rather been a case of 'cherry-picking'. Generally, the privatization of water supply and sanitation services in most developing countries is a phenomenon mainly taking place in large urban centres, particularly in non-poor urban pockets, and slum areas and poorer pockets are normally not part of PPP agreements (Neto and Tropp, 2000).

A major proportion of poor people in developing countries live in rural areas, but such areas have not been subject to PPPs in water services. 'Cherry-picking' has been a contested phenomenon that concessions have been criticized for. Concessionaires may consider certain low-income neighborhoods too costly or risky to invest in. Furthermore, providing services in poor neighborhoods can be technically and operationally more problematic than in better-off areas, with difficulties of collecting revenue and warding off vandalism causing disincentives to serve the poor (Komives, 1999; Lobina and Hall, 2003). The private sector's difficulty in reaching poor people and ensuring universal coverage in water and sanitation is made evident by the current PPP statistics. In 2007, PPPs provided water to 7 percent of the urban population in developing and emerging countries, and private financing in water and sanitation accounted for somewhat less than 10 percent of the sector's total investment (Marin, 2009; World Bank, 2007). The main problems of universal coverage include bad financial management, low funding priority, lack of staff experience and qualifications, absent or weak customer service orientation, political interference, little or no independent regulation and

Table 13.1 Regional distributions of PPPs, 1991–2010

Region	East Asia and Pacific	Europe and Central Asia	Latin America and Caribbean	Middle East and North Africa	South Asia	Sub-Saharan Africa	Total
No. of PPPs by region	402	44	224	22	13	26	731
No. as percent by region	55.0	6.0	30.6	3.0	1.8	3.6	100
Investment in USD million by region	29 617	3781	24 751	3772	355	266	62 543
Investment as % by region	47.4	6.0	39.6	6.0	0.6	0.4	100

Source: World Bank, Private Participation in Infrastructure Database (PPI), available at http://ppi.worldbank.org/explore/ppi_exploreSector.aspx?sectorID=4 (accessed 10 March 2012).

an absence of civil society consultation. Privatization was believed to be efficient and capable of solving the problems of public utilities, yet these underlying problems do not seem to be remedied by the cost recovery logic of a concession. Even pro-poor concessions such as those found in La Paz–El Alto (Bolivia) were not able by themselves to resolve any of the issues that make concessionaires reluctant to serve the poor (Komives, 1999). Research from the NGO community shows that the policy of private sector participation does not tackle the underlying causes of water utilities' failure to serve the poor in a comprehensive manner (Komives, 1999; Esguerra, 2003).

Morever, the development of PPPs in different regions shows that so far PPPs have mainly benefitted middle-income countries in Latin America and East Asia in terms of both number of projects and investment levels (Table 13.1). For example, Argentina received the highest sum of private water investment from 1990 to 2003, representing over US$8 billion, followed by the Philippines, Malaysia and Chile. These are not the countries with the lowest level of access, nor are these the countries with the poorest of the poor. This suggests that investments are more easily attracted to countries and projects that have bigger markets, are more stable and that are perceived as profitable.

In high-risk markets, such as in most low-income countries, the risks associated with infrastructure projects are too large to be absorbed by the private sector (World Bank, 2005). As a result, the privatization wave of the 1990s by-passed many developing countries. In regions where water and sanitation challenges are the greatest – South Asia and sub-Saharan Africa – PPPs have been a very marginal phenomenon, in total accounting for less than 1 percent of the total investment value of PPPs. Out of 731 PPPs between 1991 and 2010, only 3.6 percent were in sub-Saharan Africa and 1.8 percent in South Asia. Similar patterns of investments being weakly correlated to needs are also visible for overseas development assistance (ODA) that goes to water and sanitation. Middle-income countries dominate the top ten countries receiving water and sanitation aid (WaterAid, 2011).

13.4.2 Problem of Public Opposition

According to Rees (1998), water is perceived by many as a public good and a 'merit good', where public goods and services provide benefits to communities rather than individual consumers. Public health and environmental benefits from sewage transportation, treatment and safe disposal are discernible examples. In addition, a 'merit good' is considered as something that should be provided by society irrespective of whether individual consumers are willing and able to pay for it. Privatization of water supply was therefore one of the most controversial issues in water resources management. How does one allow private sector actors to make profit out of a public good such as water? The PPP trend involved arrangements that led to tariff hikes and unaffordable services with the motivation to reap back costs of investments.

This was one of the key reasons for the vast public opposition against privatization of water in Latin America during the 2000s. The Cochabamba concession involved a 15–17 percent guaranteed US dollar rate of return, with the exchange risk covered by tariffs (Pinsent Masons, 2011a). To support this PPP, the government enacted a law, which allowed the concession to cover all water resources in its area. The effect was that all actual and potential customers had to connect to the system irrespective of their ability to pay. Most notably, no public consultation was made over either the law or the concession process and prices were raised without warning by 20 percent as stipulated by the government (Pinsent Masons, 2011a). This was an example of a concession that was extremely poorly conducted and not representative of all concessions. These problems led to large-scale protests over the privatization of the city's water supply with many people injured (see, for example, Prasad, 2006). After these events, there have been several public protests and opposition to privatization in Latin America. Few Latin American governments showed any interest or political will to engage with the private sector. This is illustrated by the fact that the Inter American Development Bank (IADB) does not have an explicit PPP strategy in its water program (Krause, 2011).

In Indonesia, privatization of drinking water in the 1990s was pursued as a government policy to address its budgetary constraints due to the economic crisis. One of the first water PPPs in Indonesia was a concession contract on Batam Island by PT Aditia Tirta Batam (ABT) in 1996. In 1998, a similar scheme was followed by the City of Jakarta by entering into a 25-year concession agreement through direct appointment with two private foreign firms, namely PT Palyja of France and PT Aetra, formerly PT TPJ of UK (Kruha, 2010). Currently, concessions in Indonesia are no longer required to conduct consultations with Parliament or the governor as they have been given authority for automatic tariff increases every six months.[1] This in turn has led to frequent public opposition and dissatisfaction with unaffordable tariffs (Kruha, 2010).

Concessions do not have an automatic incentive to be pro-poor. Nevertheless, they do provide expertise and capital that many governments require. With public opposition and a government unable to set and enforce pro-poor goals and performance targets, international companies may be reluctant to engage in 20–30-year investments in water concessions, where there are risks to recovering their costs. The next section looks at the problem of risk and why private sector entities themselves are reluctant to continue with concessions.

13.4.3 Problem of Risk

Most concessions in the water and sanitation sector are granted the exclusive right to provide services in their concession areas. Exclusivity is often justified on the basis that it reduces revenue risk and thus increases the attractiveness of projects with high capital costs and long amortization periods (Komives, 1999; Lobina and Hall, 2003). Hence, a company's motive behind engaging in a concession is based on exclusivity and profit.

Currency risk

Following the common practice in most infrastructure development projects, large international water companies have made investments in the water sector in strong international currencies. Revenues, on the other hand, are collected in weak, local currencies. In parts of the world where most international water contracts have been signed (Asia, Latin America), there has been considerable exchange rate instability, which has jeopardized profitability (Marin, 2009). For example, the national concessions in Argentina had the loans in US dollars and the risk mechanisms for currency guarantees were not working, which led to the international company running at a loss. This led to a disincentive to make large investments in foreign currencies, which were needed for buying the material required for building infrastructure.

From the viewpoint of currency risk, concessions have worked better in the Philippines, since the material needed was produced domestically and therefore foreign exchange was not as much of an issue. Thus companies face less risk than they would under exchange rate instability (Catley-Carlson, 2011). When water companies are subjected to taking currency risk, they tend to perceive it as a less viable business proposition and thus do not venture into long-term infrastructure contracts such as concessions.

Default risk

In 1995, Veolia was awarded a concession for water and sewerage services to the city of Tucuman, Argentina and raised tariffs accordingly. The governor encouraged residents not to pay their bills and had the concession rescinded. At the end of the process, Veolia had to cancel the contract in 1998 at a loss of approximately US$44.7 million (Pinsent Masons, 2011a).

As Krause (2011) notes, the problems with concessions were that they involved very high risks and high costs. Political, social and economic conditions can change rapidly and there is thus always the risk that the governments concerned could change the rules of the game and thus affect profitability. Where the regulatory mechanisms of most concessions proved weak and insufficient to deal with economic and political crises, such as in Argentina, we also find the largest number of contract terminations (Marin, 2009). Out of 731 PPPs (1991–2010), close to 12 percent were cancelled or considered to be distressed.

13.4.4 Problem of Performance

A range of performance studies has concluded that there is no automatic relationship between privatization and productivity improvements (Yarrow, 1989; Bishop and Green, 1995; Martin and Parker, 1997; Lobina and Hall, 2003). As Rees (1998) notes, 'private sector involvement cannot of itself and by itself remove many of the barriers to efficiency which impede public sector operations'. One issue is that privatization may increase productivity gains where there is competition. Without competition, there is little incentive to employ least-cost production methods, spur innovation or even provide the quantity or quality of water products for which customers are willing and able to pay (Rees, 1998).

Long contract terms combined with monopolistic tendencies therefore may not lead to the performance gains anticipated by concessions. For instance, there are a number of cases where water concessions have been awarded through one-bidder tenders, or competitive procedures in which only one company put forward a bid. In the late 1990s, Argentina, Bolivia, Chile, Colombia, Ecuador, the Czech Republic, Hungary, Poland and Romania were approached by single bids by multinational corporations (MNCs) and their subsidiaries including Suez, International Water Ltd and Thames Water (Rees, 1998; Pinsent Masons, 2011a). The infamous failed Cochabamba concessions were notably also a product of a single proposal concession, and arguably uncompetitive. In Indonesia, the privatization of PAM Yaya through concessions has not been successful in terms of investments and efficiency improvements, with problems of underinvestment, slowness of building connections, leakages and unaffordable connection charges to poor communities. Water service and quality in Jakarta remains a problem despite privatization, while tariffs are repeatedly being raised (Kruha, 2010).

13.5 NEW TRENDS IN PPPS: AWAY FROM CONCESSIONS

13.5.1 Pragmatism in PPPs: Hybrid Contracts

Given the issues relating to PPPs and their limited success discussed above, there is growing recognition that management and service contracts are potentially the most competitive form of privatization for public utilities and that they also reduce the regulatory burden of governments (Rees, 1998). Under management and service contracts, private firms are under almost continuous pressure to be efficient and competitive since contracts are let at relatively short intervals and rarely longer than five years. This also makes the terms and conditions more stable and predictable than when drafting contracts for a 20–30-year period. It also allows more flexibility in setting the terms in a more socially responsible way. In addition, only limited functions are transferred, making barriers to entry likely to be fairly low. Furthermore, this does not involve large-scale infrastructure investments that push for tariff hikes to recover costs. This also removes a considerable amount of risk from the companies previously involved in concessions.

It has also been argued that the distinction between concessions, management contracts, leases and *affermage* are becoming obsolete and that many of the most sustainable projects are a mix of these different categories (Marin, 2009; Catley-Carlson, 2011; Krause, 2011). PPPs are instead displaying pragmatism, with a large portion of the most successful financing models for long-term water PPP projects in developing countries emerging as hybrid schemes. This involves transferring commercial and operational risk to the private operator, where investment relies on a variable mix of government, private money and direct revenues from customer tariffs (Marin, 2009).

Colombia was one of the first countries to depart from standard concessions and is an interesting case of mixed arrangements or hybrid PPPs. Some contracts are purely private, some are purely public and others are using the mixed-ownership companies approach to accelerate investment (Marin, 2009; Krause, 2011). The Colombian approach shows pragmatism, where the operation of a utility is delegated to a private company in a way that is economically sustainable and allows for more affordable water tariffs. However, good management and monitoring capacity must be present to ensure that service quality, cost recovery and profitability are being achieved (Catley-Carlson, 2011; Krause, 2011). This, in turn, must be set in a strong regulatory framework with appropriate options for the partnership, in accordance with its socio-economic objectives and the interests of all citizens (UNESCO, 2005). In the peak of the concession years, most private sector actors were international water companies. The next section looks at the scope for using local companies and operators in the aforesaid hybrid PPPs.

13.5.2 Trend towards National Private Water Companies

In 2005, the five largest international companies held 75 percent of the share of water operators globally. In 2011, this had decreased to 35 percent (Mahe, 2011). Even though Veolia and Suez are still the largest water operators, there has been a noticeable rise in national private sector companies (Owen and McGranahan, 2006; Marin, 2009). By 2007, local private water operators served more than 67 million people, or more than 40 percent of the market (Marin, 2009).

National private water service companies could be defined as home-grown legal entities that have signed a formal contract with the relevant municipal or state authorities for the provision of water and related services. Since 2001, such private operators from developing countries have signed most of the new contracts. The *Pinsent Masons Water Yearbook 2011–2012* (Pinsent Masons, 2011b), shows that 112 local companies in 22 countries had signed formal PPP contracts. These national private water companies are generally larger than micro-enterprises that often supply water and sanitation services informally in many low-income countries, and smaller than international water companies (Owen and McGranahan, 2006).

Some developing countries have taken active measures to encourage the participation of home-grown local companies in the water sector. For example, in Uganda, local governments are empowered by the Local Governments Act (2000) to provide water services and receive funding from the center. In consultation with the Ministry of Water and Environment they appoint and manage private operators for urban piped water schemes through management and service contracts (Government of Uganda, 2008). Management contracts are often awarded to local Ugandan companies for operating water supplies in small towns. According to Owen and McGranahan (2006), local companies have also been awarded a significant share of contracts for water and wastewater services in Asia, and particularly in China and Malaysia.

It has been argued that local water companies may have an understanding of local culture that allows them to more easily establish a viable partnership with local authorities and better mitigate political and social risks (Owen and McGranahan, 2006; Marin, 2009). It may, therefore, not be surprising that organizations such as the United Nations Educational, Scientific and Cultural Organization (UNESCO) (2005) recommend that governments should select appropriate contractual arrangements that are compatible with their socio-economic constraints and objectives and give qualified local, national and regional enterprises the opportunity to compete for PPPs. Even the World Bank acknowledges that national companies may be more appropriate in reaching areas where the needs are considerable (Marin, 2009).

For many multinational water corporations, national private water companies, or national 'champions', are becoming a popular way of doing business. Through national business subsidiaries, they can 'go local' and hire local staff, which in itself can stimulate the local economy. As part of their corporate social responsibility, they also emphasize knowledge transfer and continuity, training people and making sure they have the capacity to manage themselves after the international partners have left (Mahe, 2011). There is thus an increasing appetite to set up concessions with national companies and local shareholders.

13.5.3 The Case of Informal Private Water Operators

Whether or not international or national companies run concession enterprises, they still have problems reaching low-income populations due to cost recovery concerns discussed above. Small private operators (carriers, standpipe managers, truckers, and so on) thus play an extremely important role in providing drinking water and complementing the services offered by concessionaire enterprises.

Private water suppliers have existed in most developing countries for a long time in the form of more or less informal water vendors selling to individual households, hospitals, public agencies, private businesses and so on. Studies of small-scale informal providers from Haiti, Mali and Burkina Faso indicate that these private operators from the informal sector play a significant role in serving the low-income communities. They generate a number of important jobs and fill the gap of public and formal private services (Collignon, 1999). Collignon (1999) found that informal private operators fill in public service gaps in the drinking water sector and earn between 21 and 84 percent of the added value of the entire value chain branch. However, currently, there is insufficient focus on the small-scale private providers by analysts and stakeholders in the water and sanitation sector, partly due to the non-registration or non-recognition of such service providers as recognized businesses. This has resulted in inadequate financing for small-scale private water operators as they are seen as credit risks by financial institutions (WaterAid Ghana, 2005).

However, there are some commendable steps towards recognizing the usefulness of these informal service providers. For instance, UNESCO (2005) recommends that governments should consider involving small-scale providers in reaching unserviced groups of households in both rural and urban areas. In Ghana, it has been proposed that this growing informal sector is directly addressed through a decentralized process of registering small-scale operators so that they carry out their activities within the legal framework (WaterAid Ghana, 2005). In Paraguay new contractual approaches have been developed to target *Aguateros* (mostly small-scale illegal water companies), which have developed piped water supply in peri-urban areas with no public funding. These *Aguateros* can now legally take part in public bidding processes and one benefit has been that *Aguateros*' performance can be tracked for transparent and accountable service (Phumpiu and Gustafsson, 2009).

This increasing emphasis on the informal sector filling the gap in water service provision may indicate a trend of broadening the meaning of 'private sector participation' to also encompass a vibrant informal sector. This in turn widens the range of actors that the state can engage with.

13.5.4 The Need for More Water User Involvement

When asked how much German development agency GTZ is involved in concessions and divestitures in the water sector, the answer was: 'We do not see that many concessions and divestitures. However, we do see development partnerships with the private sector through water users' (Thombansen, 2011). One of the greatest changes over the past 20 years since privatization and the Washington Consensus is that water providers are no longer the only private sector actors that engage strategically in water. Over the past 10–15 years there has been an increasing role for the water users.

Water users are companies that require water for their operations, manufacturing and sales. International examples include Walmart, Pepsi, Coca-Cola, SAB Miller and Levis as well as companies selling water-use products such as Unilever. Two decades ago, there was barely any focus on water users. Now they are active in several strategic alliances or partnerships with donors and development aid agencies such as GTZ, USAID, DFID and UNICEF. Examples include the Safe Drinking Water Alliance with Proctor & Gamble, CARE and Johns Hopkins to increase point-of-use water treatment; the International H_2O Collaboration with Rotary International for water and sanitation projects in the Dominican Republic, Ghana and the Philippines; and the Water and Development Alliance with Coca-Cola that address a broad range of water challenges in 22 countries (USAID, 2009).

What has changed so that water users become strategic partners in the quest for providing safe water and sanitation? One issue is a greater realization of the uncertainty and risks associated with water, particularly in relation to climate change impacts and growing water demands due to growing economies and populations. Many countries are facing growing water scarcity and significant problems of water pollution and degraded quality of water, hence water is becoming an increasingly strategic resource necessary to secure and protect long-term business operations. Corporate social responsibility and public image have become normative standards that water users are more and more paying attention to. A factory cannot use local water and dump wastewater in surrounding environments and negatively affect local populations while maintaining that they are sustainable and responsible. There has, therefore, been a greater focus on addressing water scarcity, water pollution, infrastructure and water supply, and sanitation issues in the countries they are active in (Catley-Carlson, 2011; Farrington, 2011). Attempts include reducing water used in the production chain and treating wastewater on-site.

There has also increasingly been an understanding that water users cannot tackle water risks alone. Hence, they are trying to cooperate with public sector, NGOs and intergovernmental organizations to address these areas of shared interest. SAB Miller, for instance, is cooperating with its brewery subsidiary in Tanzania to address the issue of groundwater and reduce water leakages through training with the public sector. The World

Bank lent infrastructure funds for sewerage, while SAB Miller contributed to the costs of the feasibility study and design. In cooperation with local authorities and utilities, SAB Miller is helping to produce bankable projects.

The involvement of water users in addressing water issues points towards the realization of shared risks and the need for filling in gaps. They are also supporting reforms where small and medium-sized enterprises (SMEs) enter the market and take over services that the public sector does not or cannot supply; that is, supporting the use of water providers from the informal sector. Water users have been promoting the intermediate step of pro-poor water kiosks operated by SMEs and private operators. Arguably, private sector participation (PSP) with water users can also be seen as a form of PPP, albeit with a different dynamic. Regardless, this may be seen as evidence of pragmatism where water users are trying to address gaps and testing new strategies for how they can sustain their business operations through reducing shared water risks.

Companies are crucial in producing technologies that can be used to make drinking water safe in households, without the need for advanced water treatment plants. There are international companies that support local companies to manufacture liquid chloride and filter products for safe disinfection (Harvey, 2011). Safe hygiene practices are equally important to stop the spread of diseases and increased mortality rates, and the Global Public–Private Partnership for Hand Washing with Soap focuses on institution-building and advocacy to increase the practice of hand washing with soap across the globe. Key private sector actors include Unilever and Proctor & Gamble, which see it in their interest to expand their market as well as to promote their corporate social responsibility (CSR) image. UNICEF has found the campaign very useful as it can tap into private sector skills in marketing to facilitate behavioral change and increase the adoption of safe hand washing (Harvey, 2011).

Thus, it is evident that there is increased engagement of water users to address water resources risks and water services shortcomings. Many such initiatives have had positive development impacts but they have also received criticism. Firstly, some argue that the private sector is using international development and bilateral and multilateral organizations for 'green-washing', or to only expand markets. However, many multilaterals and companies are building in safeguards to avoid this type of criticism. For example, in the UNICEF case, companies cannot promote their own brands in UNICEF campaigns on the ground. For many companies their engagement is guided by their CSR policies as well as motivated by long-term business plans. In fact, it can be argued that private sector CSR engagement is even more effective if it is made part and parcel of long- and short-term business operations and planning. Secondly, some have raised concerns that such partnerships illustrate the increasing 'corporatization' of development, where both bilateral and multilateral donors are now seeking to attract funding from private sector actors as their own financing is running dry (Doane, 2011).

The private sector at present contributes in the range of US$3.5 billion in ODA. Companies are currently investing more than US$150 billion directly in emerging markets and contributing another US$3 billion in charity and in-kind contributions. At this level, multinational companies would rank as an Organisation for Economic Co-operation and Development (OECD) significant overseas development provider (Business Civic Leadership Center and Corporate Citizenship, 2009). Hence, exploring private sector funding is an increasingly attractive funding opportunity for many bilateral and multilateral organizations as well as various types of NGOs. Lastly, although greater private sector involvement is a pragmatic approach that attempts to involve all actors involved with water, there are concerns of an increased view of poor groups of people as 'consumers' and 'entrepreneurs'. This in turn can shift the responsibility of 'development' onto only the poor people themselves without taking into account structural problems of systematic social inequalities and injustices. In many ways it may also discourage the large task of promoting and enabling development by the local authorities and government.

13.6 IMPORTANCE OF BILATERAL AND MULTILATERAL AGENCIES IN THE CHANGING PERCEPTIONS AND ROLES OF THE PRIVATE SECTOR

Since the early 1990s, much has happened within bilateral and multilateral systems and how they work with the private sector. According to the International Finance Corporation (IFC) (2008), between the early 1990s and 2007 there has been tenfold growth – from around US$4 billion to US$40 billion per year – in multilateral finance to the private sector.[2]

Almost all UN agencies currently have designated Private Sector Focal Points, which work with the private sector in a variety of ways, from fundraising to strategic partnerships. There is also a system-wide UN Private Sector Focal Point Network that provides frameworks for developing operational guidelines for working with the private sector, and making the case for strategic private sector engagement in the work of the UN. United Nations entities – the Global Compact and the United Nations Office for

Partnerships – have been created solely to engage with the private sector. These entities are responsible for developing new partnerships between the UN and the private sector under the framework of reaching the MDGs. Many different UN agencies also have their own programs on PPPs as a way of pursuing MDG achievement and poverty reduction. For example, the United Nation Development Programme (UNDP) has worked with the global facility Public–Private Partnership for the Urban Environment (PPPUE) as a mechanism for developing countries to obtain support in their efforts to define, promote and implement public–private partnerships to reduce poverty by increasing the access of the urban poor to basic services.

Some of these multilateral initiatives can be seen as alternative governance models of PPPs in the form of large concessions. For example, UNDP-PPPUE considers communities to have a direct role to play in PPP arrangements as beneficiaries, expressing the price the community would pay for an acceptable level of service, and an indirect role to play in shaping policy for the urban environment. In a small PPP, the community may even be able to take the role of the private partner.[3] Hence, it is important to point out that the PPP agenda is broad and that many multilaterals also have considered it important to counterbalance some of the more large-scale and top–down PPPs with bottom–up PPPs in collaboration with communities.

Moreover, there have been considerable bilateral and multilateral investments in capacity development and technical support to developing countries to promote PPPs in water as well as other areas such as energy. For example, the Public–Private Infrastructure Advisory Facility (PPIAF) of the World Bank partners with its donors, regional development banks, and other institutions to deliver technical assistance to developing country governments. The PPIAF is a multi-donor technical assistance facility, financed by 17 multilateral and bilateral donors. Many international development agencies, such as Deutsche Gesellschaft für Internationale Zusammenarbeit (GIZ), USAID, DFID, and many more, all have some kind of private sector engagement policy and/or office.

Increased private sector engagement, as well as closer relations between the private sector and most bilateral and multilateral organizations, has not come without criticism. A donor perspective to enable and promote private sector involvement can reduce policy space for developing countries to pursue specific national development goals. For example, the indicators of the IFC Doing Business Report have been criticised as they favor labor policies that make it easier to hire and fire the workforce, which can conflict with social issues such as job security. Moreover, indicators of lowering taxes to attract investments could be in conflict with government aims at increasing public income for expanding public services through increased tax revenues (Davies, 2011). This is particularly relevant for expanding water services to poor people and where massive public investments are required. Bilateral and multilateral donor conditionality for tax-cutting policies can thus be counterproductive for assuring universal access to water and sanitation.

13.7 CONCLUSION: PPP TRENDS AND THE NEED FOR BROADENING THE WATER GOVERNANCE AGENDA

13.7.1 Trends in PPPs

PPPs have not met the expectations of reaching poor people and significantly increasing investment in water and sanitation services, as expected by many bilateral and multilateral agencies. As a result there is a shift from promoting PPPs in the form of long-term and top–down concessions towards a more pragmatic view of PPPs emerging as a part of the Post-Washington Consensus. In recent years the number of PPPs has decreased worldwide, with the overall investment in PPPs significantly declining. Moreover, bilateral and multilateral organizations are now looking at alternatives to the traditional PPPs. These include concessions and management contracts for private sector participation in development projects with shorter time horizons and reduced risks. However, private sector competition, knowledge and entrepreneurship are still perceived as vital for development, and governments should provide both enabling environments as well as effective regulatory structures for their effective engagement.

In our study of development for this chapter we have identified three discernible trends:

1. A trend towards management and service contracts. The private sector will continue to play an important role in parts of the water services value chain, but decreasingly so in managing the whole services value chain. There is now increasing reference to shorter-term service and management contracts (2–5 years) between governments and the national private sector. Some examples of this can be seen in Uganda and Kenya, where the national small-scale private sector engages in management and service contracts with local governments in smaller towns. These can be straightforward performance contracts of private sector maintenance and operations, but where the investment risk is carried by the public sector.

2. An increasing role for the national local private sector as well as for informal water vendors in many developing countries. In many new PPPs private sector partners are made up of national corporations that in some cases are subsidiaries to multinational corporations. The informal private sector remains critical to cater for the poor. Although there are emerging examples on how the public sector can engage constructively with the informal private sector, this remains highly problematic due to high prices, unregulated water quality, dynamics of slum politics and local politicians having financial stakes in private informal water companies.
3. Water is an increasingly strategic future resource for private sector water users. Some discernible trends are that private sector water users increasingly perceive water as a strategic resource that is critical to manage, both as part of their CSR policies and, perhaps more importantly, as part of their long-term business plans. Will water be available for future productions and how can it be secured? As a result, many multinationals are currently developing corporate water policies and long-term water risk assessments and analyses. In addition to the private sector, multinational corporations in particular have become a significant donor and support water development programs through grants to multilateral organizations and NGOs.

13.7.2 Future Prospects: Need for Broadening the Water Governance Agenda to Include Stronger Consumer and Citizen Perspectives

Past experience has witnessed that the 'developmental state' was not able to deliver basic water services to people. Similarly the trend of privatization, such as through PPP concessions, has hitherto largely turned out to be a rather marginal phenomenon in relation to reaching the poor and increasing water investments. For the private sector it has turned out to be a very risky and non-viable business proposition to target poor populations as well as most low-income countries. The major water governance transition in many countries has not been from public to private operation of water utilities but rather devolution of public water service responsibilities from central to local levels (Baietti et al., 2006). As a consequence, other types of governance architectures have not received the same amount of attention, such as how to make public utilities more efficient; how to ensure greater consumer participation and involvement; or how to make devolution of water services work (Tropp, 2007).

Our analysis of trends in PPPs reveals that water services provisioning has strongly focused on the role of government as enabler and regulator, and the private sector as the service provider. However, little attention has been paid to the roles and interests of water services consumers and citizens. A recent study concludes that consumer pressure on water utilities and the development of redress and complaints mechanisms can have several positive effects on service improvements (Baietti et al., 2006). Moreover, multilateral organizations, such as UNDP, UN-HABITAT and the Water and Sanitation Programme of the World Bank (WSP), along with many NGOs, are increasingly working with consumer and citizen perspectives and rights-based approaches. Despite the debate on participation in water services, many consumers, the poorer in particular, lack the voice and mechanisms for redress when water services are not working. There are currently a growing number of examples, such as water watch groups in Kenya and Zambia, where the water regulator works with consumer interests to maintain checks and balances on private and public water service providers. The role of consumers has so far been an untapped potential and should be more systematically explored as a way of improving water services. The United Nations (2010) has formally acknowledged water and sanitation as a human right. To what extent the right to water and sanitation agenda will impact on a stronger consumer and citizen focus, and whether it will make interfaces between consumers, government and the private sector more mainstream in national water development, remains to be seen.

It is critical that bilateral and multilateral organizations avoid another development panacea in promoting 'one-size fits all' solutions and instead open up for multiple water governance architectures that systematically also strengthen consumer and citizen participation and interest, particularly for poor people, in relation to national water regulators as well as to private and public service providers. The governance roads that bilateral and multilateral organizations choose matter, and will guide future international and national development agendas and investments.

NOTES

1. According to the Regulatory Body, the Jakarta water tariff is one of the highest in Southeast Asia outside Singapore and Hong Kong. Since 2004 an Automatic Tariff Adjustment has been applied every six months. Based on this mechanism, the private operator has the right to propose a tariff increase once every six months to the Regulatory Body for submission and approval by the Governor. During the water service undertaking by the private partner there have been nine tariff hikes and the last was in December 2008 (Kruha, 2010).
2. There is, however, little in-depth or comparative analysis of the evolving relationship between overseas development assistance (ODA) and the private sector. An analysis of ten international development agencies, focusing on their relationship with multinational corporations, shows that agencies are at various stages of developing their

initiatives and that financial and other commitments are difficult to quantify due to lack of, or differences in, reporting (Business Civic Leadership Center and Corporate Citizenship, 2009).

3. See PPPUE Toolkit for developing pro-poor public–private partnerships, available at http://www.margraf-publishers.com/UNDP/PPPUE/.

REFERENCES

Amjad, U. (2003), 'Re-conceptualizing public–private partnerships in the water sector as a coevolving system of innovation', paper presented at the Complexity, Ethics and Creativity Conference, London School of Economics, 17–18 September, available at http://www.psych.lse.ac.uk/complexity/Conference/Amjad.pdf (accessed 10 October 2011).

Annan, K. (1998), 'The United Nations and the private sector: a new era', *Natural Resources Forum*, **22**(2), 73–4.

Baietti, A., W. Kingdom and M. Van Ginneken (2006), 'Characteristics of well-performing public water utilities, water supply and sanitation', Working Notes: No. 9, Washington, DC: World Bank.

Bishop, M. and M. Green (1995), 'Privatization and recession – the miracle tested', CRI Discussion Paper 10, London: CIPFA.

Business Civic Leadership Center (BCLC) and Corporate Citizenship (2009), 'Partnering for global development: the evolving links between business and international development agencies'.

Checchi, D., M. Florio and J. Carrera (2009), 'Privatization discontent and utility reform in Latin America', *Journal of Development Studies*, **45**(3), 333–50.

Collignon, B. (1999), 'The potential and the limits of private water providers: independent sellers in Francophone Africa', UNDP–World Bank, Water and Sanitation Program.

Davies, P. (2011), 'The role of the private sector in the context of aid effectiveness, consultative findings document', Final Report, 2 February, report commissioned by the OECD Development Assistance Committee.

Economic Commission for the Latin America and the Caribbean of the UN (ECLAC) (1995), 'Policies to improve linkages with the global economy', Santiago: United Nations.

Esguerra, J. (2003), 'New rules, new roles: does PSP benefit the poor? The corporate muddle of Manila's water concessions', WaterAid and Tearfund.

Faulkner, J. (1997), 'Engaging the private sector through public–private partnerships', in *Bridges to Sustainability*, Yale Bulletin Series, No. 101, New Haven, CT: Yale University Press.

Government of Uganda (2008), 'Water and sanitation sector performance report', Kampala: Ministry of Water and Environment.

Guislain, P. (1997), *The Privatization Challenge: A Strategic, Legal and Institutional Analysis of International Experience*, World Bank Regional and Sectoral Studies, Washington, DC: World Bank.

Gwynne, R. and C. Kay (2000), 'Views from the periphery: futures of neoliberalism in Latin America', *Third World Quarterly*, **21**, 141–56.

Hall, D., E. Lobina and R. de la Motte (2005), 'Public resistance to privatization in water and energy', *Development in Practice*, **15**(3–4), 286–301.

Harvey, D. (2011), *A Brief History of Neoliberalism*, New York: Oxford University Press.
International Finance Corporation (IFC) (2008), *2008 Annual Report – Creating Opportunity*.
Johnson, C. (1982), *MITI and the Japanese Miracle: the Growth of Industrial Policy 1925–1975*, Stanford, CA: Stanford University Press.
Jolly, R., L. Emmerij, D. Ghai, and F. Lapeyre (2004), *UN Contributions to Development Thinking and Practice*, Bloomington, IN: Indiana University Press.
Jolly, R., L. Emmerij and T.G. Weiss (2009), *UN Ideas that Changed the World*, Bloomington, IN: Indiana University Press.
Khan, M. (2004), 'State failure in developing countries and institutional reform strategies', in B. Tungodden, N. Stern and I. Kolstad (eds), *Toward Pro-Poor Policies: Aid, Institutions and Globalization*, New York and Washington, DC: OUP–World Bank, pp. 165–95.
Kirkpatrick, C., D. Parker and Y.-F. Zhang (2004), 'State versus private sector provision of water services in Africa: an empirical analysis', paper presented at 3rd International Conference, Pro-Poor Regulation and Competition: Issues, Policies and Practices, Cape Town, South Africa, 7–9 September.
Komives, K. (1999), 'Pro-poor water and sewer concessions: early lessons from Bolivia', World Bank Policy Research Working Pape No. 2243, Washington, DC: World Bank.
Kruha (2010), 'Contribution to the public consultation on the issue of human rights obligations in the context of private sector participation in the provision of water and sanitation services'.
Labonne, B. (1998), 'Public–private partnerships in natural resources management', *Natural Resources Forum*, **22**(2), 75–6.
Lobina, E. and D. Hall (2003), 'Problems with private water concessions: a review of experience', Public Services International Research Unit (PSIRU), University of Greenwich, London.
Marin, P. (2009), *Public–Private Partnerships for Urban Water Utilities: A Review of Experiences in Developing Countries*, Washington, DC: World Bank.
Martin, S. and Parker, D. (1997), *The Impact of Privatization: Ownership and Corporate Performance in the UK*, London: Routledge.
Mkandawire, T. (2004), 'Good governance: the itinerary of an idea', *Development and Cooperation*, **31**(October), 10.
Moreau-Le Golvan, Y. and P. Bréant (2007), 'Organisation and financing models of the drinking water sector: review of available information on trends and changes', Report Techneau 07, April, TECHNEAU, Integrated Project by the European Commission under the Sixth Framework Programme.
Neto, F. and H. Tropp (2000), 'Water supply and sanitation services for all: global progress during the 1990s', *Natural Resources Forum*, **24**(3), 225–35.
Owen, D.L. and G. McGranahan (2006), 'Local water companies and the urban poor', Human Settlements Discussion Paper Series, International Institute for Environment and Development, London, UK.
Phumpiu, P. and J.E. Gustafsson (2009), 'When are partnerships a viable tool for development? Institutions and partnerships for water and sanitation service in Latin America', *Water Resources Management*, **23**(1), 19–38.
Pinsent Masons (2011a), 'Under the hood: what do the Vietnamese PPP regulations say?' PPPs in Vietnam: Series 2, August.

Pinsent Masons (2011b), *Pinsent Masons Water Yearbook 2011–2012*, London: Pinsent Masons.
Polanyi, K. (1944), *Great Transformation*, Boston, MA: Beacon Press.
Prasad, N. (2006), 'Privatization results: private sector participation in water services after 15 years', *Development Policy Review*, **24**(6), 669–92.
Rees, J.A. (1998), 'Regulation and private participation in the water and sanitation sector', *Natural Resources Forum*, **22**(2), 73–4.
Rodrik, D. (2006), 'Goodbye Washington Consensus, hello Washington Confusion? A review of the World Bank's economic growth in the 1990s: learning from a decade of reform', *Journal of Economic Literature*, **44**, 969–83.
Saad-Filbo, A. (2010), 'Growth, poverty and equality: from Washington Consensus to inclusive growth', UNDESA Working Paper, No. 100, New York: United Nations Department for Economic and Social Affairs.
Sida (2004), 'Policy guidelines for Sida's support to private sector development', Stockholm.
Tropp, H. (2007), 'Water governance: trends and needs for new capacity development', *Water Policy*, **9**(S2), 19–30.
United Nations (2002), 'World Summit on Sustainable Development, Johannesburg Plan of Implementation (JPoI)', A/CONF.199/20, United Nations, New York.
United Nations (2010), 'The human right to water and sanitation', resolution adopted by the General Assembly, A/RES/64/292, 3 August, United Nations, New York.
United Nations Economic and Social Council (UNESCO), Economic Commission for Africa (2005), 'Public private partnerships for service delivery: water supply and sanitation', 3rd meeting of the Committee on Human Development and Civil Society, Addis Ababa, Ethiopia.
USAID (2009), 'Public private partnerships: doing business through partnership', available at http://www.sida.se/Global/Aboutpercent20Sida/Konferenser/Greenpercent20Solutions/1.percent20USAIDpercent20Greenpercent20Weekpercent20Financingpercent20Panelpercent20-percent20Final.pdf (accessed 25 September 2011).
WaterAid (2011), 'Off-track, off target: why investment in water, sanitation and hygiene is not reaching those who need it most', WaterAid Policy Report.
WaterAid Ghana (2005), 'Small scale private sector participation in the rural water supply R8335, Phase 3 analysis and dissemination', final country report for Ghana, available at http://practicalaction.org/docs/consulting/FinalCountry_Report_Ghana.pdf (accessed 25 September 2011).
World Bank (2005), *Economic Growth in the 1990s: Learning from a Decade of Reform*, Washington, DC: World Bank.
World Bank (2007), 'Emerging public–private partnerships in irrigation development and management', Water Sector Board Discussion Paper Series, Paper No 10, May.
Yarrow, G. (1989), 'Privatization and economic performance in Britain', *Carnegie-Rochester Conference Series on Public Policy*, **31**, 303–44.
Zhong, L., A.P.J. Mol and T. Fu (2008), 'Public–private partnerships in China's urban water sector', *Environmental Management*, **41**(6), 863–77.

Informants Interviewed

Nine informants in their individual capacities and not as representatives of their organizations were interviewed for the purpose of developing this chapter. Any views expressed by them do not necessarily reflect their organizations. The informants were used to get a deeper understanding of the topic and as a way to get some expert reflections on PPPs and current trends. The authors are grateful for the time they took to answer questions. The list of the informants is provided below.

Catley-Carlson, M. (2011), Interview with Margaret Catley-Carlson, 25 August 2011, World Water Week 2011, Stockholm, Sweden. She has held many international board positions at global policy, science and international finance organizations, such as Board Member of UNSGAB on water and sanitation and Chair of Global Water Partnership and Suez Environment Advisory Board. She has also held several other positions, such as Deputy Director of UNICEF and President of the Canadian International Development Agency, CIDA.

Doane, D. (2011), Presentation on the Crisis in International Development, International Development Conference, 8 September 2011, London School of Economics, UK.

Farrington, R. (2011), Interview with Robin Farrington, Africa team for Cooperation with the Private sector GIZ; 24 August 2011, World Water Week 2011, Stockholm, Sweden.

Harvey, P. (2011), Interview with Peter Harvey, Senior Adviser – Water Supply and Quality, United Nations Children's Fund (UNICEF), 24 August 2011, World Water Week 2011, Stockholm, Sweden.

Krause, M. (2011), Interview with Matthias Krause, Senior Water and Sanitation Specialist, Inter American Development Bank (IADB), 25 August 2011, World Water Week 2011, Stockholm, Sweden.

Mahe, P.A. (2011), Interview with Pierre Allain Mahe, Project Manager at the Sustainable Development Department, Veolia Environnement, 24 August 2011, World Water Week 2011, Stockholm, Sweden.

Meagher, K. (2010), 'Good governance and the Post-Washington Consensus', Lecture at the London School of Economics, 16 January 2010, London.

Palanisami, K. (2011), Interview with K. Palanisami, Director of IWMI-TATA Policy Program, 22 August 2011, World Water Week 2011, Stockholm, Sweden.

Thombansen, C. (2011), Interview with Conrad Thombansen, Senior Adviser for Water sector at GIZ, 24 August 2011, World Water Week 2011, Stockholm, Sweden.

14. Re-municipalization of water utilities: back to the public fold

Asanga Gunawansa and Vanessa Garcia

14.1 INTRODUCTION

On 31 October 2011, the seven-billionth human being was born on this planet (UNFPA, 2011). More people live on earth today than have ever lived before since the beginning of time. Since the end of World War II, the world's population has almost quadrupled in size, sharing the same resources that we have had for centuries. This squeeze on our resources is being felt in every sector, including the water sector.

Water is an essential resource for life on earth and it is generally recognized that people have a fundamental right of access to clean and consumable water. However, water as a natural resource is not fairly distributed across the globe. Furthermore, only a part of natural water resources in the world can be contained and utilized. Whilst some countries and cities might be fortunate to have water in abundance, in other places water security is a key impediment to sustainable development.

According to the UN (2010), water use increased sixfold during the twentieth century, more than twice the rate of population growth. In some developing countries, including China and India, and also developed countries, including some parts of the United States, groundwater is being consumed faster than it is being replenished, and as a result groundwater tables are said to be falling steadily. Thus, managing water resources and developing facilities for sustainable management of water to cater to the needs of the growing population has become a fundamental challenge for all countries. Ensuring that water is available for household use, industrial use and agricultural use is a challenge faced by most countries in the world, especially developing-country governments with financial, technological and managerial constraints to providing efficient and effective water services to the people.

For centuries, water infrastructure was traditionally developed and governed by public sector entities. Traditionally, monarchs, rich landowners and various forms of collective governance typically fronted the high costs of developing water infrastructure within their domiciles. However, as governance mechanisms changed over time, and monarchies came to be replaced by public entities such as municipalities, they took over this role. Post-industrialization, as populations grew and urbanization occurred at a rapid pace, public entities were simply unable to keep up with the demand for quality water and the cost of infrastructure development and maintenance. Over time, they turned to the private sector to attract much-needed finance, technological and managerial expertise.

Starting in Western Europe as early as the late 1800s and increasingly in the late 1900s, private sector engagement in developing, managing and operating utilities slowly started to become the norm. The trend peaked in the 1980s and 1990s, particularly as developing countries eagerly grasped their new-found independence and moved into an era of tremendous growth and development. The history of private sector participation in the water sector has been outlined in brief in Chapter 2 in this volume and discussed in greater detail in Chapter 15. Although private sector participation in utilities peaked in the 1980s and 1990s, some governments soon learned that privatization was a mixed blessing, as a result forcing the public sector to resume responsibility of the utilities in a trend that is now referred to as 're-municipalization'.

Re-municipalization is sometimes defined as the transfer of water services from private companies to municipal authorities (Pigeon et al., 2012). However, this definition in our view is too narrow, as taking water governance back into the public fold can occur at any level of governance, including local, municipal, regional and national. In the circumstances, for the purpose of this chapter, we use the term 're-municipalization of water' to refer to the transfer of water utilities and related assets, and their control and management, from private entities back to public sector entities.

There are several reasons for re-municipalization of water supply, which was hitherto in private hands. For example, in Potsdam (Germany), the change was the result of concerns relating to Eurowasser's projected future price increases. In Grenoble (France), citizens' movements against overpricing of water played an important role (Hachfeld et al., 2009). In some countries, legal developments have excluded the private sector from the provision of water supply in urban areas. For instance, the Uruguayan Constitution prohibits any form of private sector participation in the water sector. The Netherlands also enacted a new water law in 2005 that allows only publicly owned companies to provide drinking water services (Hall et al., 2004). In 2007, a new water law was proposed in Italy that supports the establishment of progressive public models of water management (TNI, 2009).

This chapter examines the growing trend of re-municipalization with particular focus on its history, its enabling factors and its impact. An attempt has also been made to deduce a set of best practices based on an analysis of case studies from three uniquely different countries: France, Malaysia, and Trinidad and Tobago.

14.2 LIMITED SUCCESS WITH PRIVATIZATION

As discussed in the first two chapters in this volume, the term 'privatization' refers to the shift in control from the public to the private sector, through a transfer of ownership or management responsibility for water supply infrastructure. The 2002 'New Economy of Water' report (Gleick et al., 2002) stated that water can be privatized in many ways, for example, through 'the transfer of the responsibility to operate a water delivery or treatment system, a more complete transfer of system ownership and operation responsibilities, or even the sale of publicly owned water rights to private companies'. The report also highlighted that combinations of the aforementioned also exist.

Privatization of water supply systems has existed for many years. Widespread privatization efforts grew in the late twentieth century, especially when the public sector found it unviable to utilize precarious public finance and limited technology and expertise for managing a scarce resource to cater to the demands of a growing population, and when international financial institutions such as the World Bank and the International Monetary Fund required developing countries seeking assistance to deregulate, abolish subsidies, and even sell much of their water systems and infrastructure to private investors.

The management, operation and in some cases the ownership of public water systems were transferred to private companies in an attempt to increase efficiency of service, lower prices and reduce corruption (World Savvy Monitor, 2009). The impetus by the international lending agencies has resulted in critics such as Michael Goldman arguing, 'these days, an indebted country cannot borrow capital from the World Bank or IMF without a domestic water privatization policy as a precondition', and that this model of development was 'forcing many countries to commodify their water resources, rather than using their expertise in the public sector to acknowledge water as a universal human right and an essential public service' (Goldman, 2005). Despite this impetus, however, the attempt at privatization essentially failed to live up to expectations and by 2006 many countries renegotiated, re-municipalized or terminated private water contracts.

14.3 KEY REASONS FOR RE-MUNICIPALIZATION

As stated earlier, privatization was often introduced as a means of ensuring equitable provision of water and adequate investment in water utilities. However, in many instances across the world, these attempts and measures at exposing public water systems to market forces never paid off. Desperate populations and/or their governments often returned to municipalization to solve the problem. Further, despite the hype about privatization of water, it is interesting to note that the real percentage of the world's population serviced by private sector water operators is very low. For example, according to the World Bank's Public–Private Infrastructure Advisory Facility (PPIAF), although the population served by private operators has grown from 6 million in 1991 to over 164 million in the year 2000, as of 2009, only about 7 percent of the urban population in the developing countries is served by private water operators (Marin, 2009).

In multiple cases privatization resulted in little or no improvement in service, exorbitant increases in water tariffs, labour cuts, lack of or reduction in investments by private companies, lack of transparency and accountability, and allegations of corruption. A 2003 assessment of the distributional impact of privatization revealed that the popular perception was that privatization was widely and increasingly unpopular, largely because it appeared to be 'fundamentally unfair, both in conception and execution' (Birdsall and Nellis, 2002). The numerous issues and problems were, therefore, detrimental to the sustainability of many privatized water systems and disgruntled actors in societies across the world have attempted to halt or been successful in halting the privatization process or taking utilities back into the public fold.

These reasons aside, however, it is important to consider that the peculiar characteristics of water almost mandate a high degree of natural monopoly, including the high capital and sunk costs involved in infrastructure development, the natural availability of water resources, its absolute necessity in every aspect of life on earth, as well as the perception that public provision is the best way to guarantee universal access (Mehta, 2003). This makes the management of water infrastructure without profit-driven motives very challenging, often making it impossible for private entities to manage these projects.

The following paragraphs elaborate specific reasons why water utilities were wrested from under the control of the private sector and moved towards re-municipalization.

14.3.1 Public Protests

One of the primary reasons for a return of water to public control was concerted public protests against private sector involvement in water governance. Rapidly growing opposition to privatization that involved multiple powerful societal actors such as trade unions, consumers, water professionals, environmentalists, political groupings, human rights and community organizations caused governments to bend to their will. For example, in Brazil, in the mid-1990s, there was a multi-actor long-term campaign against private sector engagement and commercialization of water when Frente Nacional pelo Saneamento Ambiental (National Front for Environmental Sanitation) brought together 17 civil society organizations, including unions, managers, professional associations, non-governmental organizations (NGOs) involved in urban reform and consumer groups, in defense of public management for sanitation and treating water as a public good and basic human right (Hall et al., 2005; Filho, 2002).

Opposition to privatization was mainly founded on the 'economic implications of privatizing water resources, the risks to ecosystems, the power of private operators, foreign control over a fundamental natural resource, inequities of access to water, the perceived loss of sovereignty and the exclusion of communities from decisions about their own resources' (Gleick et al., 2002). Protests emerged in several countries such as Bolivia, Italy and the Philippines, and in numerous conferences with the aim of ensuring that water returned to public hands, and that the poorest of the poor had access to safe, affordable water (Corral, 2007). Protests varied in methodology: in Grenoble (France), Rio de Janiero (Brazil) and Nkonkobe (South Africa), resistance against privatization was showcased through legal actions against the private holding companies; in Cochabamba (Bolivia), massive public protests in the streets famously brought an end to privatization mechanisms; and in Mauritius, Poland and Panama, the public voted to end privatization of their water systems at the ballot box during local or national elections. Table 14.1 outlines some of the cases in which public protests led to rejection of private sector engagement in the water sector during the period 1994–2004.

Of this list of protests and others like them, one of the most significant was the protest in Cochabamba (Bolivia), the effects of which were felt across the world. This is because following the Cochabamba incident, civil society and public and private sectors began to seriously evaluate the nature of their involvement within the water sector. Even multilateral institutions such as the World Bank and the Asian Development Bank that had hitherto strongly supported the move towards privatization began to

Table 14.1 Opposition to and rejection of water privatization, 1994–2004

Country	Province/City	Year	Decision mechanism	Active groups
Argentina	Buenos Aires	2002	Government decision	Trade unions, consumers and citizen groups
Argentina	Tucuman	1998	Local government election	Consumers and citizen groups, opposition political parties
Bolivia	Cochabamba	2000	Government decision	Trade unions, consumers or citizen groups, local business entities and NGOs
Brazil	All	2002	National election	Trade unions, consumers and citizen groups, local business entities, environmentalists, NGOs and opposition political parties
Brazil	Rio de Janeiro	1999	Court ruling	Trade unions, consumers or citizen groups, (local) business, environmentalists, NGOs and opposition political parties
Canada	Montreal	1999	Municipal government decision	Environmentalists, trade unions and opposition political parties
France	Grenoble	2001	Municipal government decision following election	Environmentalists, consumers and citizen groups, opposition political parties
Germany	Potsdam	2000	Municipal government decision	Consumers and citizen groups, opposition political parties

Germany	Munich	1998	Municipal government decision	Trade unions and citizen groups
Honduras	Honduras	1995	Government decision	Trade unions and NGOs
Hungary	Debrecen	1995	Municipal government decision	Trade unions and NGOs
Mauritius	All	2000	National election	Consumers and citizen groups, environmentalists, trade unions
Panama	All	1999	National election	Unions, consumers or citizen groups
Paraguay	All	2002	Government decision	Trade unions
Poland	Poznan	2002	Municipal government decision	Opposition political parties
Poland	Łódź	1994	Local government election	Trade unions
Sri Lanka	All	Ongoing	–	Trade unions, consumer and citizen groups
South Africa	Nkonkobe	2002	Court ruling	Trade unions
Sweden	Malmo	1995	Municipal government decision	Trade unions, consumer and citizen groups
Thailand	All	2002	Government decision	Trade unions
Trinidad	All	1999	Government decision	Trade unions
USA	Atlanta	2003	Municipal government decision	Trade unions, consumer and citizen groups
USA	Birmingham	2000	Municipal government decision	Trade unions, consumer and citizen groups
USA	Washington DC	1996	Municipal government decision	NGOs

Sources: PSIRU Database on Privatization and Restructuring of Public Services Worldwide (2011).

change their tune. A number of factors made the case of Cochabamba so powerful, including the length of the protest period, the broad range of civil society represented amongst the protestors, and the vehement and unrelenting nature of the protests that ultimately garnered significant media attention and brought the government to its knees.

The protests spread out over several months in 2000, gaining fury and increasing media attention over time. Protestors ranged across a broad range of society including farmers, unions, consumer groups, NGOs and local businesses, making it very inclusive and representative of the majority of the population (Nickson and Vargas, 2002). The government was so overwhelmed that it finally revoked the concession contract with Aguas del Tunari (owned by the United States' transnational company Bechtel along with Italy's Edison and Spain's Abengoa corporations).

The Cochabamba water riots also laid the foundation for an important international resolution. On July 28, 2010, Bolivia's representative presented a resolution to the 64th UN General Assembly entitled 'The human right to water and sanitation'. He claimed that the human right to water 'had not been fully recognized, despite references to it in various instruments, including the Convention on the Elimination of All Forms of Racial Discrimination, the Convention on the Elimination of All Forms of Discrimination Against Women and others' (United Nations, 2010). The resolution was approved by a vote of 122 in favor to none against, with 41 abstentions, and it is seen as a significant victory for activists in the global anti-water privatization movement who have spent the last decade organizing and lobbying.

The aforesaid resolution has now confirmed the value system and ethics established under the public trust doctrine, a principle proposed in AD 528 when Justinian added to Roman law the words: 'By the law of nature these things are common to all mankind, the air, running water, the sea and consequently the shores of the sea' (Dowie, 2005). This doctrine was reiterated in English law and passed into the common law of the United States (US). It has been applied in the seminal US case *Illinois Central Railroad v. Illinois* (1892), and more recently in the Californian case of *National Audubon Society v. Superior Court* (1983), where the Supreme Court held that the saline Mono Lake was a public trust resource of the state of California.

Further, the 2010 United Nations General Assembly Declaration has also confirmed that the human right to water is also found in several other international conventions such as the Universal Declaration of Human Rights (1948), which implicitly recognized a right to water as part of a right to an adequate standard of living, and the International Covenant on Civil and Political Rights (1966), which recognized water as a key resource of subsistence of all living forms. The United Nations Committee on Economic, Cultural and Social Rights (2002) also recognized that countries are obligated to progressively ensure that economic, social and cultural rights of individuals, inclusive of labor rights, the right to health, the right to education and the right to an adequate standard of living, are realized (United Nations, 2011).

14.3.2 Diverging Private Sector and Public Sector Goals

Re-municipalization has often become the default *modus operandi* when privatization attempts have failed. Unfortunately, disagreements between the private operators and public sector managers who share responsibility for water utilities are all too common; often a result of differing priorities between a focus on cost recovery versus public accessibility. While the people expect the governments to provide water free or heavily subsidize it and most governments consider it one of their public welfare duties to make water affordable to the people, the key aim of private sector entities which engage in water governance is to maximize their profits. As Hall (2001) concluded, '[private] companies will simply leave the poor without water rather than incur losses'. This face-off between private sector goals and public sector goals could be categorized as 'privatization versus equity', which is in contravention of public sector ambitions, including pressures to increase tariffs, disconnection of non-payers, termination of local unionized workers and the perception that private interests are being served (Hall et al., 2005).

There are many examples from both developed and developing countries of termination of contracts for water infrastructure development and/or provision of water services as a result of conflicting aims and interests of the public sector and private sector players involved. For example, in the USA, the city councils of Atlanta and Montara determined to terminate the contracts as a result of differences they had with the private parties concerned. Atlanta privatized its waterworks system in 1997, by entering into a 20-year contract with United Water, which at that time was considered to be the largest and longest privatization of infrastructure in US history. However, in 2003, merely four years after the effective date of the contract (1999), the parties dissolved the arrangement with the city taking back the operation of the system. Factors leading to the takeover included differences regarding the performance of the water system, contract payments or change orders, and the status of the system before the contract was entered into (Segal, 2003).

In 2003, the residents of Montara bought back the municipal water system from American Water, then owned by RWE of Germany, after suffering poor service and some of the highest rates in California. One of the key reasons behind the takeover concerned the pricing of water. The purchase was financed by a property tax hike of about US$159 per year for every US$100 000 of assessed home value. Water and sanitation services are now administered by the Municipal Water Board, which includes community representatives (Food & Water Watch, 2008).

In 1993, the city of Szeged in Hungary, with approximately 180 000 inhabitants, was the first municipality to engage the private sector for provision of water services by selling off 49 percent of the water utility, Szegedi Vizmu, to Vivendi along with a 15-year concession contract to supply water. However, following disputes relating to a variety of issues, ranging from the extent of services provided to their pricing, the concession contract was terminated in 1999 and the operation of the water company was taken over by the utility (Lobina, 2001).

Similar examples can be cited from developing countries. For example, in Argentina, the government followed the International Monetary Fund (IMF) recommendations and privatized 71 percent of urban water services by 1999. However, many of the companies which invested in water infrastructure had disputes with the government concerning the failure to fulfill government contractual obligations and representations made to lure investors before the Argentine financial crisis of 2001–02, when Argentina withdrew from the IMF program. After the crisis and devaluation of the currency, the Suez-owned Aguas Argentinas company, which had the contract for water services in Buenos Aires, demanded steep price increases from the regulator. However, the Argentine government refused such increases on the basis that water should remain affordable. Eventually, the contract was terminated in 2004 when the company failed to reach agreement with the government about price levels (Hall and Lobina, 2007).

In Antalya, Turkey, a water utility contract was awarded to a Suez subsidiary in 1996. The privatization was not supported by the city council but imposed by the World Bank as the condition of a loan. Following the signing of the contract, Suez demanded that water prices should be more than doubled, and that sewerage prices should be increased by over ten times. When the city council refused, the Suez subsidiary declared bankruptcy and abandoned the contract in 2002. The service was taken over by the municipal water company and interestingly the prices decreased gradually after that (World Bank, 2005; Cinar, 2007).

Although not terminated as yet, the case of water utility privatization in Selangor, Malaysia can be cited as a good example of an ongoing conflict between public and private sector players concerning water utility management. In the federal capital territory of Kuala Lumpur and the state of Selangor, a 26-year build–operate–transfer (BOT) contract for an

1120 megaliters per day water treatment plant was signed with Puncak Niaga Holdings (PNH) in 1994 as part of the Selangor River Water Supply Scheme. The contract required PNH to invest US$760 million in the project. The state government provided a revenue guarantee to make the project financially viable. Subsequently, in 2004, without competitive bidding, a 30-year concession contract was signed with Syarikat Bekalan Air Selangor Sdn Bhd (SYABAS), a subsidiary of PNH. This concession required an investment commitment of US$2.5 billion and was supported by fixed government payments to SYABAS for services as the tariffs from end-users, which were fixed at the time, were considered too low to make the concession financially viable. By 2010, both contracts were listed as 'distressed' following demands by the state government, one of the public sector contracting parties (the other being the federal government) for their cancellation (World Bank, 2011). The key reasons for the dispute are said to be underperformance by the company, failure to recover the operating costs and the failure of the regulator to increase the tariff (Le Strat, 2011). Although the state of Selangor offered to buy back the water concessions with the aim of re-municipalization, in May 2011, the federal government bailed out the debts (around RM6.5 billion) of the privatized water companies in Selangor. This bail-out has hindered the Selangor state government's hope of striking a deal with these cash-strapped companies in an attempt to re-municipalize the water systems (Word Press, 2011).

14.3.3 Lack of Good Governance Practices

One of the downsides of shifting to privatization is that there could be room for irregularities in the procurement process, regulation of services and the supervision of private sector management. Further, adhering to illegal and unethical practices during the project life by the key stakeholders could adversely affect privatization. Not surprisingly, such problems in privatized systems could lead to re-municipalization.

Corruption can take many forms, from petty to grand corruption. Petty corruption in the water sector includes extortion of bribes for water connections and water use licenses. Further, poor water quality monitoring and low levels of sanctioning water pollution can also be related to corruption. Different forms of grand corruption include collusion, policy capture by the elite, embezzlement of government assets and funds, bribery in international deals, bid rigging and nepotism (UNDP, 2011).

One of the most infamous cases of corruption involving procurement of private sector participation in water is that of Grenoble (France), where water privatization was introduced in 1987 when a contract was awarded to Suez Corporation. Following allegations concerning transparency in the procurement process and corrupt practices in awarding the contract, the city's mayor at the time, Alain Carignon, a close friend of then President Jacques Chirac, was sentenced to jail in 1994 for accepting illegal campaign contributions and €3 million as bribe (UNDP, 2011; Godoy, 2003). Two Suez executives were also convicted on related charges. In 1996, Carignon's appeal was dismissed by the Court of Appeal. A second court case found that the Suez subsidiary responsible for managing water services in Grenoble had been overcharging the customers through fraudulent accounting. In particular, it was exposed that the private company had inflated the price charged to consumers and engaged in dishonest accounting by: (1) charging for 'entry fees'; (2) manipulating the indexation of prices; (3) practicing fictitious cost accounting; and (4) failing to include profits from exclusive works contracts (Hall and Lobina, 2001).

As a result, the company was ordered to pay back all the water charges imposed between 1990 and 1998 (Godoy, 2003). The re-municipalization of Grenoble's water resulted in the effective reduction of water prices and the city now enjoys one of the cheapest water rates in France. Raymond Avrillier, an anti-water privatization advocate and manager of the municipal water utility of Grenoble, argues that the success of the Grenoble campaign hinged on access to information and an independent analysis of the role of the private sector, which are essential for good governance (Avrillier, 2005).

There have been other instances of corrupt practices concerning water governance in France. For example, in Angoulème, a former mayor was jailed in 1997 for two years, with another two years suspended, for taking bribes from companies bidding for contracts, including Générale des Eaux. Some executives of Générale des Eaux were also convicted of bribing the mayor of St-Denis (Ile de la Réunion) to obtain the water concession (UNDP, 2011).

Camden County, New Jersey, one of the poorest cities in the USA, serves as another example of corruption involved in procuring private sector engagement in water. In 2000, Camden privatized its water services under a 20-year contract with United Water, a subsidiary of Suez Environment (Katz, 2009). In 2009, following a special audit of the water contract conducted by the city, it was revealed that: (1) the city had paid US$2.2 million to United Water in connection with an amendment to the contract, even though that amendment was never legally approved by elected officials; (2) the company had charged the city administrative fees not specified in the contract, even though the work was subcontracted to other companies; and (3) various payments had been made to the company without proper authorization. Three mayors of the city have been jailed for corruption.

Another classic example of failure with regard to good governance in the water sector comes from Potsdam (Germany). In 1997, the city council of Potsdam sold 49 percent of the shares in the public utility to Eurawasser, a subsidiary of Suez and the German Corporation Tyssen, for the sum of DM167 million (Hachfeld, 2008). The complicated details of the privatization were contained in a total of 13 individual contracts, which were subjected to confidentiality. One of the most important aspects of the transaction was the fact that despite its minor share of 49 percent, the power in the company was de facto conveyed to Eurawasser. The process of ending the privatization began, however, when Peter Pfaffhausen, manager of the municipal electricity company, was appointed Vice-Director of the water company. Pfaffhausen was an active board member who meticulously observed and voiced dissatisfaction with any board decisions which were unfavorable to the city. Such decisions included the manner in which Eurawasser recorded expenses: Eurawasser was recording expenses not as operating costs but as investments, since investments had to be paid by the city alone. Further, the city found out that the original purchase price for the privatization had actually not been paid by Eurawasser, as required under the contract. Instead, the DM167 million purchase price had been paid by a bank which, in return, got an entitlement from Eurawasser for the revenue from the water charges of the company over a period of more than 20 years with a value of about DM 400 million.

As a result, the city secretly re-examined the purchase agreements and prepared a re-municipalization plan (Hachfeld, 2008). The city, armed with this information, in a shareholders' meeting successfully obtained representation rights for the city's managing director and with that was able to buy back the utility. Eurawasser was taken completely by surprise and after several discussions the company and the city resolved their dispute in 2001 via an out-of-court settlement. From the content of this settlement agreement only few aspects are known. It is reported that Eurawasser received a last tranche of service orders and received financial compensation estimated at DM25 million. According to the city authorities, in addition to the irregularities in the procurement, the main reason for the re-municipalization in Potsdam was their concern as to future price developments. It is said that as projected by Eurawasser, water prices could have increased by over 100 percent had the city not taken the initiative to re-municipalize the utility (Schellenberger, 2001).

Lack of good governance in the private sector's engagement in the water sector is not limited to developed countries. A notable example comes from Lesotho and concerns the Highlands Water Project, one of the largest water infrastructure projects in the world (UNDP, 2011). Based on a treaty signed by the governments of South Africa and Lesotho in 1986, the project comprises a system of several large dams and tunnels, mostly located in Lesotho's highlands, to augment the water supply to the Johannesburg metropolitan region of South Africa and generate electricity for Lesotho in the process. The project was overseen by the Lesotho Highlands Development Authority (LHDA), a parastatal organization. During an internal audit in 1993 ordered by the minister of natural resources of Lesotho, irregularities were found in declarations by the LHDA's chief executive. Further, the investigations led to the discovery of bank accounts for the chief executive in Switzerland and South Africa and illegal payments of more than US$6 million in bribes to those accounts by local representatives of international construction firms during the bidding process (UNDP, 2011). In 2002, the chief executive of the project was sentenced to 15 years of imprisonment.

Overall, the poor and marginalized groups in society are most affected by corrupt practices in developing countries. According to Hall (1999), 'effective action against corruption has to involve effective sanctions by developing countries against multinationals which engage in corrupt practices; greater political transparency to remove the secrecy under which corruption flourishes; and resistance to the uncritical extension of privatization'.

14.3.4 Change in Government

Changes in government can be very disruptive to public infrastructure management, as competing agendas and ideological stands mean budgetary cuts and lack of support for development and management. Not surprisingly, in many countries, systems that were being managed privately or semi-privately began to fall apart and were re-municipalized. This has been particularly obvious in the French cities of Paris and Varages as well as in Tucuman, Argentina; Brazil; Mauritius; and Panama. As shown in Table 14.1, there have been instances where privatized water projects have been re-municipalized following election of new governments.

In Paris, for example, party changes in 2001 coupled with a new left-wing ideology saw the reconsideration of privatization of the city's water services. This new thinking was founded on efficiency, improved services and capable management, as the private sector entities operating water services were found to be lacking in these aspects. The discussion progressed until 2010, when the water utility returned to public provision, one of the re-municipalization schemes that has been characterized as successful. Prior to the return, a thorough assessment was made and upon re-municipalization a clear management framework that was capable of overcoming potential obstacles was implemented. Under this new management framework it is reported that priority is given to research, capacity building of personnel and the empowerment of municipal bodies (Le Strat, 2010).

In some cities referendums have been used as a tool to earn public support to re-municipalize privatized water projects. For example, in the city of New Orleans (USA), it was decided that any future proposals to privatize water would have to be put to a referendum. This prospect alone led Suez to announce that it would not seek a concession that would be subject to such a vote. Such initiatives have been taken in other countries too. For example, in Hamburg (Germany), referendums have been used as a tactic to oppose water privatization (Hall et al., 2005).

14.3.5 Unfulfilled Contractual Obligations

Failure to fulfill contractual obligations is also one of the key reasons for re-municipalization of water utilities. Such failures include substandard services, construction defects, management failures, non-fulfillment of investment undertakings, breach of payments to the government, if any, and failure to comply with regulatory requirements. Such failures have led to a steady decline in public support for the involvement of private sector entities in water service delivery, especially given that in democratic societies, public sector entities who procure private sector engagement in such projects are ultimately answerable to the general public who give them the political mandate to govern.

For example, in Latin America, there has been a steady decline in public support for the involvement of private firms in water service delivery. In Buenos Aires province (Argentina), in 1999, the management of water and wastewater treatment services, which covered 74 towns from 46 municipalities, was contracted out to Azurix under a 30-year contract (CEO & TI, 2008; Ortiz, 2011). By 2002, failure to meet most of the contractual obligations led to mounting public dissatisfaction with Azurix, as it had failed to honor commitments towards service quality and capital investment, and refused to pay fines for polluting nearby rivers (CEO & TI, 2008).

In Trinidad and Tobago, as explained in detail in section 14.4.2, the targets specified in the contract to be met by the private sector manager of the water utility, Severn Trent, never materialized. As a result the planned privatization did not proceed beyond the 'probationary' 3–5 year interim management contract awarded to the company. The interim management contract in fact provided the government with the opportunity to gather information for future risk allocation, to develop a stable tariff structure and to access the private operator's credibility in providing quality service and fair tariffs (Nankani, 1997).

In some instances private companies, when faced with the reality that the target rate of return would not be achieved or other realities which would affect profitability, would disregard the terms of the contract and instead seek to increase prices through constant renegotiation or clever interpretation of water contracts (Hall, 2001). For example, the Siza Water Company (SWC), which became the first private company to manage a water and wastewater utility in South Africa in 1999, under a groundbreaking 30-year concession contract, refused to pay a ZAR3.6 million lease payment due to the municipality of KwaDukuza and instead demanded that prices be increased by 15 percent and that investment be decreased from ZAR25 million to ZAR10 million so as to restore profitability. It justified this increase on the basis that Siza's revenues fell short because housing developments and, by extension, demand for water had fallen far short of projections (Hall, 2001). This contract, however, was not terminated and the project was not re-municipalized, as Siza recovered following a renegotiation of the terms of the contract in 2001 that allowed a reduction of the technical assistance advisory fee and execution of a series of measures to improve performance (USAID, 2005).

The World Bank has reiterated that whether a contractual alternative, be it in management contracts or full concessions, 'performs better than full provision by state-owned enterprises depends in particular on whether performance risk is effectively shifted from taxpayers to the private shareholders of the company that enters into a concession-type arrangement' (World Bank, 2002). Performance risk is defined as the 'risks associated with the commercial provision of water supply . . . such as failing to meet the agreed service targets or facing declining revenues as a result of decreasing consumption' (Hall and Lobina, 2003). In other words, the success of private sector engagement in the provision of water services depends on the extent to which the private sector entities concerned are capable of providing efficient services to the end-users as expected under the contracts. If they fail to fulfill this fundamental obligation, then the public entities that have awarded contracts to them would be compelled to consider public demand for reclaiming water utilities back into the public fold.

Furthermore, Hall and Lobina (2003) have pointed out that many technical mechanisms such as 'international arbitration, political risk insurance and guarantees' have transferred political risk from the private sector managers to the public in an attempt to allow private companies the opportunity to perform efficiently without government interference. This research has concluded however that empirical evidence has proven that risk allocation in private water concession departs in practice from what is laid out in theory, as in reality performance risk is shifted away from the private sector, leaving local taxpayers and consumers to bear the full costs of distorted incentives. There is merit in this logic as neutral dispute resolution mechanisms such as international arbitration, and instruments such as political risk insurance and guarantees of payment, would provide comfort to investors against breach of obligations by the governments and the end-users, as often the end-users are at the receiving end when contractual obligations are breached by the private sector entities engaged in the provision of water services. This is because resolution of disputes, as well as reclaiming utilities back into the public fold or renegotiation or re-contracting with private sector entities, takes time, during which the end-users may not get adequate services. Table 14.2 shows selected privatized water projects that have been terminated following contractual disputes.

14.3.6 Legal Reforms

Many countries across the world have been campaigning to regain public control of their water supplies, while others have taken steps to reinforce the essentiality of water in their constitutions and national laws. In some instances these initiatives have resulted in the passing of new laws or amendment of existing laws by the legislature enabling the re-municipalization of water utilities and/or prevention of privatization of water utilities.

For example, in the Netherlands, in 2004, the Water Supply Act[1] was passed, which states *inter alia* that 'drinking water services to consumers may only be provided by a "qualified legal person"'. Furthermore the Act, under Article 1(f), defined 'qualified legal person' as follows:

1. Public legal person which includes state, province, municipality or water board
2. Public limited company or private limited liability company which meets the following criteria:
   i. The shares of the company are held directly or indirectly by public legal persons and
   ii. The company has not committed to share its controlling rights with others than a public legal person or company as defined in the Act
3. Joint ventures whose members must comply with conditions set under 2 above.

Additionally, in accordance with Article 1(g), only existing water companies, which sold water on September 1, 2000, as well as their legal successors, controlled by a qualified legal person, were permitted to provide

Table 14.2 Water contracts terminated and re-municipalized

Country	Province/ city	Initial contract duration	Private company[a]	Termination year[b]	Details
Argentina	Buenos Aires city	1993–2023	Suez Group, France (46.3%), Veolia Group France, Anglian Water PLC, Aguas de Barcelona S.A., Spain (23%), Local partners	2006	Although service coverage increased in initial years, there was significant rise in tariffs and the concessionaire failed to meet investment commitments and failed to expand connections to poor suburbs. Due to failure to re-negotiate tariff hikes after the Argentine financial crisis, the private stakeholders decided to terminate the contract. On 30 July 2010, ICSID decided that the Argentinian government unfairly refused to allow the concessionaires to raise tariffs during the period of economic turbulence following the devaluation of the Argentine peso in 2001. The effect of the devaluation eventually made the concessions so unprofitable that the European operators requested their termination.[c]
Argentina	Greater Buenos Aires province (60 municipalities)	1999–2029	Azurix (Enron), USA (100%)	2002	Azurix withdrew from the contract accusing the regional government of 'serious breaches'. An ICSID Tribunal rendered an award in 2006 condemning Argentina to pay USD 165 million instead of the USD 565 million claimed by Azurix. In 2007, an ICSID tribunal dismissed in its entirety the

					application by Argentina for annulment of the USD 165 million decision and Argentina was ordered to bear all expenses incurred by ICSID in connection with the proceedings, including the fees and expenses of its members.[d]
Argentina	Buenos Aires province (7 municipalities)	2000–2030	Impregilo SpA, Italy (43%), ACS Group (Actividades de Construccion y Servicios), Spain (27%), Aguas de Bilbao Bizkaia, Spain (20%)	2006	Impregilo filed a compensation claim before ICSID and the case concluded in May 2011. The arbitral tribunal declared that although the Argentine Republic did not violate the contract by nationalizing the Impregilo's investments, it failed to treat it in a fair and equitable manner. As such, the government should compensate Impreglio USD 21 million along with a compound interest of 6% until payment is made.[e]
Argentina	Mendoza	1993–	South Water Argentina (32%) Saur, France (32%)	2009	In 2010 it was reported that the city council, which took over the company in 2009 on a 'temporary' basis, completed its financial, technical and environmental audit. This audit revealed that SAUR was deficient in all areas analyzed and could be sued for approximately USD 768 million in damages.

Table 14.2 (continued)

Country	Province/ city	Initial contract duration	Private company[a]	Termination year[b]	Details
Argentina	Santa Fe	1995–2025	Suez, France (51.7%), Aguas de Barcelona S.A., Spain (10.89%), Interagua – Servicio Integral de Agua S.A., Spain (14.92%), Banco de Galicia y Buenos Aires S.A., Spain (12.5%) Aguas Provinciales de Santa Fe's employees, Argentina (10%)	2006	After failing to sell its share to private sector buyers, Suez withdrew from the contract. On 30 July 2010, the arbitrators found Argentina liable for breaching the fair and equitable treatment obligations of investment protection treaties with the UK, France and Spain.[f]
Argentina	Tucuman	1995–2025	Veolia Environnement, France (90%)	1997	Terminated after consumers stopped paying bills due to serious operational failures. An ICSID tribunal rendered an award in 2007 condemning Argentina to pay USD 105 million instead of the USD 375 million

					claimed by Vivendi. The case is still pending due to appeal and no payment made.
Belize	National	2001	Cascal/Biwater, UK (45%) Nuon, Netherlands (45%)	2005	Renationalized by buying shares in BWS from Cascal/Biwater. Government then floated BWS for sale to Belize citizens only. It is reported that the Belizean government owns 80% of BWS, the Social Security Board owns 10% with the rest being owned by minority shareholders.
Bolivia	Cochabamba	1999–2039	International Water Ltd. (55%) [Bechtel (US) and Edison (Italy)], Riverstar International (25%) Four Bolivian companies (20%)	2000	The concession was terminated following social unrest. Bechtel filed a compensation claim in front of ICSID demanding payment of USD 50 million. Due to pressure from an international civil society campaign, Bechtel decided to abandon the claim to protect its reputation. The claim was formally settled for a token payment of 2 Bolivianos (USD 0.30).
Bolivia	La Paz/El Alto	1997–2027	Banco Mercantil, Bolivia (22%) Suez, France (55%)	2007	The contract achieved sizeable improvement in access to poor families. Tariff reforms, prior to concession, significantly lowered monthly bills for small households while increasing for larger ones. However, low consumption levels in El Alto significantly lowered revenues for the company, forcing the government to increase connection fees for new customers. These coupled with protests from people outside the concession area led

Table 14.2 (continued)

Country	Province/ city	Initial contract duration	Private company[a]	Termination year[b]	Details
					to early termination of the contract after amicable negotiations, which required the Bolivian government to assume responsibility for loans, amounting to USD 9.6 million, and pay USD 5.5 million as compensation to Suez.
Central African Republic	Bangui	1991–2006	Saur, France (51%)	2001	The SAUR contract was terminated 5 years before its expiry. No reports of compensation paid.
China	Shenyang city, Liaoning province	1997–2017	NWS Holdings Limited, China (19%) Suez, France (19%)	2000	Due to high price of bulk water and huge losses to state-owned company due to high guaranteed returns, the contract was terminated and re-sold to state-owned company.[g]
Colombia	Bogota	1998–2018	Suez, France (100%)	2004	The Salitre wastewater BOT was terminated after the city council calculated that the project was charging ten times too much, and that it was worth paying USD 80 million to buy out the contract.
France	Grenoble	1989–	Suez	2000	The concession contract was awarded illegally, against the protests of environmental and civil groups, as the then mayor was significantly bribed. Concessionaire failed to make required

					investments and earned extensive profits. When the extent of corruption by the mayor was revealed in 1998, court decisions led to termination of the contract.[h]
Germany	Potsdam	1998–	Eurawasser (Suez)	2000	The Potsdam city council terminated the contract in June 2000, when the municipality found that prices were increasing in a way customers found untenable. More precisely, Eurawasser projected tariff increases from DM 8.80 cm^3 in 2000 to DM 16.40 cm^3 in 2017 as opposed to the city council's estimate of increases up to DM 11.95 cm^3 in 2017. The termination led to a legal dispute between the parties to the contract over payment of compensation, which was settled in 2001 with the agreement that the city council would pay Eurawasser an undisclosed sum and that in the future Potsdam Water Works would contract Eurawasser to carry out consultancy work.[i]
Hungary	Pécs	1995–2020	Suez	2009	In 2004 the mayor of Pécs announced that they were considering the re-nationalization of the water works, mainly because the price of water was increasing and investments were not forthcoming. However, Suez won a court case claiming it was entitled to massive compensation for lost future profits, which the city could not afford, and so forced the council to continue the contract. In 2009, the

Table 14.2 (continued)

Country	Province/ city	Initial contract duration	Private company[a]	Termination year[b]	Details
					second attempt made to re-municipalize the water works was successful and in March 2010, a Hungarian arbitration court ruled that the city council of Pécs was legally entitled to terminate its water service contract with Suez and take over the running of the service itself.[j]
Mali	Bamako	2000–2020	Energie du Mali (EDM) (60%) SAUR (65%)	2005	Conflicts emerged between the parties, mostly due to poor contract design with respect to tariff setting formulae, which led to a vicious circle of mutual accusations and uncooperative behavior. The concessionaire was being criticized for its insufficient performance and it was unable to raise long-term finance to fund its investment obligations due to a series of tariff cuts. As a result, the concessionaire slowed down investments significantly and in 2005, the company was renationalized without paying any compensation.[k]
Philippines	Manila West Zone	1997–2027	Maynilad Water Services Inc., consortium of Suez and Benpres Holdings	2003	Price hikes, failure to extend water connections to poor areas and non-fulfillment of investment obligations, led to takeover of the company by the public water authority MWSS. A new contract was awarded to a

					consortium of DM Consunji Inc., Philippines (42%), First Pacific, Hong Kong (42%) and SUEZ (16%) in 2006, with significantly smaller investment commitments (USD 504 million) than the first concession.
South Africa	Amahthali (Stutterheim)	1993–2003	Suez, France (50%) Everite, South Africa (50%)	2003	In July 2003, the Amathole District Municipality (ADM) took over the functioning of water services for all local municipalities (except Buffalo City Municipality). The ADM is responsible for the provision of effective, efficient and sustainable water services for all.
South Africa	Nkonkobe (Fort Beaufort)	1995–2005	Suez, France (50%) Everite, South Africa (50%)	2001	The city council won a court ruling invalidating the contract. In July 2003, the Local Government Municipal Structure's Act, 117 of 1998, declared the Amathole District Municipality as the Water Services Authority for the Nkonkobe Municipality.
Tanzania	Dar-es-Salaam	2003–2013	Biwater, UK (25.5%) Gauff Engineers, Germany (25.5%) Superdoll, Tanzania (49%)	2005	The country's high dependence on international donors forced the government to award a 10-year lease to City Water Services (CWS) in 2003; however, due to the private company's poor preparation, difficult relations with staff and shareholder in-fighting, the government abruptly terminated the contract in 2005.[1] Biwater filed two compensation claims before two different international arbitration tribunals and lost both. It lost the case filed in front of the UK

Table 14.2 (continued)

Country	Province/ city	Initial contract duration	Private company[a]	Termination year[b]	Details
					High Court under UNCITRAL rules (on grounds of early termination of the contract) and was condemned to pay GBP 3 million in damages, which Biwater refused to pay. It also lost the case filed in front of an ICSID tribunal on grounds of expropriation. The ICSID tribunal 'found that while Tanzania's government had violated its bilateral investment treaty with UK, the company is not entitled to compensation because the value of its investment in CWC was nil at the time of the expropriation'.
Turkey	Antalya	1996–2006	Enka, Turkey (50%) Suez, France (50%)	2002	Suez declared bankruptcy and terminated the contract when the municipal council refused to agree to Suez's demands for water prices to be doubled and sewerage prices to be increased 10 times. After the water supply was taken over by the municipal authorities prices actually declined slightly.[m]
Uruguay	Barra de Maldonado	1993–2018	AgBar, Spain (60%) STA Ingenieros (30%)	2006	Suez left after reaching an amicable settlement with the government and selling its shares. Obras Sanitarias del Estado (OSE) is now responsible for water services in Maldonado.

Uruguay	Maldonado Department	2000–2030	Aguas Bilbao Bizkaia, Spain (20%) Bilbao Bizkaia Kuxta BBK (31%) Iberdrola SA, Spain (49%)	2005	Urbaser and Aguas de Bilbao filed a compensation claim in front of ICSID and the case is still pending. No payment has been made.

Notes:
a. Private Participation in Infrastructure (PPI) Project Database, available at http://ppi.worldbank.org/explore/Report.aspx.
b. Ibid.
c. http://www.globalwaterintel.com/archive/11/8/general/suez-seeks-12bn-damages-argentina.html.
d. ICSID Case No. ARB/01/12, available at http://icsid.worldbank.org/ICSID/FrontServlet?requestType=CasesRH&actionVal=showDoc&docId=DC1171_En&caseId=C5.
e. ICSID Case No. ARB/07/17, available at http://icsid.worldbank.org/ICSID/FrontServlet?requestType=CasesRH&actionVal=showDoc&docId=DC2171_En&caseId=C109.
f. http://www.iareporter.com/articles/20100818_9.
g. Manthan Adhyayan Kendra (2010).
h. Hachfeld (2008).
i. Hall and Lobina (2003).
j. http://isslerhall.org/drupal/content/court-rules-favour-pecs-decision-terminate-suez-water-contract.
k. Fall et al. (2009).
l. Pigeon et al. (2012).
m. http://www.herinst.org/BusinessManagedDemocracy/government/privatisation/pricesWater.html.
n. Water Supply Act of 2005.

Source: Updated from Hall et al. (2010).

water services to consumers. It must be noted that the Act refers only to drinking water. The Act further prohibited the private provision or delivery of water to consumers and the creation of legal transactions which directly or indirectly allowed a water company to be controlled by someone other than a qualified legal person (Articles 3(j) and 3(m)). Furthermore, Article 3(n) prevented changes in company statutes, the transfer of shares or the transfer of property which permitted a third party to become a share owner in a water company or owner of works needed for drinking water production. Hence, privately owned water companies are unambiguously prohibited from providing drinking water services to the citizens of the Netherlands and this places the responsibility of drinking water squarely on the shoulders of the state.

A constitutional amendment in the Republic of Uruguay is another example. Uruguay is known as the pioneer which enshrined the right to water in its constitution and provided inspiration for many other Latin American countries to follow suit. This right was proclaimed by a referendum held in 2004, parallel to its national elections, whereby 62.75 percent of Uruguayans voted to reform the constitution to make water a national asset for the public good (Ortiz, 2011). This referendum included the statement that 'the public service of sewerage and the public service of water supplying for the human consumption, will be served exclusively and directly by state legal persons' (Hall et al., 2004).

Section 1(b) of the referendum also declared that the users and civil society should be permitted to take part in 'all instances of planning, management and control of water resources' and it highlighted that any 'authorization, concession or permission' that contravened the public right to water or the users' right to participate in planning would be 'stopped with no consequences'.

Following the successful victory in the election and the referendum, the constitution of the country was amended. Article 47 of Uruguay's constitution now declares that the 'access to drinking water and access to sanitation constitute fundamental human rights'. Furthermore it states under Article 47(1)(c) that priority will be given to the provision of drinking water for the population in Uruguay's National Water and Sanitation policies, and under Article 1(d) that economic grounds cannot override social grounds. Thus, the said amendments not only ensure that water is kept safe for public use by prohibiting exploitation by private companies but also enshrined the right of public participation in the planning, management and allocation of water for all.

In Malaysia too, changes introduced to the laws in 2005–06 have affected the future of private sector engagement in water sector. In the early 1990s several states of Malaysia embarked on a policy of private sector participation in water and sanitation as discussed in the case study in section 14.4.2 below. Prior to the aforesaid legal reforms, 'water' was a subject under the administration of state government agencies that were responsible for operation and maintenance, and the development of water supply infrastructure. Traditionally, the federal government agencies were responsible for research, planning and development of water resources. However, in order to avoid fragmented water resources management and to ensure better coordination amongst various stakeholders in the federal and state governments in the management of water, the National Water Services Industry Commission Bill and Water Services Industry Act were amended in 2006, bringing the central management of water-related services in 13 states in Malaysia under the federal government (Mokhtar and Tan, 2008). To this end, the National Water Services Industry Commission (SPAN) was established under the provision of the Water Services Commission Act as the top body for water governance. The state governments were required to corporatize their state water authorities, which were to be regulated by SPAN as the central regulatory body. These reforms allowed the federal government to seize all assets previously owned by local water operators, public and private alike, and to fast-track asset development with government funding. In the policy statements which were issued introducing these legal reforms, the government made it clear that its intention was not to embark on new private sector contracts for water provision (PAAB, 2008).

14.4 RE-MUNICIPALIZATION: CASE STUDIES

As we have noted thus far, most re-municipalization schemes happened when the private sector failed to ensure equitable provision of water and proper investment by shareholders into water utilities. However, is it proper to entirely eliminate private sector involvement? Or is it possible for both public-sector and private operators to work together? To examine these questions further, a comparative study of three unique re-municipalization cases – Paris (France), Malaysia, and Trinidad and Tobago – has been conducted.

14.4.1 Paris (France)

Until 1984, the water systems in Paris were entirely run by the city, with only the responsibility for billing being outsourced to Veolia. Privatization of the water sector in Paris was conducted in two stages under the administration of Jacques Chirac, the then mayor of the city. The first step, in

1984, involved awarding a 25-year contract for water supply and billing services to two private companies, Veolia and Suez, for the right bank and left bank of the River Seine, respectively. The second step, in 1987, concerned delegation of water production and distribution responsibilities to a semi-public company, Société Anonyme de Gestion des Eaux de Paris (SAGEP), which was owned by the city (70 percent), Veolia (14 percent), Suez (14 percent) and Caisse des Dépôts et Consignations (CDC), a public national investment bank (2 percent) (Pigeon et al., 2012). It is important to note that these contracts were awarded without a formal procurement procedure and lacked specific and challenging performance targets (Le Strat, 2010); moreover, no assessment of the water network's condition was done prior to the privatization. Interestingly, it has been found that the then chief executive officer (CEO) of Suez, Jérôme Monod, was a co-founder of Chirac's party (Pigeon et al., 2012).

Following the privatization, water tariffs increased significantly, as much as 265 percent between 1985 and 2009, and although the private companies had a declared profit of 6–7 percent, it was much higher in reality (Pigeon et al., 2012). However, due to confidentiality of information of the private companies, the city authorities lacked reliable data on the technical and financial aspects of water supply, and regulatory action was hindered by the fact that the companies were not legally breaching the contract obligations as the contract hardly had any special requirements (Pigeon et al., 2012). Things started to change in 2001, when Bertrand Delanoë, the newly elected mayor, decided to regain control of the water systems of Paris and in 2002, a specialized unit was set up within the municipalities' services, with the responsibility of policing the activities of private operators. In 2003, a number of negotiations were carried out between Paris and the three operators with the objective of rectifying the inadequacies of the contracts. The renegotiations included the implementation of an investment plan of €153 million to be carried out by Veolia and Suez without any impact on water tariffs, as well as the adoption of detailed objectives to improve the water network efficiency (Le Strat, 2010).

In order to strengthen their position in the face of these powerful water giants, the politicians launched a series of technical and financial audits and expert consultations to identify the consequences and potential of terminating the contracts. These investigations revealed that, given the legal and technical complexities, it would be best to wait for the contracts to expire in 2009. However, steps towards re-municipalization started in 2007, when Veolia and Suez were obliged to sell their shares in SAGEP to CDC, which were in turn purchased by the city in 2008. Another key event was that, in 2007, during his campaign for his second term as the mayor of Paris, Delanoë promised that water systems would be taken back to the public sector if he was re-elected. His argument was that public water management can be more efficient, and it enables the city to manage water as a common good rather than being treated as a private commodity. Following his re-election, the final decision was made in November 2008, when the Paris City Council voted for the entire water system, from resource protection to end-user, to be operated by one unified public entity with effect from January 2010 (Pigeon et al., 2012).

In late 2009, the water production and distribution activities were transferred to the public utility, Eau de Paris, and SAGEP was dissolved. The transition from private to public management has been an enormously challenging task, especially with respect to customer databases and information systems (Pigeon et al., 2012). Eau de Paris is now a semi-autonomous body with a separate budget and legal status, with Anne Le Strat as the president of the board as well as the vice-mayor for water, sanitation and canals. To allow effective regulation of this public water utility, a Memorandum of Understanding (MoU) has been signed between Eau de Paris and the municipality, and numerous performance indicators have been introduced in the MoU to guarantee service standards and efficiency (Le Strat, 2010). In the very first year, the city had a savings of €35 million and 8 percent reduction in tariff. According to Anne Le Strat, the successful re-municipalization initiative in Paris is providing food for thought for other municipalities as well, and with 75 percent of delegated water services contracts in France up for renewal by 2015, councils are mulling over whether to consider in-house management (Le Strat, 2011).

14.4.2 Malaysia

Malaysia consists of 13 states and three federal territories and since its independence from British rule in 1957, water supply has been the responsibility of the state governments, while the federal government provided interest-free loans for capital investments and tariff subsidies. During the late 1980s, in order to meet the demand of the fast-growing industries, several states, especially the richer and more populated ones, started to grant long-term concessions to private operators (Pigeon et al., 2012). In 1992, Johor became the first state to award a 20-year BOT contract for a water treatment plant to a consortium led by the Malaysian company Kembangan Dinamik with a minority share by the French firm Suez. In the federal capital territory of Kuala Lumpur and the state of Selangor, a 26-year BOT contract for a water treatment plant was signed with Puncak Niaga Holdings in 1994 as part of the

Selangor River Water Supply Scheme, and in 2004, a 30-year concession was awarded without competitive bidding to Syarikat Bekalan Air Selangor Sdn Bhd (SYABAS), a subsidiary of Puncak Niaga Holdings. Also during the 1990s, PPP contracts were awarded in other states, including Sabah, Penang, Kuala Lumpur, Kendah, Kelantan and Negeri Sembilan. In 2002, Penang became the first state to privatize its water supply through an outright sale of assets for the entire network (Hoque at al., 2012).

The 1997 Asian financial crisis, coupled with the El Niño droughts in 1997–98, revealed the declining environmental and economic status of the water sector, characterized by decreasing surface water quality and quantity, technical inefficiencies, indebtedness of the states and non-fulfillment of contract obligations by the private sector. All these triggered initiatives for an entire sector reform and in 2004, the newly created Ministry of Energy, Water and Communications hired the business consultant KPMG to propose ways to tackle the challenges. The proposal consisted of bringing together ownership and management of water systems under the federal government, and corporatizing the public utilities to enable a transition to full-cost recovery. The first step of the reform was a constitutional amendment approved in January 2005, which allowed water services to become a shared responsibility between the states and the federal government. This was followed by enactment of two important laws in 2006: (1) the Water Service Industry Act, which outlined details of the reform and established a national water asset holding company called Pengurusan Aset Air Berhad (PAAB); and (2) the National Water Services Commission Act, which led to the establishment of an independent national regulator called Suruhanjaya Perkhidmatan Air Negar (SPAN). PAAB was tasked with the responsibility of raising funds from the market with government credit rating and buying all water assets in the country from existing operators, which would in turn be awarded as lease contracts for the same utilities. This reform was touted by many as a unique national experiment that involved a departure from long-term concessions to a water sector run by the public authorities using business models (Pigeon et al., 2012).

By mid-2011, only five states (Melaka, Negeri Sembilam, Perlis, Johor and Pulau Pinang) had transferred their assets and liabilities to PAAB, and based on data provided by SPAN, technical efficiency has improved in three of them, with a decrease in unaccounted-for water (UfW), particularly in Johor, where the previous operator was a private company. However, it has been observed that the intended 'cheap funding' for the water sector, raised by PAAS on international bond markets at lower interest rates, turned into a money-making business for PAAB, as it lends this money to the state operators at different higher interest rates, based on their capacity to reimburse. One of the central tenets of this reform was 'tariff depoliticization', as tariff setting has now moved from the state government, which was reluctant to raise tariffs, to the federal government, where consensus building is necessary with the cooperation of NGOs and consumers. Although it is too early to assess the full outcomes of the reform, some positive aspects include greater commitment of the national government to water services and the likelihood of improvement in service coverage and efficiency due to new public capital investments. On the downside, exclusion of water resource management issues from the reforms may lead to long-term environmental impacts (Pigeon et al., 2012).

14.4.3 Trinidad and Tobago

Although many re-municipalization cases proved costly for the state, as they were required to make compensation payments to the private company, Trinidad and Tobago is one of the exceptions to the aforementioned premise. Trinidad and Tobago is a twin-island state in the Caribbean just 11 km off the coast of Venezuela. The Water and Sewerage Authority (WASA) is charged with the responsibility of providing water and wastewater services for the islands.

In 1994, WASA reported that it was suffering from operational inefficiencies; 55 percent of the population was experiencing an interrupted supply; there was 50 percent UfW and a 16:18 ratio of staff per 1000 water connections (Mycoo, 2005). Furthermore highly subsidized water tariffs coupled with excessive accounts receivables and high operating costs handicapped WASA's ability to provide an efficient service. As a result, the Trinidad and Tobago government slated WASA for privatization but not before the completion of a feasibility study by an established task force (Environmental and Development Division, 1998).

In 1994, the government awarded a 3–5-year interim management contract to Severn Trent for the first phase of its water privatization program so as to allow the government the opportunity to gather information for future risk allocation, to develop a stable tariff structure and to assess the private operator's credibility in providing quality service and fair tariffs (Nankani, 1997). Thereafter, they could have been eligible for a 20–30-year contract.

When Severn Trent took over the operations of WASA, it did not manage to fulfill the targets stipulated in the five-year management contract, and the results of the privatization published by the Trinidad Union (NUGFW) found that (Hall, 1999):

1. Less than 30 percent of customers had a constant supply of water and less than 60 percent received a piped supply for more than half the time (12 hours).
2. 30 percent of the population had access to the sewerage service, which is similar to the percentage before Severn Trent took over.
3. WASA had a net operating deficit of US$420 million, which was actually an increasing amount when compared to previous years.
4. Leaks were rampant throughout the whole system but the operator responded by laying off staff instead.
5. A US$450 million loan was made to WASA by Trinidad and Tobago Water Finance Ltd (Severn Trent), which would take 25 years to return.
6. Meters were installed in all homes to reduce water consumption (up to 40 percent), but this led to increase in tariffs of 16 percent per annum until charges were high enough for the private sector to make enough profit, estimated to be sometime in 2009.

The Trinidad and Tobago government, dissatisfied with Severn Trent's performance, decided not to renew the contract after expiry and took WASA back to the public fold. Water management contracts, due to the large amount of investment and the scale of the project, are often long term. However, the Trinidad and Tobago case highlighted the importance of conducting all necessary research prior to fully entering into an agreement. Targets specified in the contract to be met by Severn Trent never materialized; however, due to the 'probationary' 3–5-year interim contract adopted by Trinidad and Tobago, there was a smooth return of the utility to state control without any exorbitant cost to the state as is normally observed. The interim management contract provided the government with the opportunity to gather information for future risk allocation, to develop a stable tariff structure and to assess the private operator's credibility in providing quality service and fair tariffs (Nankani, 1997).

14.5 SCOPE FOR BETTER MANAGEMENT OF RE-MUNICIPALIZED FACILITIES

Given that a key reason behind the privatization of water utilities is the efficiency of private sector management, an important question that needs to be answered is whether there is scope for better management of re-municipalized facilities. We answer this question in the affirmative. Our answer is based on lessons garnered from two re-municipalized entities, namely the Departamento Municipal do Agua e Esgoto (DMAE), the public water utility of Porto Alegre (Brazil), which was re-municipalized back in 1904; and the recently re-municipalized system of Paris (France).

In this section, we seek to outline the areas where there are opportunities for better management in re-municipalized systems. However, it should be noted that in some instances these recommendations can also be applied to privately managed water utilities and other alternative forms of water management, such as public–private and public–public partnerships.

The first prospect for better management is in the area of public finance. This recommendation is premised on the fact that the goal of the privatization model was to achieve financial targets and returns on investment for shareholders, whereas the focus of re-municipalized models would be different as there would be an internalization of revenue. It has been reported that in France, since there are no dividends to be paid, all the revenue from the provision of water services is reinvested in the service and there are now competitive public calls for tender, whereas previously jobs would be handed to subsidiaries of the private company that would in turn overcharge for the goods and services provided (Le Strat, 2010). As a result of these strategies, the Paris utility that was re-municipalized in 2010 has already reported initial profits in the sum of €35 million (Le Strat, 2010). Porto Alegre also practices 100 percent reinvestment in the services and this approach to finance has allowed for investment in better technology, better wages for staff and more efficient services (Maltz, 2005).

The second opportunity for better management comes in the area of capacity development, which is defined as 'the process by which individuals, groups, organizations, institutions and countries develop their abilities, individually and collectively, to perform functions, solve problems and achieve objectives' (UNDP, 2001). Re-municipalized companies therefore need to prioritize their re-empowerment and must endow their utilities with the expertise and capacity necessary to guarantee an adequate provision of water services. According to the United Nations Development Programme (UNDP), one aspect of successful capacity development is the ability to solve problems (UNDP, 2001). For example, in Porto Alegre, now a successful public water utility, when workers were faced with technical difficulties from an obstruction of pipes caused by invasion of the golden mussel, the utility invested heavily in research so as to combat the challenge and as a result Porto Alegre is now recognized as a leading consultant on this issue (Maltz, 2005).

Better management of utilities can be achieved when public managers adhere to the pillars of good governance, namely accountability, transparency, rule of law and participation, all of which must be supported by the foundation of a strong civil society (Schiavo-Campo and McFerson, 2008). Re-municipalization in many instances occurred as a result of the support and actions of citizens and a strong civil society. It is therefore now imperative that focus be directed toward ensuring accountability, transparency, rule of law and participation.

The aforementioned 'pillars' have been implemented in Porto Alegre and Paris. For instance, Porto Alegre is fully accountable to the people and all transactions are executed in an open and transparent manner (Maltz, 2005). In France, a Management Framework has been developed which includes many technical, social, financial and other indicators, and the use of this clearly outlined framework has enabled managers to track implementation, evaluate investments and personnel, research and provide reports on the company, and receive feedback (Le Strat, 2010).

The most important scope for better management is observed in the area of public participation. Following the dissatisfaction of the people in Cochabamba, Bolivia and in Italy, with each referendum the importance of community engagement and stakeholder participation in the management of public utilities is underscored. As nations seek to achieve the shift towards a green economy, the success and sustainability of any water system needs the buy-in, support and input from people. In Porto Alegre, DMAE rests upon an integral deliberative council that enables citizens to exert influence and participate in the functioning of their own public water system, including a 'social audit' process involving citizen oversight and participation in budgeting for actual water works. The system features a participatory budgeting mechanism, whereby 16 regions are consulted, votes are taken and input is gathered concerning areas for improvement and expansion. These ideas are studied for feasibility before being integrated into the following year's budget (Maltz, 2005).

Utilities that cannot utilize the participatory budget approach are advised to strengthen their information gathering and sharing mechanisms. Paris is now engaged in better follow-up of individual consumption and customer expectation and the administrators try to ensure improved information of the services (Le Strat, 2010). Citizens' feedback is also important and there must be ongoing follow-up with users. In this vein Paris has established a citizens' control mechanism via the Municipal Water Watch, and this group enables users to evaluate the provision of services and the submission of feedback (Le Strat, 2010).

Scope for better management is also afforded in the formulation of policy and legal frameworks to guide the re-municipalized system. The water policies and legislation should include provisions on the right to safe drinking water, equitable provision of water and environmental protection. Additionally the roles and responsibilities of local actors coupled with realistic performance indicators and penalties for non-performance should also be included. The need for regulation must also be emphasized under re-municipalized systems so as to avoid a repetition of practices observed under private management and to ensure that there is consistent delivery of service obligations, 'efficient' pricing, conservation of water, professionalism from managers and the financial viability of the utility.

It is recommended, therefore, that a successfully managed water system in the twenty-first century is one which takes into account the needs of the people, the environment and even future generations. So by implementing policies and legislation, ensuring access to information, encouraging feedback and engaging all stakeholders, solutions for technical problems can be derived and innovative ways to better manage re-municipalized entities can be achieved.

14.6 CONCLUSION

In this chapter we have discussed the key reasons for re-municipalization and have argued that, despite the perception amongst many that privatization is an effective mechanism for efficient water sector management, it has not always yielded the expected results. Further, we have argued that re-municipalized water utilities could be effective in efficient management, provided that the public sector entities taking charge of them upgrade themselves, technically as well as financially, in addition to improving their managerial skills. This would need the introduction of public funds to finance the development, schemes for providing the necessary technical and managerial training, and laws and regulations which would facilitate efficient operation and management of the services. It should be noted that public water operators are serving more than 90 percent of the world's major cities.

Despite the success observed with re-municipalized systems and the potential for improvement, there is still room for meaningful engagement of the private sector in the provision of water utility services, especially in light of the high cost attached to the development of water infrastructure. However, in order to ensure the success of such partnerships, it is important that the roles and functions of public sector and private sector entities are clearly outlined, decisions are taken in a transparent environment, and accountability and performance mechanisms are installed. A successful model is also one which encourages public participation, and is instrumental in the development of fair pricing structures and providing legitimacy to the decisions taken by the regulators and the service providers. Chapter 15 in this volume includes a detailed discussion on the crucial parameters required for successful engagement of the private sector in water governance.

NOTE

1. Water Supply Act of 2005.

REFERENCES

Avrillier, R. (2005), 'A return to the source: re-municipalization of water services in Grenoble, France' in B. Balanya, B. Brennan, O. Hoedeman and S. Kishimoto (eds), *Reclaiming Public Water*, Amsterdam: Transnational Institute and Corporate Europe Observatory, pp. 63–72.

Birdsall, N. and J. Nellis (2002), 'Winners and losers: assessing the distributional impact of privatization', Working Paper, Washington, DC: Center for Global Development.

Cinar, T. (2007), 'Trend in infrastructure development and investments related to urban water and sewerage services on Turkey', quoted in David Hall and Emanuale Lobina (2009), 'The private sector in water in 2009', PSIRU, available at www.psiru.org.

Corporate Europe Observatory and Transnational Institute (CEO & TI) (2008), Water Re-municipalization Tracker, available at http://www.re-municipalization.org/cases#Buenos_Aires_Province (accessed 20 October 2011).

Corral, V.P. (2007), 'Water privatization and ADB, its impacts and responses from peoples' movements', Forum Report, Public Services International Research Unit (PSIRU).

Dowie, M. (2005), 'In law we trust', available at http://www.orionmagazine.org/index.php/articles/article/122/ (accessed 14 November 2011).

Environmental and Development Division (1998), 'Report on the Second Workshop on Private Participation in Water Supply and Sanitation Utilities in the Americas', Workshop Report, Economic Commission for Latin America and the Caribbean.

Fall, M., P. Marin, A. Locussol and R.Verspyck (2009), 'Water P-Notes', Issue 38, World Bank, available at http://water.worldbank.org/water/publications/public-private-partnerships-reform-urban-water-utilities-western-and-central-africa (accessed 14 October 2011).

Filho, A. De O. (2002), 'Brazil: struggle against the privatization of water', Inter-American Water Conference, San Jose, Costa Rica, 8–10 July, available at http://www.observatoriosocial.org.br/servpub/relatorios/resistprivating.pdf (accessed 10 March 2012).

Food & Water Watch (2008), 'Faulty pipes: why public funding, not privatization, is the answer for US Water Systems', available at http://documents.foodandwaterwatch.org/doc/FaultyPipesSeptember2008-web.pdf (accessed 10 March 2012).

Gleick, P., G. Wolff, E.L. Chalechi and R. Reyes (2002), 'The new economy of water: the risks and benefits of globalization and privatization of fresh water', Oakland, CA: Pacific Institute for Studies on Development, Environment and Security.

Godoy, J. (2003), 'Water and power: the French connection, Centre for Public

Integrity', available at http://projects.publicintegrity.org/water/report.aspx?aid=47 (accessed 10 March 2012).
Goldman, M. (2005), *Imperial Nature: The World Bank and Struggles for Justice in the Age of Globalization*, New Haven, CT: Yale University Press.
Hachfeld, D. (2008), 'The re-municipalization of water – some reflections on the cases of Potsdam and Grenoble', available at http://www.wasser-in-buergerhand.de/untersuchungen/david_hachfeld_remunicipalisation_of_water_potsdam_grenoble.pdf (accessed 11 April 2012).
Hachfeld, D., P. Terhorst and O. Hoedeman (eds) (2009), 'Progressive public water management in Europe: in search of exemplary cases', Reclaiming Public Water Discussion Paper, Transnational Institute and Corporate Europe Observatory, available at http://www.waterjustice.org/uploads/attachments/Progressivepercent20publicpercent20 waterpercent20managementpercent20inpercent20Europe.pdf (accessed 12 December 2011).
Hall, D. (1999), 'Privatisation, multinationals, and corruption', *Development in Practice*, **9**(5), 539–56.
Hall, D. (2001), 'Water privatisation – global problems, global resistance', Public Service International Research Unit (PSIRU), available at http://www.psiru.org/reports/water-privatisation-%E2%80%93-global-problems-global-resistance.
Hall, D. and E. Lobina (2001), 'Private to public: international lessons of water re-municipalization in Grenoble, France', Public Service International Research Unit (PSIRU), available at www.psiru.org/reports/2001-08-W-Grenoble.doc (accessed 12 December 2011).
Hall, D. and E. Lobina (2003), 'Problems with private water concessions: a review of experience', Public Service International Research Unit (PSIRU), available at www.psiru.org/reports/2003-06-W-over.doc (accessed 12 December 2011).
Hall, D. and E. Lobina (2007), 'Water privatisation and restructuring in Latin America', Public Service International Research Unit (PSIRU), available at http://gala.gre.ac.uk/2940/1/2007-09-W-Latam.pdf (accessed 12 May 2012).
Hall, D., E. Lobina and V. Corral (2010), 'Replacing failed water contracts', Public Service International Research Unit (PSIRU), available at www.psiru.org/reports/2010-01-W-Jakarta.doc (accessed 25 October 2011).
Hall, D., E. Lobina and R. De la Motte (2004), 'Making water privatization illegal: new laws in Netherlands and Uruguay', Public Service International Research Unit (PSIRU), available at www.psiru.org/reports/2004-11-W-crim.doc (accessed 12 December 2011).
Hall, D., E. Lobina and R. De la Motte (2005), 'Public resistance to privatization in water and energy', *Development in Practice*, **15**(3–4), 286–301.
Hoque, S.F., L. Bhullar and A. Gunawansa (2012), 'Private–public partnerships in urban water supply sector: a study of the regional trends', Lee Kuan Yew School of Public Policy Research Paper No. LKYSPP12-04-IWP.
Katz, M. (2009), 'State faults Camden water management', *Philadelphia Daily News*, 16 December, available at http://article.wn.com/view/2009/12/16/State_faults_Camden_water_management/ (accessed 26 April 2012).
Le Strat, A. (2010), 'Paris: local authorities regain control of water management', available at http://www.tni.org/article/paris-local-authorities-regain-control-water-management (accessed 25 October 2011).
Le Strat, A. (2011), 'Paris defends re-municipalisation record', *Global Water Intelligence*, **12**(5), available at http://www.globalwaterintel.com/archive/12/5/general/paris-defends-re-municipalisation-record.html (accessed 12 May 2012).

Lobina, E. (2001), 'Water privatisation and restructuring in Central and Eastern Europe', Public Services International Research Unit, University of Greenwich, available at http://www.psiru.org/reports/2001-11-w-cee.doc (accessed 12 May 2012).
Maltz, H. (2005), 'Porto Alegre's water: public and for all', in B. Balanyá, B. Brennan, O. Hoedeman, S. Kishimoto and P. Terhorst (eds), *Reclaiming Public Water*, Amsterdam: Transnational Institute (TNI) & Corporate Europe Observatory (CEO), pp. 29–36.
Manthan Adhyayan Kendra (2010), Database of failed privatisation projects in water supply and sanitation, available at http://www.indiawaterportal.org/sites/indiawaterportal.org/files/List_Failed%20Privatisation%20Projects_Water%20Supply%20&%20Sanitation_Manthan%20Adhyayan%20Kendra_2010.pdf (accessed 16 April 2012).
Marin, P. (2009), *Public–Private Partnerships for Urban Water Utilities: A Review of Experiences in Developing Countries*, Washington, DC: Public–Private Infrastructure Advisory Facility, World Bank.
Mehta, L. (2003), 'Problems of publicness and access rights: Perspectives from the water domain', in I. Kaul, P. Conceicao, K. Le Goulven and R.U. Mendoza (eds), *Providing Global Public Goods. Managing Globalization*, New York: Oxford University Press, pp. 556–70.
Mokhtar, M.B. and K.W. Tan (2008), 'Integrated water resources management in Malaysia: an effective institutional framework', Institute for Environment and Development (LESTARI), National University of Malaysia, available at http://www.inweb.gr/twm4/abs/TANpercent20Kokpercent20Weng.pdf (accessed 16 April 2012).
Mycoo, M. (2005), 'Shifting paradigms in water provisioning: a Trinidad case study', *Water Resources Development*, **21**, 509–23.
Nankani, H. (1997), 'Testing the waters – a phased approach to a water concession in Trinidad and Tobago', Note No. 103, Public Policy for the Private Sector, Washington, DC: World Bank.
Nickson, A. and C. Vargas (2002), 'The limitations of water regulation: the failure of the Cochabamba concession in Bolivia', *Bulletin of Latin American Research*, **21**(1), 99–120.
Ortiz, G. (2011), *Inter Press Service*, 21 March, available at http://ipsnews.net/news.asp?idnews=54935 (accessed 15 November 2011).
Pengurusan Aset Air Berhad (PAAB) (2008), 'Changes under the national water services industry restructuring initiatives', available at http://www.paab.my/reform-in-water-services-industry/changes-under-the-national-water-services-industry-restructuring-initiatives (accessed 16 April 2012).
Pigeon, M., D.A. MacDonald, O. Hoedman and S.O. Kishimoto (2012), *Re-municipalization: Putting Water Back into Public Hands*, Amsterdam: Transnational Institute.
Schellenberger, V.R. (2001), 'Tenders in the volume of 150 million marks – Eurawasser aiming for strong revenue growth, Berliner Zeitung, available at http://www.berliner-zeitung.de/archiv/ausschreibungen-im-volumen-von-150-millionen-mark-eurawasser-peilt-kraeftigen-umsatzsprung-an,10810590,9919906.html (accessed 16 April 2012).
Schiavo-Campo, S. and H. McFerson (2008), *Public Management in Global Perspective*, London: M.E. Sharpe.
Segal, G. (2003), 'What can we learn from Atlanta's water privatization?', Reason

Foundation, available at http://reason.org/news/show/122661.html (accessed 25 October 2011).

Transnational Institute (TNI) (2009), 'Water: a common good, public management and alternatives', available at http://www.tni.org/multimedia/water-common-good-public-management-and-alternatives (accessed 13 October 2011).

United Nations (2010), 'General Assembly Department of Information', 28 July, available at http://www.un.org/News/Press/docs/2010/ga10967.doc.htm (accessed 14 November 2011).

United Nations Development Programme (UNDP) (2001), 'A UNDP policy document – glossary of key terms', available at http://mirror.undp.org/magnet/policy/glossary.htm (accessed 1 November 2011).

United Nations Development Programme (UNDP) (2011), 'Fighting corruption in the water sector – methods, tools and good practices', available at http://www.undp.org.tt/News/UNODC/Anticorruption%20Methods%20and%20Tools%20in%20Water%20Lo%20Res.pdf (accessed 14 April 2012).

United Nations Population Fund (UNFPA) (2011), 'As world passes 7 billion milestone, UN urges action to meet key challenges', available at http://www.un.org/apps/news/story.asp?NewsID=40257 (accessed 10 December 2011).

USAID (2005), 'Case studies of bankable water and sewerage utilities. Volume II: compendium of case studies', available at http://pdf.usaid.gov/pdf_docs/PNADE148.pdf (accessed 11 November 2011).

Word Press (2011), 'For a better Malaysia', available at http://fabm.wordpress.com/2011/05/25/bn-govt-bailout-for-selangor-privatised-water-companies/ (accessed 6 November 2011).

World Bank (2002), 'Private Sector Development Strategy: Directions for the World Bank Group', Washington, DC: World Bank.

World Bank (2005), 'Turkey – Antalya water supply project', available at http://go.worldbank.org/BR6RJOFCC0 (accessed 6 November 2011).

World Bank (2011), 'Private participation in infrastructure database', available at http://ppi.worldbank.org/explore/Report.aspx (accessed 10 March 2012).

World Savvy Monitor (2009), 'Privatization of water', available at http://worldsavvy.org/monitor/index.php?option=com_content&view=article&id=715&Itemid=1202 (accessed 1 September 2011).

15. Good governance of water: the final analysis

Sonia Ferdous Hoque and Asanga Gunawansa

15.1 INTRODUCTION

The importance of appropriate governance structures for efficient provision of water supply and sanitation services to the rapidly growing urban population in developed and developing countries has been highlighted throughout this book. As explained in Chapter 1, water governance architectures are not only tied to either public or private sector, but have evolved to include a range of arrangements between these two extremes. Chapter 1 also provides a detailed overview of the evolution of urban water governance architectures over the last two centuries in relation to the global economic and political situation and dominant international ideologies. Although much focus has been given to the privatization wave of the 1980s and 1990s, as substantiated by the mass of literature on this topic and the active involvement of international and multilateral organizations, in Chapter 2 Perard has pointed out that private participation in the urban water sector is neither a new phenomenon nor a new mode for service delivery. Historical analysis reveals that the involvement of the private sector in water supply has fluctuated over long periods of time and has shown spatial variation.

As discussed in Chapter 13, the emergence of long-term concession contracts and other forms of public–private partnerships (PPPs) in the water sector during the 1980s and 1990s has largely been shaped by the influence of bilateral and multilateral organizations which, taking advantage of their leverage as creditors, played an important role in guiding the development policies of developing countries. Most of these arrangements were made with the expectation that the private sector could rescue the ailing public utilities by injecting private capital and incorporating management and technical efficiencies. However, not all projects in the water sector that were developed with private sector engagement have had the expected success. For example, countries have garnered important experiences from some highly publicized failures of PPPs, such as those in Buenos Aires and

Tucuman (Argentina), Cochabamba and La Paz–El Alto (Bolivia) and from the cases of re-municipalization in Grenoble and Paris (France), Dar es Salaam (Tanzania) and Malaysia. On the other hand, relatively successful PPPs in the water sector in countries such as the Philippines (Manila) and Senegal and the privatization of the water industry in England and Wales offer lessons on successful partnering with, or engagement of, the private sector in water governance.

Rather than engaging the private sector as a quick-fix solution to existing water challenges, some countries, as exemplified by China, India and Chile, have been undergoing years of sectoral reforms to create a conducive environment for private sector participation. Meanwhile, in some countries, public water utilities have also shown considerable success in achieving goals of financial sustainability, technical efficiency and universal access by injecting changes into their administrative structures. As exemplified by Singapore's Public Utilities Board (PUB) and the Phnom Penh Water Supply Authority (PPWSA) analyzed in Chapter 3, good governance of water does not necessarily require private sector engagement.

Throughout this book, the authors have discussed the successes achieved and the challenges faced by private and public water utilities in different countries, including the United Kingdom (UK), China, Spain, France, India, Saudi Arabia, Singapore, the Philippines, Malaysia, Cambodia, Australia and others. This chapter aims to provide a snapshot of these different modes of service delivery in the water sector, based on the issues discussed in the preceding chapters, along with findings from other studies. Section 15.2 examines the ideological notions usually set forward by proponents of the public or private sector, which have been briefly outlined in Chapter 4, in relation to the outcomes perceived so far. Further, in this final chapter, an attempt has been made to outline the crucial parameters required for successful management both by the public and the private sector, using examples of good practice from selected utilities.

15.2 PUBLIC VERSUS PRIVATE PROVISION OF WATER SERVICES: RATIONALE AND OUTCOMES

There have been long-continued debates about whether water and sanitation services should be provided by the public sector or the private sector and the arguments most often mobilized by proponents of the public sector are based on the innate characteristics of water being a public good which leads to a natural monopoly and is linked to principles of basic human rights (Budds and McGranahan, 2003; Gunawansa and Hoque,

2012). On the other hand, supporters of privatization argue that water is an economic good subject to full cost pricing and that the private sector can mobilize funds and result in greater efficiency due to increased competition. As argued in Chapter 8, often the choice between public and private delivery is made ideologically rather than rationally. The following paragraphs examine the extent to which these rationales are justified based on the outcomes observed in the sector so far.

15.2.1 Nature of Good

Gleick et al. (2002) argue that water can be both a social and an economic good, and the challenge is to achieve the right balance between the two theoretical perspectives. Access to clean water is fundamental to human survival and critical to reducing prevalence of water-borne diseases and as stated in General Comment No. 15 of the United Nations Committee on Economic, Cultural and Social Rights (OHCHR, 2002), 'The human right to water entitles everyone to sufficient, safe, acceptable, physically accessible and affordable water for personal and domestic uses'. However, the latter statement is often misinterpreted and results in the assumption that it is the duty of the public sector to provide water at subsidized rates or free of cost, especially to the poor. For example, as mentioned in Chapter 7, due to the influence of Islamic principles which prioritize social equity and the 'right to thirst', governments in the Middle East and North Africa (MENA) region are often reluctant to raise water tariffs to cost recovery levels as this would impact religious and political sentiments. Similarly, as mentioned in Chapter 6, the 1978 Spanish constitution provided the legal basis for public management of water resources, based on the notion that water is a free natural resource. However, water management principles have evolved in these countries over the years to incorporate cost-effective mechanisms.

On the other hand, the Dublin Statement (Principle 4) (1992), which states that 'water has an economic value in all its competing uses and should be recognized as an economic good', often results in the argument that water should be treated as a commodity and hence, the private sector which can opt for full cost recovery is the best option to manage this scarce resource. Unfortunately, while a lot of focus has been given in literature to this first line of the Dublin Principles, the consecutive sentences are often overlooked, stating that:

> within this principle, it is vital to recognize first the basic right of all human beings to have access to clean water and sanitation at an affordable price. Past failure to recognize the economic value of water has led to wasteful and

> environmentally damaging uses of the resource. Managing water as an economic good is an important way of achieving efficient and equitable use, and of encouraging conservation and protection of water resources . . .

Hence, the economic approach does not inevitably lead to management of water as a commodity in all aspects, but merely states that efficient management of water is essential to prevent wastage of this important resource so that the basic human right can be fulfilled. As exemplified by Gleick et al. (2002), water pricing that subsidizes the fixed-charge portion (for the physical water connection) of water rates but imposes a volumetric charge (for actual water used) that reflects the highest-value use of water, treats each unit of water consumed as a commodity but treats the piped connection itself as a social good.

15.2.2 Competition

It is true that piped water and sewerage networks resemble natural monopolies as these involve huge capital investment, and multiple networks competing for the same consumers are totally unjustified. Unbundling of functions often proves to be difficult and hence, competition in the urban water supply sector is restricted to 'competition for the market' rather than 'competition within the market' (Budds and McGranahan, 2003). However, in case of large contracts such as concessions, 'competition for the market' is also restricted as only a handful of companies will be capable enough to handle such big projects. Moreover, as the preparation of a bid incurs substantial costs, private operators often opt for 'opportunistic bidding' quoting unrealistic tariff reduction. This strategy is useful in winning the bid, provided that early renegotiations are possible to increase the tariffs (Bakker, 2010). Many governments often agree to new demands by the private firm given the high political and financial costs of cancelling a contract and initiating a new bidding process (Davis, 2005). One of the most cited examples related to this issue is that of Manila Water and Maynilad Water, which won the concessions for the East and West Zones of Manila, respectively, by quoting only 26 percent and 57 percent of existing tariffs, respectively, in their bids. Ultimately both companies set out to renegotiate with the regulator, and despite initial resistance, tariff increases were approved and the costs were passed on to the consumers (Esguerra, 2003). In Chapter 5, Chong mentioned that under the 1993 Anti-Corruption Law (Loi Sapin) in France, local authorities are required to advertise their willingness to contract out the provision of local public services and to organize an auction for interested candidates. Thus, by increasing the potential number of bidders, the law can lead to more competition during the auction process, which will make it more efficient.

Another type of competition that can be implemented in the water sector is yardstick competition or comparative performance, innovated by the British system of economic regulation. Under yardstick competition, the regulator collects data from the utilities under its jurisdiction and uses them to set benchmarks for all utilities based on the performance of the most efficient utilities (Berg, 1997; Cowan, 1997). Comparative performance has been used to evaluate the performance of companies operating in different parts of cities (Haarmeyer and Mody, 1998). As mentioned in Chapter 3, the main intentions of the International Finance Corporation (IFC) and the Metropolitan Waterworks and Sewerage System (MWSS) in dividing Manila city into two concession areas was to promote competition. The same rationale was put forward during the privatization of the water industry in the UK, discussed in Chapter 4. However, such regulation works better the greater the correlation between the exogenous shocks (or cost drivers) that affect the costs of the firms (Cowan, 1997). For example, the UK Office of Water Services (OFWAT)'s calculation of individual price caps for the water companies suggests that each company operates under different conditions, making OFWAT less successful in developing robust measures of relative performance (Berg, 1997). Moreover, when very different initial price bids result in large price differences in different jurisdictions, as in Manila, the regulator may be under political pressure to make adjustments that are not economically justified or risk reducing the acceptability of the process (Haarmeyer and Mody, 1998).

15.2.3 Efficiency

Improvement in operational efficiency is one of the major rationales for supporting private sector participation in water supply. The rationale is based on the economic notion that firms operating under competitive conditions will try to maximize profits by minimizing operation costs. However, this is not often the case as competition in the water sector is very limited and the government regulates private firms. Based on a review of 100 studies that compared the efficiency of state-owned and private companies, Yeaple and Moskowitz (1995) concluded that 'in competitive environments, public enterprises rarely outperform private enterprises and generally perform worse. In highly regulated industries, the public sector's performance is relatively better, occasionally outperforming the private sector.' Empirical studies involving the measurement and comparison of performance of utilities are plagued by lack of data, omitted variables, endogeneity and selection bias. Based on a review of 20 studies,

Gunatilake and Jose (2008) concluded that studies using econometric estimations of cost functions, stochastic frontier production function analysis and data envelopment analysis do not provide conclusive evidence of higher production or technical efficiency among privately run water utilities. Estache et al. (2005) mentioned that 'in general, there is no statistically significant difference between the efficiency performance of public and private operators in this (water) sector', while Hall and Lobina (2005) concluded that 'there is no systematic intrinsic advantage to private sector operation in terms of efficiency'. Based on a review of case studies, Marin (2009) observed that private operations have been effective in efficiency improvement in three dimensions, namely, water loss reduction, bill collection and labour productivity. Similarly, Chapter 4 mentions that financial and productivity analysis of the UK water industry pre- and post-privatization by different authors has shown that privatization has resulted in higher profits but few efficiency gains.

15.2.4 Access

Right to water should ideally include a variety of dimensions such as access to water, affordability, ownership, delivery and participation in decision-making processes (Gunawansa and Hoque, 2012). As far as access to water is concerned, one of the important expectations is that the private sector, with greater access to financial resources, will expand access to water to the poor. However, whether there have been significant improvements in coverage in the contract areas and whether these improvements can solely be attributed to the private companies are debatable issues. Bakker (2010) and Gunatilake and Jose (2008) have mentioned that private operators, being driven by the profit motive, indulge in 'cherry picking', that is, they serve the wealthier urban customers and ignore the poor and rural consumers. For example, in Cartagena (Colombia) and La Paz–El Alto (Bolivia), low-income settlements on the periphery were excluded as they were defined as being outside the cities' limits, and sparsely populated rural areas were excluded from the renegotiated contract in Côte d'Ivoire. In Córdoba and Buenos Aires, operators excluded poor households within the contract area on the grounds that they did not have legal land tenure (Budds and McGranahan, 2003).

Some projects have used innovative methods to make service expansion to poor communities profitable (Colin and Lockwood, 2002). For example, in the Buenos Aires (Argentina) concession, although the original contract allowed the company to finance new connections by charging US$600, the contract was renegotiated and rather than charging the poor customers only, the network extensions were financed through a solidarity tax on all consumers – the Universal Service and Environmental Improvement fee (Hall and Lobina, 2006). In La Paz–El Alto (Bolivia), the private company used techniques such as involvement of community groups, the use of micro-credit schemes and voluntary labor by the inhabitants to make connections, and the use of the shallow 'condominial' sewerage system (Hall and Lobina, 2006).

15.2.5 Affordability

While recognizing the right to water allows some to argue for free provision of water supply by the public sector, the cost involved in development and provision of water facilities supports the argument that pricing of water is inevitable (Gunawansa and Hoque, 2012). Thus, treating water as a commodity, especially in circumstances in which private sector entities have invested and participated in developing water infrastructure, is not unusual.

While the private sector is expected to make investments in upgrading and expanding infrastructure, it should also be understood that private companies will need to recoup these investments through revenues earned from monthly tariffs paid by consumers. In most cases, due to political reasons, the tariffs charged by the public sector are very low, as a result of which public utilities cannot even recover their operating costs. Hence, when contracts are awarded to private companies, tariff increases are likely to follow, as has been well substantiated in the case of many PPP contracts (Gunatilake and Jose, 2008; Davis, 2005). In Chapter 5, an analysis of PPPs in France shows that when price is considered as a measure of performance, both simple statistical analysis and econometric analysis point to negative performance of PPPs. The author calculated that the average price premium associated with a PPP for water services is about €15 per 120 m^3, accounting for various factors that may influence water prices.

Sometimes, in order to create a win–win situation for both the private company and the poor consumers, governments advise or oblige the companies to implement an increasing block tariff (IBT) structure, based on the assumption that poor households consuming less water will pay a lower monthly charge while the company can earn revenues from higher charges paid by rich households. As illustrated in Chapter 7, this policy has been adopted by a majority of the countries in the MENA region. But often the situation is not as simple; as noted by Boland and Whittington (2000), poor households often do not have individual connections and hence share a single connection between several households, which ends up in a higher consumption block and hence attracts a higher tariff. Moreover, such IBT structures encourage private companies to focus on wealthier communities which pay higher charges.

15.2.6 Finance

As discussed in Chapter 1, during the 1980s and 1990s, private sector participation was touted as the most effective means of meeting the water supply and sanitation needs of developing countries, as the private sector can mobilize funds for investment and thus free the limited public funds for other important purposes. According to Davis (2005), although most of the PPPs failed to achieve the investment levels committed in the contracts, private sector participation did accelerate capital investments in the water and sanitation sector. However, it has been observed that foreign investors have mostly been interested in large markets with low risks and, hence, private finance often failed to reach the low-income countries where it was required more.

According to PPIAF (2011), during the period 1991–2011, of the total private sector investments of US$64 billion, the highest proportion went to East Asia and the Pacific region (47 percent) and to Latin America (41 percent) whereas the poor and densely populated countries in South Asia and sub-Saharan Africa together received only about 1 percent. This is because PPPs in the water sector in South Asia started only after the mid-2000s, while those in Africa have been mainly management contracts that required no investment from the private companies. However, this gap has largely been filled by official development assistance (ODA) for the water sector and of the total US$19 billion of ODA disbursed during the period 2003–08, 29 percent went to sub-Saharan Africa and 18 percent to South and Central Asia (OECD–DAC, 2010).

A recent development in the realm of financing of water infrastructure projects is the injection of funds by corporatized public entities and the creation of special public funds to meet the development cost of water infrastructure. For example, as mentioned in Chapter 9, in China, domestic government-affiliated water companies are the key investors in water infrastructure development in the country. In India, as mentioned in Chapter 8, two funds have been established, the Indian Infrastructure Project Development Fund (IIPDF) to provide funds to meet project development expenses and the Viability Gap Fund (VGF) to provide grants to projects in order to encourage private sector investment.

15.3 SUCCESSFUL PUBLIC WATER UTILITIES: LESSONS LEARNT

Based on experience over the last few decades, it is now evident that successful water governance is not dependent on whether the assets are owned and managed by the private or the public sector. Both forms of governance can be successful and the ingredients for success vary from one case to another. This section discusses some of the important factors that may contribute to the success of a public water utility, with examples of relevant cases.

15.3.1 Political Commitment and Leadership

Water and sanitation should be explicitly integrated with economic development and public health goals in local and national policy documents in order to secure long-term commitment from government leaders. Having a secure political commitment ensures that water and sanitation development will have consistent budget allocations and specific, enforceable regulations for drinking water and wastewater quality. The link between water security and economic growth, industrialization and urbanization can be especially useful in elevating water projects on political agendas (Kremak and Lol, 2007). An ideal example of the importance of political commitment in water sector reforms is Uganda, where water was identified as a priority in the national poverty reduction strategy in the mid-1990s. The 1999 water policy developed an investment plan aiming at 100 percent water coverage by 2015. Budget allocations to water increased from 0.5 percent of public expenditure in 1997 to 2.8 percent in 2002 and coverage levels increased from 39 percent in 1996 to 51 percent in 2003, which translated to an additional 5.3 million people gaining access to water (UNDP, 2006). As mentioned in Chapter 1, the Dublin Statement as well as the UN Millennium Development Goals make a strong case for political commitment for water sector reforms.

In Singapore, strong political will has enabled the island nation to pull together resources from different national agencies in order to increase and diversify its water resources, so that the country becomes self-sufficient in water by the time the water import contract with Malaysia expires in 2061. According to Tan et al. (2009), the setting up of the Water Planning Unit under the Prime Minister's Office in 1971 to study the scope and feasibility of new conventional sources and unconventional sources (water reuse and desalination) led to the Water Master Plan in 1972. Similarly, as highlighted in Chapter 3, under the leadership of Ek Sonn Chan, the Phnom Penh Water Supply Authority (PPWSA) in Cambodia has been able to turn itself from a dysfunctional utility to one of the most highly acclaimed and well-run public utilities in the world.

15.3.2 Autonomy

Numerous investigations of public sector enterprises point to undue government interference as the root cause of performance problems. An early study of public enterprises by Hanson (1959) observes that 'the problem is to give them adequate freedom rather than to bring them under closer supervision'. Hoffer (1995)'s in-depth analysis of four water supply companies led him to conclude that 'autonomy is the key determining variable for effective urban water supply'. The various governance mechanisms and case studies considered in this book illustrate that absolute autonomy for the public sector, when it lacks the financial, technical and managerial capacity for efficiency and effectiveness in water governance, could adversely affect the water sector. On the other hand, it has also been established that when autonomy in water governance is supported with necessary institutional reforms, financial, technical and managerial capacity, public entities can govern the water sector efficiently and effectively. The case of Singapore offers an ideal example.

A number of countries have embarked on reforms aiming to improve the performance of public sector organizations by enhancing their autonomy. Reforms have tended to focus on changing the legal status of organizations embedded in the public sector, based on the assumption that it would create an arm's-length relationship between organizations delivering public services and oversight departments. It is expected that granting statutory autonomy would enhance accountability and transparency as well as reduce government interference in the day-to-day business of service delivery.

Although there is widespread agreement among both researchers and practitioners that autonomy affects organizational effectiveness, many questions remain to be answered regarding the nature of the autonomy–performance relationship. Braadbaart et al. (2007) developed a managerial autonomy construct and applied it to a sample of 22 water utilities in developing countries. By associating the resulting autonomy scores with a number of key performance indicators, the authors found that the degree of autonomy was statistically significant for six out of eight key performance indicators: service coverage, revenue water, water tariff, bill collection ratio, labor productivity and quality of financial reporting, which confirmed their hypothesis that autonomy does matter for the performance of water utilities. As mentioned in Chapter 1, good leadership by the top management, a strong institutional culture, a degree of professionalism of the staff and the level of support from the ministries and international donors and lending agencies are the prerequisites of an autonomized model.

Based on their analysis of ten public utilities, Baeitti et al. (2006) mentioned that most of the utilities required approvals for major financing decisions, some required outside approvals for compensation and salary scales, and all required prior authorization to set or revise tariff levels. Although the freedom of internal decision-making did not severely affect the performance of the utilities, the authors advocated that certain decisions such as procurement of small amounts of goods and services, recruitment of qualified staff or disconnecting households due to non-payment of bills, should lie with the utility manager in order to avoid bureaucratic delays and efficiency losses. For example, in both the National Water and Sewerage Corporation (NWSC) and PUB, the process of tariff adjustment is initially proposed by the utility, followed by review by the relevant ministries and approval by the cabinet. PUB determined its own salary scales using government salaries as a guide and staff salaries are comparable to the private sector, which allows the utility to attract competent staff. However, NWSC struggles to attract qualified staff because its salary structure is based on the projected income and the obligations that the NWSC has to meet, and hence, remuneration levels are not as competitive as those of some other corporate bodies in the country.

15.3.3 Performance Incentives

It is basic human nature to respond to performance incentives, which can therefore be utilized to enhance labor productivity and efficiency of the organization. For example, in 2004, Singapore's PUB moved from tenure-based fixed incremental pay increases to a system of performance-based merit increases, which was introduced first for executives and then negotiated with the PUB Employees' Union. Similarly, in Uganda, key staff members of the regional utilities are paid a management fee including a fixed fee (100 percent of uncontrollable and 75 percent of controllable monthly costs), performance fee (25 percent of controllable costs provided all minimum performance standards are met), and incentive fee (extra payments for exceeding minimum performance standards) (USAID–ARD, 2006).

Misalignment of such incentives can also lead to inefficiencies. For instance, during the 1980s and early 1990s, in the PPWSA, the employees received free supply of water. However, after Ek Sonn Chan's appointment as the General Director of the PPWSA, this practice was stopped and the staff were obliged to install meters and pay monthly bills on time like other citizens (Biswas and Tortajada, 2010). As discussed in Chapter 3, in order to motivate staff, the PPSWA has significantly increased staff, incorporated a profit-sharing system that monitors corruption practices by all the employees, and created a solidarity fund financed by employees that provides interest-free loans to poor employees. The PPWSA evaluates its employees four times a year and may provide financial incentives as well as take disciplinary actions such as delays in salary increases or even salary deductions.

As mentioned in Chapter 1, to increase its operational efficiency, Office national de l'Eau et de l'Assainissement (ONEA), the public water utility of Burkina Faso, designed an innovative approach based on the concept of a performance-based service contract. Professional operators were contracted to manage ONEA's commercial and finance departments while at the same time completing a series of pre-identified tasks (including setting up new accounting and customer management systems). The operator would be paid a fixed monthly fee for management services along with a bonus or penalty based on its achievement of contractual targets, plus a fixed price for each specified output. These professional operators coached ONEA staff in updating the customer register and identifying illegal customers, put in place new working practices to improve meter reading, established a meter repair workshop, and advised on improving customer service. Under their guidance, ONEA's performance improved markedly in terms of bill collection, non-revenue water, total number of household connections, staff ratio and water production capacity (Marin et al., 2010).

15.3.4 Innovative Mechanisms

Instead of adhering to ideological debates and formulae, utilities sometimes implement innovative strategies that are pragmatic and also suitable for local context. For example, in Haiphong, Vietnam, the crisis of public outcry against poor service delivery inspired the Haiphong Provincial Water Supply Company (HPWSC) to implement its innovative 'Phuong or Ward Model', which focused on overhauling water supply services of a whole ward at a time. This decentralized service delivery allowed the utility to achieve impressive results: during 1993–99, UfW declined from more than 70 percent to about 32 percent, metered connections increased from 0 to 81 000, employees per 1000 connections declined from 30 to 7.4 and, most importantly, HPWSC went from a loss of 3.1 billion dong (D) in 1993 to a profit of D6.9 billion in 1999 (Baietti et al., 2006).

As discussed in Chapter 3 and Chapter 10, given land shortage and scarcity of conventional water resources in Singapore, PUB has managed to overcome its dependence on imported water from Malaysia and become self-sufficient in water by implementing innovative projects such as NEWater, a project that recycles greywater. Although consumers had initial reservations concerning the use of recycled water, the government organized a massive public awareness campaign about the potability of NEWater (Tian, 2010). Moreover, Singapore is one of the few countries which collect rainwater through a comprehensive network of drains, canals, rivers and stormwater collection ponds before it is channeled to its 17 reservoirs for storage (PUB, 2011). Moreover, as mentioned in Chapter 10, in the design–build–own–operate (DBOO) contracts for the desalination plants in Singapore, the private companies were given flexibility to propose innovative and cost-effective means to meet contractual obligations, such as coupling desalination with on-site power generation for the second desalination project in Tuas and the installation of reverse osmosis membranes on top of the micro-filtration feed tank in the Ulu Pandan NEWater factory.

15.3.5 Consumer Orientation

Consumers can often play an important role in well-intentioned public utilities by voicing their needs and their inclusion in decision-making can instill trust and help the utility to build a healthy relation with them. This can often counter the general perception of politicians that consumers are not willing to pay and hence tariffs must be kept low despite the negative impact on service quality (Baietti et al., 2006). Consumers need to be well informed about the billing and collection system so that they are more aware about the cost of different services. Utilities should also establish a timely and effective complaint process and periodically survey customers on their satisfaction levels and interests in improved services.

In Uganda, regular strategic alliance meetings are held by NWSC with different customer segments and annual customer surveys are held to gather the customers' views regarding areas where the NWSC still needs improvement. In HPWSC, customer surveys are normally used to obtain opinions from the customers for new projects, such as expansion of the network to new areas. In Singapore, PUB actively seeks the opinions of its customers through customer satisfaction surveys held every three years, regular dialogue sessions and feedback forms. In the UK, too, periodic consumer surveys are used for restructuring water tariffs.

In order to get useful feedback on the impacts of reforms brought about by the Water Act of 2002 in Kenya, the Ministry of Water and Irrigation introduced the Citizen Report Card (CRC) initiative, a social accountability tool that was pioneered in Bangalore, India. Modeled on the private sector practice of conducting client satisfaction surveys, CRCs solicit user perceptions on quality, efficiency and adequacy of the various public services that are funded by taxpayers. Qualitative user opinions are aggregated to create a 'score card' that rates the performance of service providers. The findings present a quantitative measure of overall satisfaction and perceived levels of corruption among an array of other indicators. They are a useful medium through which citizens can credibly and collectively signal to agencies about their performance and pressure for change (WSP, 2011).

The 'participatory budgeting' system in Porto Alegre has been discussed in Chapter 14. Similarly, as mentioned in Chapter 6, the introduction of a participatory model in EMACSA, the autonomous public utility in Cordoba (Spain), allowed civil society representatives, trade unions and opposition political parties to participate in the organization's decision-making processes. This participation has been a key element in guaranteeing financially, socially and environmentally efficient management.

15.3.6 Affordability

In order to assist poor people to gain access to domestic connections, the PPWSA in Cambodia established a revolving fund, which financed 6708 water connections (World Bank, 2006). Poor households are entitled to receive subsidies of 30, 50, 70 or 100 percent of the connection fee, depending upon their financial condition, which is jointly evaluated by a committee of the utility and local communities. This approach to reach the poor was developed after several years of experimentation with unsuccessful approaches. For example, the reselling of water to the poor through 'community representatives', an approach that works well in Manila, had failed in Phnom Penh because the community representatives resold water at much higher prices than agreed (Biswas and Tortajada, 2010).

In Porto Alegre, Brazil, the Departamento Municipal do Agua e Esgoto (DMAE) implements cross-subsidization in tariffs for water. Lower-income citizens are entitled to $10m^3$ of water per month but pay for only $4m^3$, while tariffs rise steeply for those who use between 20 and $1000m^3$ per month. This 'social tariff' translates into a system whereby the wealthy help to subsidize the utility's reinvestment of tariff monies into improvements in the system itself: such improvements are 70 percent financed through tariffs on usage (Maltz, 2005).

In Singapore, low-income households, which have difficulty in paying their water bills, receive assistance in the form of Utilities Save (U-Save) rebates. This involves placing a fixed amount of cash in the household's utility account, which can be drawn at any time to pay any utility bill, including water; moreover, if the cash is not used up for one month, it stays in the account and can be used for subsequent months (Tan et al., 2009). Thus, rather than having a low per unit charge for the first block under an IBT structure, which is enjoyed by all consumers irrespective of their income levels, this targeted subsidy achieves three purposes: assisting the poor, spreading the message that conservation is important even at low consumption levels, and also earning revenues from all consumers who can afford to pay.

As stated in Chapter 7, although the majority of countries in the MENA region no longer rely on Sharia law, Islamic principles have a significant influence on water resources management, particularly the emphasis on affordability. The authors argue that the widespread subsidies given by the governments of these oil-rich nations, as a means of maintaining political stability, make no sense from an economic or environmental point of view. As in many parts of the world, the poor and marginalized communities in MENA countries are often not connected to municipal networks and, due to their reliance on private providers, often pay much higher rates than publicly supplied consumers.

15.3.7 Financial Sustainability

It is essential for public water utilities to achieve self-sufficiency in financing operational costs and, if possible, earn profits to allow necessary investments. However, the urge of politicians to hold back tariff increases based on social welfare justifications often makes cost recovery difficult in developing countries. Baietti et al. (2006) argue that subsidies for operational expenses increase dependency of the utility and, even if such support is necessary, proper phase-out plans should be devised. Moreover, rather than providing a continuous flow of funds, subsidies can support discrete one-time activities such as offsetting connection fees for the poor or constructing pipelines to expand coverage into new areas.

In order to earn higher revenues and cover operational costs, the PPWSA proposed to increase tariffs in three steps over a period of seven years, along with improvements in service delivery. It believed that, to avoid social unrest, consumers first needed to experience a better quality of reliable service from the utility before any tariff increase. The PPWSA also conducted a socio-economic survey which collected information on how much consumers were paying for water from other vendors, and what were likely to be their reactions if these vendors were replaced by supply from the PPWSA. This survey showed the willingness and capability of the consumers to pay a higher tariff for a vastly improved service (Biswas and Tortajada, 2010). Apart from convincing the general public, the PPWSA also urged the army and senior government officials, who have traditionally received water free of cost, to install meters. While tariffs were increased to cover costs on the one hand, on the other hand the decline in unaccounted-for water (UfW) and increase in bill collection also increased revenues. As of 2008, the PPWSA's operating revenue was US$21.9 million, out of which it made a profit of US$7.3 million (Das et al., 2010).

Even when public utilities can achieve operational efficiency by incorporating changes in their management and institutional structure and become financially self-sufficient in terms of operation and maintenance (O&M) costs, large-scale investments are often required to build infrastructure such as dams, reservoirs and pipe networks. Funding from external sources becomes necessary at this point and donors are sometimes willing to pay only when the utility demonstrates good performance. For example, external cooperation played a major role in bringing about the turnaround of the Phnom Penh water utility. During the first years the utility relied on grants, while after its financial situation improved and it became an administratively and financially separate entity, it also took loans. Japan (US$85 million in grants), France (US$21 million in grants and US$14 million in loans), the World Bank (US$29 million in loans), the Asian Development Bank (US$13 million in loans) and the United Nations Environment Programme (UNEP) provided financial and technical assistance to the PPWSA. The external financial assistance totalled approximately US$165 million between 1993 and 2009 (Das et al., 2010).

In Burkina Faso, the major expansion of production capacity in Ouagadougou and associated investment in network expansion required donor funds, and since ONEA was able to demonstrate positive results within its first decade of corporatization, donors were ready to fund its innovative performance-based service contract, which was supported by an investment program of about €200 million financed by a pool of donors (including the European Investment Bank, Agence Française de Développement, KfW, and the African Development Bank) (Marin et al., 2010).

15.3.8 Accountability

As mentioned in Chapter 1, in many developing nations the governance of the water sector as a whole is in a state of confusion and dysfunction with little responsiveness or accountability to citizens. For example, in China, discussed in Chapter 9, the fragmentation and frequent overlap of responsibilities and authority, between central and local level, as well as at the local level, has resulted in disorganized management, inefficiency and unclear administrative accountability.

Accountability to multiple stakeholders, such as the owners, independent regulators, external auditors and even to consumers, is important to ensure that the utility provides a good quality of service and maintains a healthy financial statement. Utility managers often have to report to their board of directors about their level of progress with regard to well-defined performance indicators. For example, the Haiphong Provincial Water Supply Company (HPWSC) submits annual reports on quantifiable performance targets to its owner, the Haiphong People's Committee (HPPC), while financial reports are submitted monthly and quarterly to the Transportation and Urban Public Works Service (TUPWS), the local Tax Department and the local Finance Department. The utility also has to provide annual reports to international lending agencies concerning revenue, profit, loss and contribution to the city budget. The only independent regulator is the Haiphong Preventive Medical Centre, which is responsible for maintaining the quality of water samples and even if they fail to meet national standards, the regulator has no authority to penalize the utility (Baietti et al., 2006).

The Rawalpindi Water and Sanitation Agency (RWASA) in Pakistan established two water districts within its service area, each equipped with complaint centres, as an attempt to decentralize services and to make it more convenient for consumers to interact with the relevant utility staff. Complaint centres, an integral part of RWASA's Customer Relationship Management system, are served by designated technical and engineering staff. They deal with complaints related to water supply, sewerage and billing; and according to records, about 90 percent of complaints are satisfactorily rectified within 24 hours (WSP, 2011).

15.4 SUCCESSFUL PUBLIC–PRIVATE PARTNERSHIPS: LESSONS LEARNT

There is a plethora of literature that elucidates the crucial parameters for a successful PPP project in the water sector (for example, Marin, 2009; Gunatilake and Jose, 2008; Gentry and Auyuan, 2000; Kanakoudis et al., 2007). Based on the experiences of PPPs in different countries, authors have come up with long lists of success factors such as political backing; public acceptance; pro-poor contract designs; sectoral reforms in terms of laws, policies and pricing; viable regulatory system; guarantee for public funds or loans; increasing competition through integration of domestic players; ensuring profit of the public company; transparency in bidding; accountability in contract enforcement and so on. This section aims to integrate these factors into broader topics and discuss some of the most important determinants of success, with examples of relevant PPP projects in the water sector.

15.4.1 Conducive Policy and Institutional Environment

According to Gunatilake and Jose (2008), private sector participation in the water sector has made greater progress in high- and middle-income countries, due to an investment climate that is conducive to private capital and reduces the risks associated with operation. Indeed, PPPs are more likely to produce positive results when they are part of a wider reform that supports financial viability, ensures social responsibility and enhances accountability for performance, as exemplified in the case of Chile, Cote d'Ivoire and Senegal (Marin, 2009). As discussed in Chapter 10 of this book, recently PPPs have been used successfully in Singapore to develop wastewater treatment and desalination plants. The success of these projects is largely due to the presence of a clear policy framework and strong institutional capability that ensure a transparent and fair tender process. Rather than being a dormant agency in the partnership, PUB shares its experience and technical knowledge that it has gained from the previous projects with its new private partners. Similarly, as discussed in Chapter 1, the privatization of water in Chile during 1998–2005 was preceded by a decade of institutional and policy reforms during which public utilities were strengthened, tariffs were increased and a system of subsidies for poor households was introduced to ensure affordability. Chapter 6 provides an overview of the paradigm shift in the legal framework and water policy in Spain. The 1978 Spanish constitution provided the legal basis of public management of water resources, based on the notion that water is a free natural resource. In 1985, the enactment of the Water Act represented a step forward in configuring the water management system by granting financial and institutional autonomy to the river basin authorities. This was reformed in 1999 to increase the role of the market, by permitting holders of water rights to exchange them and allowing public corporations to build water infrastructure by signing financial agreements.

15.4.2 Incorporation of Social Goals

Engaging the private sector in delivery of water and sanitation services does not necessarily relieve the government of its responsibilities of ensuring that basic public needs are met by all sections of the society. During the initial years of operation, the tariff revenues gained by the private operator may not be sufficient to expand network coverage to previously unserved areas. Moreover, if low tariff rates are implemented to support poor households, the private operator may indulge in 'cherry picking' and be reluctant to expand access to these communities. Thus, in order to establish new connections in poorer areas, without affecting affordability of the poor households, government often needs to inject public funding through some form of subsidy scheme.

Senegal in West Africa is often touted as an example of having a successful reform process in its water and sanitation sector. The reformation of the water sector in Senegal started in 1996, when the government, under instructions of the World Bank, introduced limited private sector participation through a ten-year *affermage* contract, with a possibility of moving to a concession model at the end of the contract. The reform consisted of splitting the existing national utility, the Societe Nationale d'Exploitation des Eaux du Senegal (SONEES), into the following entities: (1) a public asset-holding company, Société Nationale des Eaux du Sénégal (SONES), which was in charge of managing and investing in water service assets in the 56 urban centers previously served by the public utility; and (2) a private operator, Sénégalaise des Eaux (SDE), mostly owned by SAUR International, for operating and maintaining water services in the same urban centers (Fall et al., 2009). Since a majority of the population in Senegal is below the poverty line (54 percent of the total 9.2 million in 2002), improvements to water supply and sanitation for the poor were identified as a priority by the government, bilateral donors and the World Bank. The subsidies targeted at the poor were of three types: (1) subsidized connections through the social connection program, financed by government funds as well as the water sector project of the World Bank; (2) construction of standpipes (managed by private operators recruited by SDE in consultation with the local community) in areas where there are people without private connections, and supply of water to these standpipes at low rates; and (3) subsidized consumption at low levels of consumption, financed through a cross-subsidy between customer categories and delivered through an IBT structure, with a subsidized 'social tariff' for levels of consumption below 20 m^3 in a 60-day period (Brocklehurst and Janssens, 2004).

Consumers are usually opposed to paying higher tariffs unless they witness a satisfactory level of service delivery. However, since the latter takes considerable time to develop, some contracts delink revenues collected from consumer tariffs from the remuneration of the private company in order to achieve the social goal of providing wider access to water whilst maintaining affordability. In such cases, the government often buys bulk water from the private company in exchange for a fixed fee or pays a volumetric fee to the company that covers its O&M costs. This allows the government to adjust tariff rates at a slower pace depending on the local social and economic structure. For example, as discussed in Chapter 10, in order to avoid interaction between the private companies and end-users, the PPP projects awarded by PUB in Singapore involve a 'Water Purchase Agreement', which specifies the amount of water to be purchased by PUB over the lifetime of the PPP as well as the payment mechanisms.

15.4.3 Effective Risk Allocation

PPP is an effective approach to enhance project productivity by introducing management efficiency and creative skills from business practice, and reducing governmental involvement by using private sectors in the provision of public services (Shen et al., 2006). A PPP contract apportions the project-associated risks mainly to the public and private sectors. Details on the different types of risks facing companies in the water sector are discussed in Chapter 12. Best practice suggests an optimal rather than a maximum risk transfer from the public to the private sector. A problem may however occur if risk is inappropriately transferred; the government may pay a premium that is higher than necessary or jeopardize the long-term sustainability of a PPP arrangement.

Performance risks are associated with the commercial provision of water supply and sanitation services, such as failing to meet the agreed service targets or facing declining revenues as a result of decreasing consumption. Moreover, the real value of revenues from operations in a transition or developing country may be eroded by currency devaluation, as experienced by the private companies in Argentina during the 1990s. However, according to Hall and Lobina (2003), international companies can be virtually risk-free if government guarantees are provided. For example, in Cochabamba (Bolivia), Aguas del Tunari, a subsidiary of International Water Limited (IWL), was guaranteed a 15 percent real return for 40 years (Lobina, 2000). There are a number of cases where prices have been forced up within a year or two of a concession being started, where the assumption in the original contract about the level of demand turned out to be too high and the real level of demand was then used as a justification for large price rises.

In order to reduce currency risk, water multinationals seek to index tariffs to the value of the US dollar. For example, in Manila, the devaluation of the Philippine peso (PHP) from US$1 = PHP26 in December 1996 to US$1 = PHP50 by 2000 due to the Asian financial crisis of 1997 led to renegotiation of the water concessions in 2001, which introduced the Foreign Currency Differential Adjustment mechanism. The latter allowed the operators to automatically pass all the costs of currency devaluation on to consumers on a quarterly basis. As a result, prices charged by Maynilad increased from PHP4.96/cm^3 in 1997–98 to PHP15.46/cm^3 in 2002, while prices charged by Manila Water Company rose from PHP2.32/cm^3 to PHP6.75/cm^3 during the same time period (Esguerra, 2003).

Political risks include risks faced by a commercial water operator due to undue governmental interference, such as in the case of breach of contract, expropriation, restrictions on currency transfer, or war and civil disturbance. Sometimes, unfulfilled contractual obligations or tariff hikes by private operators force governments to terminate the contracts, which have often resulted in international arbitration and heavy compensation being paid by the government to the companies (see Table 14.2 in Chapter 14). For instance, as highlighted in Chapter 9, the absence of a 'uniform supervisory legal system' in China, which has led to inconsistent legal practice at the local level, and the frequent 'changes in laws', as exemplified by the revocation of the guaranteed fixed rate of return for foreign investment by the State Council in 2002, have proved to be major concerns for foreign investors.

15.4.4 Contracts with Realistic Targets

As discussed in section 15.4.2, private companies often indulge in 'opportunistic bidding' and often quote unrealistic performance targets, such as large reduction in tariffs and significant increase in coverage. This falsification allows them to win the contracts initially, but results in a series of renegotiations when financial constraints arise during practical operation. Hence, in many cases, the private operators failed to meet their contract obligations, resulting in public protests and early termination of contracts.

Unrealistic targets also arise due to inaccuracy of baseline information against which performance can be monitored (Marin, 2009). For example, in the *affermage* contract in Senegal, the original technical efficiency targets were very ambitious, and required the operator to raise efficiency from an estimated 73 percent in 1995 to 85 percent in 2002, thus reducing UfW from 27 percent to 15 percent. However, it was later discovered that the baseline information had been inaccurate, which deterred SDE from achieving the targets spelled out in the contract. As SDE's remuneration depended on its attainment of targets, this flawed information resulted in financial loss during the first two years. However, on recognizing the inaccuracy, both SDE and SONES renegotiated the contract in 1998 by incorporating two major changes: (1) the base year technical efficiency was reduced from 73 percent to 68.5 percent, and the year in which the efficiency was to reach 85 percent was delayed from 2000 to 2002; and (2) an agreement was reached to reimburse the difference between SDE's remuneration in each year according to the old targets and according to the new targets. The renegotiation of the targets had a dramatic effect on the financial health of SDE. In this case, on the one hand, while SDE was willing to go through a renegotiation process without resorting to arbitration or threatening withdrawal, on the other hand, SONES also recognized that an operating company with poor financial status was not good for reforms. Hence, mutual cooperation between relevant parties solved the issue without charging the operator for not meeting its contractual goals (Brocklehurst and Janssens, 2004).

15.4.5 Effective Regulatory Mechanism and Contract Enforcement

Since product competition in the water sector is not feasible, the regulatory mechanism is viewed as an imperfect substitute for competition aiming at safeguarding consumers' interests while providing private companies with the incentives to invest and operate efficiently (Rees, 1998; Klein, 1996). As mentioned in Chapter 12, there are broadly three types of regulatory regimes with respect to their capacity to interfere with the allocation of property rights as well as decision rights: (1) command-and-control, in which public authorities have the power to interfere directly, through bureaus or politically dependent administrators, with the exercise of these rights; (2) regulation by contract, in which public authorities commit to delegate rights defined *ex ante* to private or semi-public operators and transfer power to monitor the respect of the commitment to external agencies or to the judiciary; and (3) regulation by general laws as in the case of full divestiture that involves full transfer of rights to private operators.

One of the constraints to regulation identified by the literature is the asymmetry of information between the regulator and the regulated operator, that is to say, the fact that regulators 'tend to be less well informed about costs and quality of water systems operations than the water company management' (Klein, 1996). Private companies often insist on maintaining commercial confidentiality and in many cases the contract itself is treated as a restricted secret document and withheld even from elected representatives. In Paris, following the award of concession to Veolia and Suez, the city council realized that it had begun to lose control over the system's knowledge and it was becoming entirely dependent on the companies for information on the network's state. Due to absolute financial opacity, the city had no idea about the actual profits made by the companies, which was unknown to all except the companies' most senior executives (Pigeon et al., 2012).

Lack of transparency and poor institutional development are key factors in determining the failure of regulation. Sometimes, regulators are susceptible to serve the interests of the private companies because of their reliance on a fixed percentage of tariffs collected by private water operators. For example, the Buenos Aires regulator ETOSS was funded by a 2.67 percent charge on tariffs collected by Aguas Argentinas (Azpiazu and Forcinito, 2002) and funding for the Santa Fe regulator ENRESS came from a 4.92 percent charge on tariffs collected by Aguas Provinciales de Santa Fe (Muñoz, 2002). Regulation and strict enforcement of contracts is often prohibited by corruption of public officials, who may accept large sums of bribe to award contracts without competitive tendering or strict obligations or for re-negotiating contracts and allowing tariff increases based on the private company's demands (Hall and Lobina, 2003). The case of Grenoble in France, discussed in Chapter 14, illustrates the importance of disclosure of information, independent assessment of the private companies and the consequences of corruption.

15.5 CONCLUSION: THE WAY FORWARD

An examination of the evolution as well as the decline of various governance architectures for water indicates a basic universal trend. There has been a paradigm shift in urban water governance from supply-driven, infrastructure development solutions towards demand-based approaches with greater emphasis on the role of institutions and economic and social instruments. The purpose of the shift is to ensure that whilst the public's right to water is maintained, supply of water to the public remains commercially viable and generates sufficient revenue for the development and maintenance of sustainable water infrastructure. The case of Australia, discussed in Chapter 11, illustrates this observation. Although public policy has largely rejected a model of extensive private provision of water or funding of water services, the public utilities providing water and wastewater services have been affected profoundly by a range of policy reforms since the 1990s intended to modernize the ways in which public utilities operate through 'corporatization models' and to introduce greater competition. There has been a major shift in the drivers for change in water governance with less emphasis on investment in supply and more attention being given to broader policy issues of integrated water planning with more extensive demand-side considerations.

Governance of water resources is no longer tied to the binary solutions of public or private sector ownership and management, as it has been sufficiently substantiated that both forms can be successful given some crucial factors are in place. In Chapter 3, based on a comparative analysis of both public and private successful water utilities in Phnom Penh, Manila and Singapore, Wong concludes that setting the right policies and strategies, coupled with changes in behavior and financial resources to implement these goals, are the tripartite keys to success, while the final piece needed to complete the puzzle is good management.

An assessment of the ideological notions put in favor of either public or private sector provision of water services reveals that while theoretical concepts can fuel significant debates, in practice, ensuring competition, accessibility, affordability or financial sustainability does not depend on the mode of service delivery. While governments recognize that appropriate tariff structures are necessary to send correct signals to consumers about the cost of treating and supplying water, they are also sympathetic to the basic needs of the poor, which is reflected by targeted innovative subsidies, both in public utilities and PPP contracts. The economic justification of ensuring efficiency through competition is often not realized in the case of water PPPs, but effective regulation and performance-based remuneration can yield efficiency gains. Moreover, it is crucial that water service providers, whether public or private, maintain good financial health. It is imperative that governments realize that even when private companies are awarded PPP contracts, some level of public funding or donor support is necessary to ensure adequate investments and to safeguard public interests.

As exemplified by the cases of the relatively successful public water utilities in Porte Alegre (Brazil), Phnom Penh (Cambodia), Singapore, Uganda, Burkina Faso and Haiphong (Vietnam), which have been discussed in this volume, it is evident that the crucial factors contributing to a well-functioning public utility include: (1) strong political commitment demonstrated through appropriate policy and legal framework; (2) a well-motivated and corruption-free workforce achieved through performance incentives; (3) greater autonomy in decision-making balanced by a strong system of external and internal accountability; (4) efficient customer feedback mechanisms and active public participation; and (5) a well-designed tariff structure to ensure cost recovery, coupled with targeted subsidies to help the poor.

Similarly, the successful PPP cases in Senegal, Chile and Manila, and unsuccessful cases in Buenos Aires, La Paz–El Alto and Cochabamba, highlight some crucial parameters that are required for a PPP to remain viable. These include, among others: (1) supportive policy and institutional environment; (2) proper contract design ensuring effective risk allocation and realistic targets; (3) effective regulatory mechanisms and contract enforcement; and (4) incorporation of social goals.

In Chapter 12 of this book, while referring to the fact that PPPs are here to stay, Ménard emphasizes that the absence of *ex ante* estimation of the comparative costs and benefits of PPP with respect to alternative solutions, or weak hypotheses and lack of rigor in such estimations when they are done, or lack of transparency about hypotheses that were used, can challenge the legitimacy of the choice of PPP as a solution to the problems facing delivery of water in the urban environment.

A key argument made in several chapters of this book is that governments are responsible for providing basic public needs and PPPs should only be used as a tool to do so. Historical analysis of the water sector shows that the choice of public or private management of water utilities depends on the dominant ideologies and the global socio-economic situation during each period. Highly publicized failures of PPPs in Latin America and re-municipalizations in France do not necessarily mean that PPPs have come to an end; however, they have important lessons for the developing countries willing to experiment with the private sector. PPPs in the water sector have gained momentum in India and China, where they are part of sectoral reforms rather than being quick-fix solutions to problems of urban water supply.

In Chapter 8, based on examples of PPPs in the states of Tamil Nadu and Karnataka (India), Mahalingam claims that the choice of project delivery mode is not based on the features of the project alone, but rather on the ability of the institutional environment to support the delivery mode. Moreover, the author proposes that governments should select PPP models that can survive within their ambient environment, and 'upgrade' them while simultaneously undertaking measures to strengthen the institutional environment. Even in countries which are reluctant to involve the private sector, governments are modernizing urban water governance by providing greater autonomy, integrating corporate management cultures, restructuring tariff rates and increasing credibility of public utilities to enable them to access market finance (refer to examples of corporatization in Uganda and Zambia in Chapter 1).

Climate change, which is one of the most pressing global issues of the twenty-first century, represents an additional burden for water utilities, the extent of which will vary based on geographical location and level of development. In the UK, drought is predicted to have a major impact on water security, with decreasing run-off in the South-East and growing populations leading to greater water scarcity (refer to Chapter 4). In Spain, increasing abstraction of groundwater for agricultural purposes, coupled with predicted reduction in precipitation and growing water consumption by the urban population, is very likely to increase the stress on scarce water resources (refer to Chapter 6). Similarly, the arid countries of the MENA region are also likely to face water security challenges due to climate change, unless steps are taken to reduce water use in agriculture, a sector that pays low tariffs and also has low contribution to gross domestic product (GDP) (refer to Chapter 7). These chapters highlight the fact that water governance mechanisms should urgently incorporate implications of climate change to achieve long-term sustainability. In light of the decreasing quantity and quality of water, coupled with rising demand, it is imperative that an integrated management approach is implemented, which promotes the coordinated development and management of resources, including water, land and vegetation, so as to equitably maximize economic and social welfare without compromising the sustainability of the ecosystem.

REFERENCES

Azpiazu, D. and K. Forcinito (2002), 'Privatisation of the water and sanitation systems in the Buenos Aires Metropolitan Area: regulatory discontinuity, corporate non-performance, extraordinary profits and distributive inequality', paper presented at the First PRINWASS Project Workshop, University of Oxford, 22623 April.

Baietti, A., W. Kingdom and M. Ginneken (2006), 'Characteristics of well performing public water utilities', Water Supply and Sanitation Working Note No. 9, World Bank.

Bakker, K. (2010), *Privatizing Water: Governance Failure and the World's Urban Water Crisis*, Singapore: National University of Singapore Press.

Berg, C. (1997), 'Water privatization and regulation in England and Wales', Public Policy for the Private Sector, Note No. 115, World Bank, available at http://rru.worldbank.org/documents/publicpolicyjournal/115vdbrg.pdf (accessed 15 April 2012).

Biswas, A. and C. Tortajada (2010), 'Water supply of Phnom Penh: an example of good governance', *International Journal of Water Resources Development*, **26**(2), 157–72.

Boland, J.J. and D. Whittington (2000), 'The political economy of water tariff design in developing countries: increasing block tariffs versus uniform price with rebate', in Ariel Dinar (ed.), *The Political Economy of Water Pricing Reforms*, New York: Oxford University Press, pp. 215–35.

Braadbaart, O., N.V. Eybergen and J. Hoffer (2007), 'Managerial autonomy: does it matter for the performance of water utilities?', *Public Administration and Development*, **27**, 111–21.

Brocklehurst, C. and J.G. Janssens (2004), 'Innovative contracts, sound relationships: urban water sector reform in Senegal', Water Supply and Sanitation Sector Board Discussion Paper Series, Paper No. 1, World Bank, available at http://www.partnershipsforwater.net/psp/tc/TC_Tools/040B_Contract%20 Senegal.pdf (accessed 25 March 2012).

Budds, J. and G. McGranahan (2003), 'Are the debates on water privatization missing the point? Experiences from Africa, Asia and Latin America', *Environment* and *Urbanization*, **15**(2), 87–114.

Colin, J. and H. Lockwood (2002), 'Making innovations work through partnerships in water and sanitation projects', Business Partners for Development Water and Sanitation Cluster, available at http://www.aguaconsult.co.uk/uploads/pdfs/agua_bpdinnovation.pdf (accessed 25 March 2012).

Cowan, S. (1997), 'Competition in the water industry', *Oxford Review of Economic Policy*, **13**(1), 83–92.
Das, B., E.S. Chan, C. Visoth, G. Pangare and R. Simpson (2010), *Sharing the Reform Process – Learning from the Phnom Penh Water Supply Authority (PPWSA)*, Bangkok, Thailand and Phnom Penh, Cambodia: International Union for Conservation of Nature and Natural Resources (IUCN) and PPWSA.
Davis, J. (2005), 'Private sector participation in the water and sanitation sector', *Annual Review of Environment and Resources*, **30**, 145–83.
Dublin Statement on Water and Sustainable Development (1992), available at http://www.wmo.int/pages/prog/hwrp/documents/english/icwedece.html (accessed 20 March 2012).
Esguerra, J. (2003), 'The corporate muddle of Manila's water concessions', WaterAid and Tearfund.
Estache, A., S. Perelman and L. Trujillo (2005), 'Infrastructure performance and reform in developing and transition economies: evidence from a survey of productivity measures', World Bank Policy Research Working Paper 3514.
Fall, M., P. Marin, A. Locussol and R. Verspyck (2009), 'Water P-Notes', Issue 38, World Bank, available at http://water.worldbank.org/water/publications/public-private-partnerships-reform-urban-water-utilities-western-and-central-africa (accessed 14 October 2011).
Gentry, B.S. and A.T. Auyuan (2000), 'Global trends in urban water supply and waste water financing and management: changing roles for the public and private sectors', Paris: Organisation for Economic Co-operation and Development.
Gleick, P.H., G. Wolff, E.L. Chalecki and R. Reyes (2002), 'The new economy of water: the risks and benefits of globalization and privatization of fresh water', Oakland, CA: Pacific Institute for Studies in Development, Environment, and Security.
Gunatilake, H. and M.F.C. San Jose (2008), 'Privatization revisited: lessons from private sector participation in water supply and sanitation in developing countries', ERD Working Paper No. 115, Manila: Asian Development Bank.
Gunawansa, A. and S.F. Hoque (2012), 'Right to water versus pricing of water: some key challenges faced by developing countries in connection with urban water supply', Lee Kuan Yew School of Public Policy Research Paper No. LKYSPP12-01 IWP, available at http://papers.ssrn.com/sol3/papers.cfm?abstract_id=1987307 (accessed 1 April 2012).
Haarmeyer, D. and A. Mody (1998), 'Competition, contracts, and regulation in water and sanitation: lessons from recent experience', available at http://www.amody.com/pdf/waterreg.pdf (accessed 14 March 2012).
Hall, D. and E. Lobina (2003), 'Problems with private water concessions: a review of experience', Public Service International Research Unit (PSIRU), available at www.psiru.org/reports/2003-06-W-over.doc (accessed 12 December 2011).
Hall, D. and E. Lobina (2005), 'The relative efficiency of public and private sector water', Public Service International Research Unit (PSIRU), University of Greenwich.
Hall, D. and E. Lobina (2006), 'Profitability and the poor: corporate strategies, innovation and sustainability', *Geoforum*, **38**, 772–85.
Hanson, A.H. (1959), *Public Enterprise and Economic Development*, London: Routledge and Kegan Paul.

Hoffer, J. (1995), 'The challenge of effective urban water supply', PhD Thesis, Twente University, Enschede.

Kanakoudis, V., A. Papotis, A. Sanopoulos and V. Gkoutzios (2007), 'Crucial parameters for PPP projects successful planning and implementation', available at www.corp.at/archive/corp2007_KANAKOUDIS.pdf (accessed 15 April 2012).

Klein, M. (1996), 'Economic regulation of water companies', Policy Research Working Paper, Private Sector Development Department, World Bank, available at http://www-wds.worldbank.org/servlet/WDSContentServer/WDSP/IB/1996/09/01/000009265_3961214162451/additional/125525322_20041117175607.pdf (accessed 15 April 2012).

Kremak, N. and L. Lol (2007), 'The history of Philadelphia's water supply and sanitation system', Philadelphia Global Water Initiative, available at http://www.pgwi.org/documents/Philalessonsinsustainability.pdf (accessed 10 November 2011).

Lobina, E. (2000), 'Cochabamba – water war', *FOCUS on the Public Services*, **7**(2), 5–10, http://www.psiru.org/reports/Cochabamba.doc.

Maltz, H. (2005), 'Porto Alegre's water: public and for all', in B. Belén, B. Brennan, O. Hoedeman, S. Kishimoto and P. Terhorst (eds), *Reclaiming Public Water*, Amsterdam: Transnational Institute (TNI) and Corporate Europe Observatory (CEO).

Marin, P. (2009), *Public–Private Partnerships for Urban Water Utilities: A Review of Experiences in Developing Countries,* Washington, DC: Public–Private Infrastructure Advisory Facility, World Bank, available at http://www.ppiaf.org/ppiaf/sites/ppiaf.org/files/FINAL-PPPsforUrbanWaterUtilities-PhMarin.pdf (accessed 10 November 2011).

Marin, P., M. Fall and H. Ouibiga (2010), 'Corporatizing a water utility: a successful case using a performance-based service contract for ONEA in Burkina Faso', Gridlines Note No. 53, Washington, DC: Public Private Infrastructure Advisory Facility, World Bank.

Muñoz, A.D. (2002), 'El agua: derecho humano o mercancia de lucro? – Suez Lyonnaise des eaux', Paper for Union de Usuarios y Consumidores.

Office of the United Nations High Commissioner for Human Rights (OHCHR) (2002), 'Committee on Economic, Social and Cultural Rights – general comments. General comment No. 15', http://daccess-dds-ny.un.org/doc/UNDOC/GEN/G03/402/29/PDF/G0340229.pdf?OpenElement (accessed 30 October 2011)

Organisation for Economic Co-operation and Development – Development Assistance Committee (OECD – DAC) (2010), 'Financing water and sanitation in developing countries: the contribution of external aid', available at http://www.oecd.org/dataoecd/34/2/45902160.pdf (accessed 13 March 2012).

Pigeon, M., D.A. MacDonald, O. Hoedman and S. Kishimoto (2012), *Remunicipalisation: Putting Water Back into Public Hands*, Amsterdam: Transnational Institute.

Public Private Infrastructure Advisory Facility (PPIAF) (2011), 'Private Participation in Infrastructure Database', available at http://ppi.worldbank.org/explore/Report.aspx (accessed 17 March 2012).

Public Utilities Board (PUB) (2011), 'Water supply', available at http://www.pub.gov.sg/general/drainageworks/Pages/default.aspx (accessed 17 March 2012).

Rees, J.A. (1998), 'Regulation and private participation in the water and sanitation sector', *Natural Resources Forum*, **22**(2), 95–105.
Shen, L.-Y., A. Platten and X.P. Deng (2006), 'Role of public private partnerships to manage risks in public sector projects in Hong Kong', *International Journal of Project Management*, **24**, 587–94.
Tan, Y.S., T.J. Lee and K. Tan (2009), *Clean, Green and Blue: Singapore's Journey towards Environmental and Water Sustainability*, Ministry of Environment and Water Resources, Singapore: ISEAS Publishing.
Tian, P.K. (2010), 'Urban water systems: factors for successful change?', *International Journal of Water Resources Development*, **26**(1), 97–109.
United Nations Development Programme (UNDP) (2006), *Human Development Report 2006 – Beyond Scarcity: Power, Poverty and the Global Water Crisis*, New York: Palgrave Macmillan.
United States Agency for International Development and Association for Rural Development (USAID – ARD) (2006), 'Good practices in public water utility corporatization', available at http://www.pwrf.info/Downloads/References/Contractualization/Good%20practices%20in%20public%20water%20utlity%20corporatization.pdf (accessed 12 February 2012).
Water and Sanitation Program (WSP) (2011), 'Promoting voice and accountability in urban water and sanitation services: lessons on introducing demand side accountability tools in Kenya', available at http://www-wds.worldbank.org/external/default/WDSContentServer/WDSP/IB/2011/07/22/000386194_20110722015951/Rendered/PDF/633470WP0Promo00Box0361513B0PUBLIC0.pdf? (accessed 20 March 2012).
World Bank (2006), 'Rehabilitating the urban water sector in Cambodia', *Water Feature Stories*, Issue 6, Water Supply and Sanitation Sector Board of the World Bank, available at http://water.worldbank.org/water/sites/worldbank.org.water/files/FS06_Cambodia.pdf (accessed 20 March 2012).
Yeaple, S. and W. Moskowitz (1995), 'The literature on privatization', Research Paper 9514, Federal Reserve Bank of New York.

Index